Industrial Transformation

Industrial Transformation

Environmental Policy Innovation in the United States and Europe

edited by Theo de Bruijn and Vicki Norberg-Bohm

The MIT Press
Cambridge, Massachusetts
London, England

MIT Press books may be purchased at special quantity discounts for business or sales promotional use. For information, please email special_sales@mitpress.mit.edu or write to Special Sales Department, The MIT Press, 5 Cambridge Center, Cambridge, MA 02142.

This book was set in Sabon on 3B2 by Asco Typesetters, Hong Kong.
Printed on Recycled Paper and bound in the United States of America.

Library of Congress Cataloging-in-Publication Data

Industrial transformation : environmental policy innovation in the United States and Europe / edited by Theo de Bruijn and Vicki Norberg-Bohm.
p. cm.
"The papers included were first presented at a workshop at the Kennedy School of Harvard University in the spring of 2001"—
Includes index.
ISBN 0-262-04228-2 (alk. paper) — ISBN 0-262-54181-5 (pbk. : alk. paper)
1. Industries—Environmental aspects—United States—Congresses.
2. Industries—Environmental aspects—Europe—Congresses. 3. Environmental policy—United States—Congresses. 4. Environmental policy—Europe—Congresses. 5. Industrial policy—United States—Congresses. 6. Industrial policy—Europe—Congresses. 7. Industrial ecology—Technological innovations—United States—Congresses. 8. Industrial ecology—Technological innovations—Europe—Congresses. 9. Sustainable development—United States—Congresses. 10. Sustainable development—Europe—Congresses.
I. Bruijn, Theo J. N. M. de. II. Norberg-Bohm, Vicki.
HC110.E5I46 2005
363.73′1—dc22 2004062537

10 9 8 7 6 5 4 3 2 1

Contents

About the Authors

Laurie K. Allen is Director of the Office of Protected Resources for the National Oceanic and Atmospheric Administration.

Per-Olof Busch is a Research Fellow at the Environmental Policy Research Unit, Berlin.

Cary Coglianese is Associate Professor of Public Policy at Harvard University's John F. Kennedy School of Government and Chair of the Regulatory Policy Program at the School's Center for Business and Government.

Theo de Bruijn is a Senior Research Associate at the Center for Clean Technology and Environmental Policy of the University of Twente, and Associate Professor of sustainable development at Saxion University for Professional Education, both in the Netherlands.

Donald A. Geffen has been a Research Associate at the Strategic Management Research Center of the Carlson School of Management. He is also a former Vice President of Alliance Capital Management and Physics Professor at the University of Minnesota.

Andrew Gouldson lectures in environmental policy at the Department of Geography and Environment at the London School of Economics. He is also Deputy Director of the newly established LSE Centre for Environmental Policy and Governance.

Mary Graham co-directs the Transparency Policy Project at the Harvard University's John F. Kennedy School of Government and is associated with the School's Taubman Center for State and Local Government and Environment and Natural Resources Program.

Peter S. Hofman is a Senior Research Associate at the Center for Clean Technology and Environmental Policy at the University of Twente.

Helge Jörgens is a Research Fellow at German Advisory Council on the Environment.

Ulrik Jørgensen is Associate Professor at the Department of Manufacturing Engineering and Management in the section for Innovation and Sustainability at the Technical University of Denmark.

Kris Lulofs is a Senior Research Associate at the Center for Clean Technology and Environmental Policy of the University of Twente in the Netherlands.

Alfred A. Marcus is Professor of Strategic Management in the Carlson School of Management at the University of Minnesota.

Robert M. Margolis is a Senior Energy Analyst in the Washington, DC, office of the National Renewable Energy Laboratory (NREL). Previously he was a member of the research faculty at Carnegie Mellon University and a research fellow at Harvard University.

Catherine Miller is a Senior Researcher with the Hampshire Research Institute.

Jennifer Nash is Director of the Regulatory Policy Program at the John F. Kennedy School of Government, Harvard University.

The late Vicki Norberg-Bohm was Director of the Energy Technology Innovation Project (ETIP) at the Belfer Center for Science and International Affairs at Harvard University's John F. Kennedy School of Government.

Bruce Paton is an Assistant Professor at San Francisco State University's business school.

Audun Ruud is Senior Research Fellow at the Program for Research and Documentation for a Sustainable Society (ProSus) at the University of Oslo, Norway.

Geerten J. I. Schrama is a Senior Research Associate at the Center for Clean Technology and Environmental Policy, University of Twente, the Netherlands.

Ken Sexton is the Bond Professor of Environmental Health Policy and Director of the Center for Environment and Health Policy in the School of Public Health at the University of Minnesota.

A Word to Readers

Just before the final manuscript of this book was handed over to The MIT Press, Vicki Norberg-Bohm passed away at the age of only 48. After a yearlong fight she lost the battle against lung cancer. Although almost all the intellectual work was done, she cannot witness the final result of our long collaboration.

The idea for this book was born years ago at one of the conferences of the Greening of Industry Network. Both interested in systemic change in production and consumption structures, she and I discussed the merits of recent innovations in environmental policies. We found that much of the debate on alternative approaches to rigid state regulatory frameworks was dominated by beliefs rather than empirical proof. Our goal from the start of the book project was to have a rigorous analysis and evaluation of the benefits and flaws of the new policy approaches. We also strongly believed in an international comparison as we discovered through our discussions that although policy approaches in different countries often seem alike, the effects differ substantially. As our prime interest was to understand the promise of the new policy approaches for systemic change, we chose to evaluate these approaches on the potential they have to set in motion a process of industrial transformation. This brought the book project to the heart of Vicki's interests. In her work she focused on the role of public policy for stimulating innovation and diffusion of environment-enhancing technologies. It was a scientific mission, but one with a strong societal impact: to understand technological change for sustainable development.

The present book brings together the results of our intentions. The papers included were first presented at a workshop at the Kennedy

School of Harvard University in the spring of 2001. Later that year Vicki invited me to spend the summer at the Kennedy School to jointly work on the book project. During that stay we not only managed to work productively; she also introduced me to her family and friends and made sure that I enjoyed my stay in Cambridge. That was characteristic of Vicki's warm personality. The papers have all been rewritten substantially. They represent the intensive and stimulating collaboration each of the authors and I had with Vicki. Vicki's capacity for always asking and rephrasing the right questions pushed us to the limits of our intellectual understanding of modern environmental policies.

Our work could not have been completed without the support of many. We are indebted to the faculty and staff of the Belfer Center for Science and International Affairs at Harvard's Kennedy School of Government and the Center for Clean Technology and Environmental Policy at the University of Twente; the US Department of Energy; the Dutch National Science Foundation NWO; Clay Morgan at The MIT Press; Kurt Fischer of the Greening of Industry Network; the participants at the spring 2001 workshop and at a workshop Vicki and I organized at the Greening of Industry Conference in Gothenburg in 2002; four anonymous reviewers; and of course our authors. Their support, especially during the past year, helped us to finalize what we started so many years ago.

We feel very sad that Vicki cannot see the final result of our challenging journey. We are grateful, though, for the opportunity we had to work with her, and we are happy that we can share many of her insights with the readers of this book.

We will remember Vicki through our work. We'll miss the stimulating and often provocative questions she posed but most of all we miss her friendship.

Theo de Bruijn
Summer 2004

1

Introduction: Toward a New Paradigm for the Transition to a Sustainable Industrial Society?

Theo de Bruijn and Vicki Norberg-Bohm

When environmental degradation emerged as a priority for government action in the early 1970s, most industrialized countries enacted media-specific legislation based on direct regulation that resulted in a set of ambient, emission, and technology standards that were enforced through permitting systems. Although direct regulation has been a powerful tool for adjusting industrial behavior, its shortcomings soon became apparent: shifting pollution from one media to another rather than eliminating pollution, constraining innovation, lax or expensive enforcement, inadequate or incorrect priority setting, and high transaction costs. Furthermore direct regulation has been criticized for being incapable of addressing the challenges of sustainability.[1]

The United States and many European countries have developed new approaches to overcome these shortcomings, including both market-based approaches (economic incentives) and the voluntary, collaborative, and information-based approaches examined in this book. This latter group of programs represents an attempt to engage industry in significant environmental improvements through dialogue, consensus-building, and voluntary action rather than through the imperatives of direct regulation or the incentives of market-based approaches. The rationale for these approaches lies not only in the shortcomings of direct regulation, but perhaps more important, in the complexity and severity of environmental problems, which necessitate a redefinition of the scope and methods of environmental policy and the roles traditionally played by government and the private sector in environmental protection.[2] "Compliance to regulation" is simply not an adequate approach to achieving sustainability. Rather, environmental policy must focus on how to use the creativity of all the actors in the production and consumption system beyond the level that has been stimulated by command-and-control strategies. Sustainability requires dialogue and consultation, collaboration, and the formation of new partnerships (Hart 1995; Hartman et al. 1999; De Bruijn and Tukker 2002). The industrial transformation is about system innovation, both technological and institutional, and goes beyond the domain of individual firms (IHDP 1999: xii).[3] The voluntary, collaborative, and information-based programs presented in this book can be seen as attempts to elicit industrial transformation.

Included in this book are evaluations of twelve such innovative programs, six from the United States and six from European countries. These and other approaches were heralded by different stakeholders to achieve a variety of policy goals, including efficiency, equity, democracy, and superior environmental performance. While the authors in this volume touch on all of these goals, the book's main focus is on the potential such programs hold for setting in motion a drive toward industrial transformation. More specifically, our standard for evaluation is whether these innovative policies can be effective in (1) stimulating beyond compliance behavior, (2) the development and diffusion of environmentally superior technologies, (3) providing opportunities and incentives for private-sector leadership in environmental protection, and finally (4) inducing change throughout the production and consumption system. The goals of this book are twofold: to understand the potential for this set of policy innovations to contribute to the industrial transformation necessary for a transition to a sustainable industrial society, and to provide guidance on the design and use of this set of policy innovations to support such a transition.

This book has good company. Over the past few years a range of titles on voluntary and collaborative approaches in environmental policy making, as well as a considerable body of literature on innovation effects of environmental policy, have been published (e.g., OECD 1999; NAPA 2000; CAVA 2001, Orts and Deketelaere 2001; Ten Brink 2002; Gunningham and Sinclair 2002). We contribute to this growing literature in different ways. First, we focus on mechanisms and programs rather than on instruments. Instruments are rarely employed in isolation. Rather, it is the combination of different instruments in programs that accounts for the effectiveness of policies. Second, where often the focus is on process rather than on outcomes, this book draws causal relationships between policy design and outcomes, within the context of the larger policy and institutional environment. Third, we look beyond the goals that the programs themselves set by evaluating the potential for industrial transformation. By including programs from the United States and Europe, this book builds on a wealth of experience and is able to draw on the diversity and similarities among voluntary, collaborative, and information-based programs developed in different contexts.

Voluntary, Collaborative, and Information-Based Approaches

At first blush, the group of programs discussed in this book are defined best by what they are not: neither direct regulation nor economic instruments. But this distinction is more than simply a negation; these programs were all conceived as an alternative or significant addition to the existing command-and-control system (which has been judged inadequate) and to environmental taxes (which often had limited support from industry or the public). Furthermore the programs were all created with high expectations of their ability to stimulate significant improvements in the effectiveness and efficiency of environmental protection. Despite the common goals, and in some cases similar basic strategies of these programs there is considerable variance among them. Perhaps the strongest dichotomy in the programs examined in this book is whether or not they are voluntary. In recent years several scholars have created taxonomies. In these, an important distinction has been made between voluntary and nonvoluntary programs (Delmas and Terlaak 2001a; Dowd, Friedman et al. 2001; Paton 1999; Ten Brink 2002). The term voluntary has been used in two distinct ways to describe this set of innovative approaches. The most common and straightforward use of the term voluntary is to describe programs that ask firms to voluntarily improve their environmental performance in exchange for benefits such as recognition, cost savings, and regulatory relief. In these programs there are no sanctions against firms that do not step forward. The term voluntary is also used in a second context to describe programs that provide a group of firms within an industry sector the opportunity to develop a voluntary agreement to reach environmental targets. This second type of voluntary program, while preserving the right of firms to choose whether to participate in this sectoral effort, provides direct regulation as an alternative if a voluntary agreement is not reached. When such approaches lead to fruition and result in a legally binding contract, they are most often called negotiated agreements. Finally, beyond these two categories, a third type of program—information disclosure—while not voluntary as it requires firms to provide information on toxic releases and other potential environmental hazards, relies on voluntary action by firms to move beyond compliance.

The second characteristic is collaboration. Over the past decade collaboration has become an increasingly more frequent element in environmental policy, and held up by many as a preferred approach for making a transition to sustainable industrial societies (Freeman 1997; Hajer 1995; Hartman et al. 2002; Mazmanian and Kraft 1999; Wallace 1995), although more critical comments can also be heard (Caldart and Ashford 1999; Driessen 1998; Poncelet 2002). All the policy innovations examined in this book depended on or resulted in increased collaboration between stakeholders. Some explicitly required collaboration in the decision-making process, bringing together private sector, government, and in many cases NGO representatives. Others created collaboration through implementation that explicitly aimed to increase the network of actors involved in environmental protection activities. Still others resulted in industry choosing to increase its interaction with stakeholders.

The third characteristic is increased information flows. All the programs examined in this book were information intensive, that is, they required increased development and dissemination of information on pollution emissions and sometimes other aspects of industrial behavior that affect the natural environment. In the information disclosure programs, increased information flows were mandated by law. Although not always mandatory, information disclosure is also a critical component for many types of voluntary and collaborative programs, as they rely on transparency in both negotiation and as part of monitoring schemes. In addition to disclosure, information development was pursued, for example, in EMS programs, to build capacity inside companies.

Table 1.1 shows that the programs in this book all contained at least two of the voluntary, collaborative, and information-based characteristics, and many contained all three, be it with different emphasis on one characteristic over another. The fact that all programs are a hybrid of different approaches, characteristics, and instruments makes the question on the mechanisms through which they work even more salient. Therefore, before moving to a description of each of the twelve programs, we discuss the potential strengths and weaknesses of this innovative set of policy mechanisms. The next section lays out the reasons one might expect these mechanisms to achieve their goals and be effective for industrial transformation. The following section examines the reasons why these new approaches might fall short of these goals. In this discussion, we draw heavily on three literatures: technology policy and management, policy sciences, and new institutionalism. However, as evidenced by the multiple theoretical perspectives in the chapters as well as the diverse literature we draw on in the introduction, understanding the sources of effectiveness requires a multidisciplinary approach. The fourth section of this chapter discusses the challenges and methods for evaluating these innovative programs and policies (and environmental policy more generally). The final section provides a more detailed introduction to the chapters in the book.

Pathways for Effectiveness

The programs in this book were established with high expectations about their ability to provide opportunities and incentives to stimulate superior environmental performance by the private sector, to encourage the development and diffusion of environmentally superior technologies, to promote leadership in environmental protection, and to involve other actors in the production and consumption system. The three big arguments for why these programs can achieve these demanding goals are discussed in this section. First, these programs may build new relationships among stakeholders, leading to better solutions for the environment. Second, the programs may engage industry in a learning process that creates the capabilities within firms to make significant environmental improvements. Third, the programs may create first movers—firms

Table 1.1
A taxonomy of voluntary, collaborative, and information-based approaches

	Voluntary		
	Strictly voluntary (no government sanctions for nonparticipation)	Voluntary with regulatory backup for nonparticipation	Mandatory information disclosure (depends on voluntary action to improve performance)
Sector based			
Dutch Target Group		×	
CSI	×		
German ELV		×	
Energy Star	×[a]		
R&D Collaboration	×[a]		
Danish CTP	×[b]		
Firm level			
Dutch EMS	×[a]		
EMAS in United Kingdom	×		
StarTrack	×		
Project XL	×		
TRI			×
Norwegian Accounting Act			×

a. Other government policies created market conditions that resulted in potential market losses for nonparticipants
b. Attempts to link with other policies that create sanctions for nonparticipants only partially effective
c. This is true only in some cases, in which firms created community advisory boards as one response to TRI
d. Not the main outcome at this point, but rather early to tell
e. Collaboration is defined here as new, more cooperative relationships between stakeholders. This sometimes involved only government and firms, and other times included additional stakeholders.

Collaboration		Information	
Requires collaboration[e]	Does not require collaboration, but stimulated increased collaboration[e]	Requires information disclosure	Successful implementation depends on information generation and disclosure
×			×
×			×
×			×
×			×
×			×
×			×
×			×
	×	×	
	×	×	
×			×
	×[c]	×	
	d	×	

willing to invest in business strategies, including managerial and technological innovations, that can lead to substantial improvements in environmental outcomes.

Building New Relationships

Innovations that aim at sustainable development are likely to require complex, collaborative settings. The programs included in this book all aim to establish new or altered relationships among various actors with a stake in environmental protection, including governments, industry, and community and environmental groups. Over the past decade all these actors have increasingly recognized that industry is not only a key contributor to environmental degradation but also can be part of the solution through the development of new processes, technologies, and products.[4] Enhancing the interactions among different relevant actors, be they firms, governments, or NGOs, holds three potential benefits.

The first potential benefit from building new relationships is an improvement in the quality of government strategies for the environment. Collaboration offers the opportunity to bring together experts from a variety of different disciplines and arenas to fashion solutions that can go beyond the limited perspectives of individual stakeholders (Gray 1989; Hartman et al. 2002). Several chapters evaluate programs that aimed to bring the benefits of close collaboration between industry and government to the development of new government policies and regulations. This collaboration may overcome one of the shortcomings of previous policy development—the fact that it has been very difficult for governments, through traditional regulatory processes, to develop regulations for industry that can provide an adequate stimulus toward sustainability. For governments it is virtually impossible to have inside knowledge of all sectors and firms. Government regulators must therefore rely on industry to provide critical information for the development of environmental policy, such as the amount of pollution, abatement costs, and technological options (Lévêque 1996). Developing meaningful regulations might be easier and more effectively done in collaboration with the target group.

The second benefit is found in increasing the steering capacity of governments by bringing in other actors in a network setting. The effectiveness of command-and-control regulation is often limited by difficulties with implementation and enforcement (Mazmanian and Sabatier 1983). Although in theory regulators set the boundaries for all firms, in practice it has proved to be very hard for regulators to effectively control all firms. Partnering with other organizations in a network setting can increase the steering capacity of governments. These partners can then act as agents of change. They can put environmental issues on the company's agenda and bring in technological knowledge, managerial expertise, and external views on the production process. Some of the programs in this book are explicitly implemented in a network setting. Many of the programs in this book, although not specifically aiming to establish networks, nonetheless facilitate their formation or strengthen existing networks through information requirements.

Third, collaboration can be used to share and reduce risk. In some cases the problem at hand is too large for individual companies to tackle, due to the costs and uncertainty involved. Governmental programs can enhance the development of networks of technological capability providing opportunities and incentives for manufacturers, suppliers, universities, national laboratories, and consultants to work together to reduce and spread risks. They do this by setting challenging goals that require collaboration and by providing funding and requiring organizational structures that increase collaboration among firms and other actors with technological capabilities.

In sum, the programs in this book may result in the building of new relationships or the altering of existing relationships. As the radical innovations needed for sustainable development require the commitment and engagement of multiple stakeholders, this is a crucial step in the process toward more sustainable enterprises. These programs can therefore develop new agents of change that will push and support firms to develop strategies that previously seemed impracticable.

Developing Environmental Capabilities in Industry

While in the past specific standards and end-of-pipe technologies prescribed the actions that industry had to take to protect the environment, the task of sustainable development requires that firms exercise considerably greater discretion. Many firms, however, lack the skills, knowledge and expertise, funds, and time to make the desired changes. In order to develop those capabilities, firms need to go through a learning process (e.g., Hart 1995; Roome and Cahill 2001). One of the potential strengths of voluntary, collaborative, and information-based approaches is their ability to engage industry in this learning process, partly through the partnerships discussed in the previous section.

The learning process toward more sustainable production takes place in different phases during which firms need to develop and implement increasingly complex and sophisticated environmental skills (Hart 1995). This is not an easy challenge: the transition to sustainable enterprises is a matter of “reinventing the company, together with its relationships with others in the sociotechnical systems in which its products and services are embedded" (Roome and Cahill 2001).

In many cases firms lack the capacity or motivation to develop the new capabilities on their own. Two of the chapters in this book examine programs aimed directly at the establishment of environmental capacity by promoting the adoption of environmental management systems (EMS). EMS set internal rules, create organizational structures, and direct resources so that the firm management can routinize environmentally superior behavior within the firm (Coglianese and Nash 2001; Welford 1998). EMS provide firms with the capability to develop an environmental strategy based on win-win opportunities, as well as to respond to external demands for improved environmental performance in a more efficient and effective way (see Coglianese 2001: 2). EMS thus may increase the likelihood of firms entering in a learning process toward sustainable enterprises.

Many of the other programs in this book, while not focused directly on EMS, nonetheless provide a stimulus to the development of new capabilities within firms. This is done first and foremost by requiring firms to collect, analyze, and sometimes disseminate new information about resource use and emissions.[5] This is most clearly the case in the information disclosure policies, but is also true for voluntary programs in which firms have to demonstrate their commitment to superior environmental performance in order to gain benefits such as regulatory flexibility and recognition. In these cases, just as is the case for EMS, new information can provide the underpinning for new action. Furthermore, by establishing long-range goals of significantly improved environmental performance, many of the programs challenge firms to re-examine their current investments and practices in light of their ability to meet these goals.

In sum, the programs in this book may evoke learning processes within firms through which they can develop environmental capabilities in order to transform themselves into more sustainable enterprises. The fact that many of the programs involve new partners, as stated in the previous section, is helpful in this respect as these partners may be better capable of engaging industry in a learning process.

Creating First Movers

The programs in this book were, often, created with high expectations about their ability to contribute to fundamental technological innovation. In this regard the ability of the programs to create niche markets for radical innovation is of special interest. The notion of radical innovation is, however, often ex post (Geels 2002: 113). A radical innovation may start small and gradually, when the actors involved recognize the potential and develop into something large. What makes an innovation into a radical one in the context of industrial transformation is its ability to change the technological trajectory in a direction that results in significant environmental improvement. Niche markets play an important role in this respect, providing steppingstones for the maturing and diffusion of radical new technology (Dosi 1982; Kemp et al. 1998; Nelson and Winter 1982). Niche markets develop as a result of actors willing to look for novel solutions. An important goal of environmental policies in the pursuit of sustainable development therefore is to motivate first movers, namely firms willing to take the risk of investments in entrepreneurial and technological innovations that substantially reduce environmental impacts. To accomplish this, these programs must either change the competitive environment of firms, or change firms' perception of their competitive environment. These innovative approaches can succeed in doing this in two ways.

First, they can provide the characteristics through which regulation is known to elicit an innovative response, including strictness, reduction in uncertainty, flexibility, and information generation.[6] Direct regulation has often fallen short of providing this set of characteristics.[7] The innovative approaches in this book have their origins, at least in part, in trying to overcome this shortfall. Some of the programs examined in this volume were built around stringent, long-term goals. Others challenged participants to develop long-term goals on a voluntary basis. In addition to providing a stringent standard, long-term goals reduce uncertainty about future markets, particularly when codified as legal requirements and combined with promises not to add additional requirements along the way. Long-term goals also provide the time frame necessary to innovate. Regulatory flexibility has been another hallmark of these innovative programs. This took a number of forms, from providing industry sectors the flexibility in choosing how to achieve publicly defined environmental goals to providing individual firms regulatory flexibility in exchange for beyond-compliance environmental performance. Regulatory flexibility can promote first movers by freeing the private sector to think creatively about how to achieve improved environmental performance, rather than responding in a more rote fashion to meeting regulations with proven and familiar technology, such as best available technology (BAT) and maximum achievable technology (MACT) (Ashford 1993; NAPA 2001; Porter and Van der Linde 1995).

Second, the programs may enhance the ability for first movers to profit from investments in superior environmental performance. Recent research on corporate strategy has identified a number of ways in which first movers can profit from investments in superior environmental performance (DeSimone and Popoff 2000; GEMI 1999; Hoffman 2000; Reinhardt 2000). The innovative programs examined in this book have in many cases enhanced the ability of firms to pursue these approaches, in the process bringing environmental policy closer to the core objectives of firms (Delmas and Terlaak 2001a; Hoffman 2000). In particular, these policy innovations have in some cases enhanced the ability of firms to pursue strategies of product differentiation, cost savings through increases in resource efficiency (e.g., pursue win–win approaches), and improved management of environmental risk.

For a firm to profit through a strategy of environmental product differentiation, it must be able to distinguish its product from that of competitors based on environmental performance, and in cases where there is an additional cost for an environmentally superior product, charge a price premium to cover the additional cost. One of the programs analyzed in this volume, the Energy Star program, enhanced the firm's ability to pursue environmental product differentiation, by creating brand recognition for energy efficient products.

Although there is considerable debate about the extent to which win–win opportunities exist, practices over the last decades make it clear that many such opportunities remain.[8] As discussed in the previous section, this set of innovative approaches enhances the ability of firms to identify such opportunities through their varied and numerous provisions for information generation. The programs provide a stronger stimulus to identify and pursue win–win opportunities to the extent that they ask firms to not only develop information about their current practices but also to make plans for future improvements.

The risks that corporations face from the environmental impacts of their production activities and products are numerous: regulatory difficulties, sour community relations, business interruption, and negative stockholder and investor reactions (GEMI 1999). The collaborative nature of these programs, combined with information disclosure and third party auditing can provide companies an opportunity to manage their risks through improved relationships with their stakeholders, including community and environmental groups, other firms in the supply chain, and the investment community.

In sum, there are numerous ways in which these innovative policy mechanisms can stimulate firms to become first movers. By creating an external environment in which firms choose to solve environmental challenges through radical innovation and by helping firms create an internal environment in which technological innovation is a profitable business strategy, the programs can initiate a process of industrial transformation.

Limitations of Innovative Policy Mechanisms

Their strengths notwithstanding, the new approaches may fail to be a force for industrial transformation because they have not succeeded in changing the competitive environment of the firm, suffer from complex implementation processes or do not fit with the dominant legislative system.

Failure to Change the Competitive Environment of Firms

Despite efforts to contribute to the firms' ability to pursue strategies of environmental product differentiation, operational efficiency, and environmental risk management, these new mechanisms may fall short of changing the competitive environment of firms in a way that will stimulate investment in the development and adoption of radical technological innovations. Below, we first discuss why multiple policies and programs may be necessary to create such a strong stimulus, and then examine the implications of this for creating first movers in general, and more specifically for strategies of environmental product differentiation.

Recent scholarship suggests that multiple policies and programs will be required to radically change industrial practices toward the environment (Baker et al. 1997; Blazeczsak et al. 1999; Jänicke et al. 2000; Jänicke and Weidner 1995; Norberg-Bohm 2000). This is because fundamental changes in production and consumption patterns may involve changing the behavior of numerous parties, and require changes to entrenched institutions and existing physical infrastructures. Thus any single program may hold only one part of what is necessary to change the competitive environment of a firm. For example, in terms of policy characteristics, flexibility without stringency or stringent goals without financial, organizational, or technological capabilities are strategies that are likely to fail. Furthermore incentives to one party, say, final manufacturers, will not be adequate if other members in the supply chain or consumers do not have the necessary incentives or capabilities. In some cases government must intervene not only to provide incentives to individual actors but also to overcome collective action problems with the creation of new institutions or support for the building of new infrastructure.

Thus these innovative programs must be looked at within the context of a broader set of environmental policies. The entire policy system has to provide a strong and coherent package of incentives, penalties, and moral persuasion in order to change the competitive environment of firms. Single programs, like the ones included in this book, may be only one element in this package.

The need for adequate incentives and penalties is particularly salient to firms that want to pursue a strategy of environmental product differentiation. Quite often there is a premium in price for new products. In the commercialization process for private goods, firms are able to charge more to lead adopters, the group of consumers willing to pay a premium for the qualities that the new good provides. However, new products whose value over existing goods is based solely on public good qualities, for example, improvements in environmental performance, may have more difficulty charging a premium in the marketplace. To the extent that an environmental product differentiation strategy depends on a price premium, there are two possibilities: a large enough group of green consumers that is willing to pay a high enough premium to bring a new, cleaner technology through the commercialization process or alternatively, environmental product differentiation strategies may need to be linked to government created niche markets, which can be done through market-based approaches, government procurement policy, and regulation. Similar issues exist for government R&D programs—their ultimate success depends on the existence of a market.

In sum, innovative policy mechanisms like the ones included in this book face the tough challenge of having to change the competitive environment of firms. This may be too large a task to ask of single innovative policies—rather it may depend on a suite of policy interventions.

Complex Implementation Processes

Program success depends on successful implementation. Long overlooked, Pressman and Wildavsky (1973) turned the attention of scholars and practitioners to the complexity of implementation, which they defined as a process of interaction between the setting of goals and actions geared to achieving them.

Policy programs are generally relatively easy to implement under the following conditions (Peters 1993): the legislator or administrator has sufficient information and a good knowledge of the causes of problems, the program holds clearly defined goals, the program aims at a small and identifiable target group, and the implementation structure is orderly. Many of the programs in this book do not fit these characteristics. Below we further discuss these issues.

First, with regard to breaking new ground, the types of programs examined in this book were developed precisely because that was their goal. The aims of these programs include shifting new responsibilities to the private sector for environmental stewardship, evoking beyond-compliance behavior, and developing and diffusing radical innovations with significant environmental improvements. Because these types of innovative programs require a great deal of change in firms, as well as other actors in the environmental policy system, successful implementation is particularly challenging (see Mazmanian and Sabatier 1981).

A second complicating factor is the lack of clear and specific goals. Programs may not set such goals, for various reasons. Most important, the general approach of the programs in this book is to give more discretion to firms. Instead of precisely demanding specific and detailed targets and prescribing certain techniques, the challenge is to get industry involved in a transformational process that will result in pro-active behavior and significant environmental improvements. Therefore the goals of the programs necessarily had to be somewhat of a general nature.

Third, implementation of the programs involves numerous actors in unfamiliar settings and roles. While establishing new relationships may be a necessary precondition for the pursuit of sustainable development, it makes the implementation process that much harder. While successful implementation usually requires the consent of a large number of participants (Pressman and Wildavsky 1973), these innovative programs tend to increase the number of participants and the complexity of obtaining consent. More specifically, some programs require complex negotiating processes among numerous public, private, and nongovernmental organizations. National legislators, local regulators, trade associations, individual firms, and interest groups. These programs, therefore, also rely on the willingness of actors to bridge mutual differences in order to reach an agreement.

In sum, given the often undefined or abstract goals, the reliance on the active participation by many actors, and the effort to create radical changes in the environmental policy system, the types of innovative programs in this book lead to complex implementation processes. This results in numerous opportunities for them to fail in the implementation phase.

Misfit with the Environmental Policy System

As previously noted, firms do not respond to a specific program in isolation; rather, their response depends largely on the environmental policy system as a whole. Thus, notwithstanding the potential advantages of the new approaches, they cannot be effective unless designed to work synergistically with the larger policy system (Jänicke and Weidner 1996). The context of a program is as relevant to its success and successful implementation as the design of the program itself. Programs that are not designed to fit with and complement the other elements of a nation's environmental policy system are likely to be less successful (De Bruijn

2003). This section first examines why the national environmental policy system is relevant to the implementation and therefore to the success of these programs. It then explores the ways in which these innovative policy mechanisms are often in conflict with the existing national environmental policy systems, thus constraining their effectiveness.

Many of the programs in this book are not only intended to evoke private sector leadership and superior environmental performance but also to *change* the existing legislative system. In almost all Western countries, direct regulation is still the dominant approach to environmental policy. All the programs in this book represent efforts to overcome the limitations this command-and-control system imposes for moving toward sustainable development, many with long-term ambitions of creating new regulatory regimes. This implies not only changing legislation but also changing organizational structures, expertise, and working routines (March and Olsen 1989). Institutions do not change easily though. Existing structures, task assignments, current procedures, and expertise limit the range of options. Change is possible, but it takes a good deal of pressure to produce that change (see Peters 1999). And the range of possibilities for that change is constrained by the existing institutional context. As a consequence, while the programs in this book might create promising alternatives to current national systems of environmental policy, it is also this existing system that constrains their design and implementation, and ultimately their effectiveness.

Although changes are hard to establish, institutions are capable of change. The programs in this book create pressures on the existing regulatory regime by developing alternative models. The chances of success depend on their "goodness of fit" (Risse 2001: 7). Goodness of fit relates to the interaction between one set of institutions, namely the existing regulatory regime, and another set, namely the innovative program or policy. In cases of a near match, new developments or programs can be easily incorporated and complied with in the existing setting. In these cases change is possible, but this change will almost by definition then be incremental and path dependent.

Programs that try to make a more radical change have a much higher chance for failure. Examples of these include innovative programs that ask government agencies to change their implementation practices and take on new roles in their relationship with industry, or ask industry to take on the responsibility for their own environmental performance, rather than just responding to government regulation. Such path-breaking programs will likely meet strong opposition from existing institutions since the required change then completely counters existing ideas, working routines, existing structures, competencies, and so on.

One of the most crucial factors concerning the (mis) fit with the existing policy system is policy style (Risse et al. 2001; Van Waarden 1995). A policy style shows itself foremost in the pattern of interaction between administrative and societal actors, which can either be formal and closed (interventionist style) or can be characterized by pragmatic bargaining, consensus, and transparency (mediating style) (Knill and Lenschow 1998). Misfits occur when a new program assumes a certain style that is contrary to the existing policy style. Consultation and collaboration, central characteristics of many of the programs in this book, might, for instance, flourish better in a more corporatist context characterized by pragmatic bargaining and consensus building between administrative and societal actors (e.g., European countries) than in a more adversarial system (e.g., the United States). In the latter, collaboration and consensus-building are not a matter of slightly adjusting current practices but rather changing the very nature of institutions. It is near impossible for single programs that do not fit the national style in environmental policies to catalyze such fundamental changes.

In sum, one would expect programs to be less successful when they are not complemented by other parts of the regulatory system and do not relate closely to those parts. Programs that are closely embedded in the total environmental policy system and are part of an ongoing process (of change) will likely be more successful. In cases with an insufficient or improper institutional capacity and especially when there is a misfit with the dominant style in environmental policies, the chances for success are limited.

An Evaluative Perspective

Evaluation research has grown as a field of inquiry and practice within the policy sciences over the last couple of decades, and encompasses a diverse set of goals and analytical approaches. In this section we place this book within the broader context of evaluation research in environmental policy. Specifically, we examine the types of evaluations undertaken in this book, the goals of these evaluations, the difficulties in providing conclusive results, and the evaluation methodology.

Policy evaluation can be defined as the judgment of the contents, processes or effects of a policy (Bressers and Hoogerwerf 1991: 22). In the environmental field, evaluation research can be divided into three types: impact, process, and efficiency (Knaap and Kim 1998). Impact evaluations focus on the results of a program, linking outcomes to activities carried out under the program. Process evaluations focus on the implementation of a policy, in order to understand why a program has or has not produced results. Efficiency evaluations examine whether a program makes good use of resources, generally through some type of cost–benefit analysis. In this volume we focus on outcome and process analysis. The authors use both qualitative and quantitative approaches to evaluation, although there is more qualitative work represented in this volume.

In terms of our outcome analysis, we have made a normative judgment to focus on a specific set of outcomes—superior environmental performance, radical technological innovation, industry leadership, and the involvement of other actors in the product-chain. These goals were often but not always articulated as part of the rationale for the programs evaluated in the book. We nonetheless believe it is relevant to look at all of the programs through this lens, as they have all been undertaken as innovative strategies for future environmental policy and have all been promoted as a way of achieving substantial environmental improvement. If these are to be a major part of the environmental policy arsenal of the future, they must contribute to a process of industrial transformation. In sum, we are taking a goal-free approach to evaluation: we are not simply taking the program goals as our impact variable, but rather, we are judging these programs against moral, political, and societal criteria (Scriven 1991).

Process evaluation was crucial to the goals of this book for two reasons. First, some programs explicitly specified process changes as a primary outcome. Second, and perhaps more important, this entire set of mechanisms are directed toward new collaborative processes for their success. As part of the process analysis the authors examine not only program design elements but also considered political factors, both internal and external to the regulatory agencies. Previous scholarship suggests that these factors can have a big impact on the policy outcomes (Rosenbaum 1998).

It is beyond the scope of this book to undertake efficiency analyses, although in many of the chapters the high transaction costs associated with the programs are discussed. Efficiency analysis is an important direction for future research, as it would contribute to understanding whether these programs are a cost-effective way of achieving environmental goals.

This book is within the policy analysis tradition of pragmatism—we are trying to provide usable knowledge to policy makers. This can best be done through formative evaluations, which examine not only the outcomes of a program, but focus on explaining why these outcomes have occurred (Scriven 1967). These can be used to provide guidance for changes in midstream as well as for the design and implementation of future programs. The programs evaluated in this book are a combination of works in progress and completed works. Certainly as an emerging paradigm for environmental policy, this group of innovative programs is a work in progress—policy makers are still trying to determine how to design these programs to be effective, and under what conditions they can be effective. Our goal is to contribute to this task by understanding why the programs examined in this volume have achieved success or failure, and to draw lessons across these twelve cases (see Patton 1978).

The authors in this volume encountered several challenges, including the long-term nature of environmental degradation, the lack of data and difficulty in measuring outcomes, and the difficulty in linking outcomes to a particular program (see Knaap and Kim 1998). Some of the programs evaluated in this volume set environmental goals for over a decade into the future, and many are still in progress. While these long-term goals are a positive development in relation to the normative interests of this book, they are a challenge for evaluators, who can often only provide process evaluations and use these to speculate about the future.

Other programs had more immediate results (e.g., the adoption of an EMS), but they still expected environmental outcomes to occur in the future, raising a similar set of concerns.

There were numerous types of data difficulties that limited the authors' ability to precisely evaluate the impact of these programs. In many cases there was simply a lack of data. In some cases there were process data available but no outcome data. An often-repeated recommendation in the chapters is the need for more monitoring.

The authors have approached these evaluation difficulties creatively and carefully, using established evaluation and social science modes of inquiry, but cannot claim to have conquered them all. These issues influenced the methodologies that were chosen by individual authors, as well as the types and strength of conclusions that each was able to reach.

Introduction to the Book

The book is divided into two parts. In the first part, the programs examined are those that aim to influence the behavior of firms by addressing an industry sector. In the second part, the programs are those that aim to influence individual firms or facilities. The grouping of programs was challenging, as programs could be grouped in one way on some characteristics, and in another on others. In the workshop that we held to present and discuss the work that now appears in this book (De Bruijn and Norberg-Bohm 2001), we divided the papers into panels based on the three defining characteristics of this book: voluntary action, collaboration, and increased information flows. Although all the programs contained a hybrid of these characteristics (see table 1.1), this distinction provided limited information about the factors leading to success. For the book we chose to divide the programs into those that were aimed at an industry sector, and those that addressed individual firms and facilities. There are a number of factors that vary along this divide—the approach and effectiveness of calls for voluntary action, the form of collaboration, the types and structure of information flows and the approach to leaders and laggards are generally different when addressing individual facilities versus industry sectors. In our concluding chapter we return to these differences.

Part I: Industry Sector Approaches

The first three chapters examine experiments in the Netherlands, Germany, and the United States. They bring together firms within an industry sector with the goal of achieving large and long-term reductions in pollution. The next three chapters, based on two US cases and one Danish case, represent government efforts to stimulate the development and deployment of cleaner technologies.

Chapter 2, by Peter Hofman and Geerten Schrama, examines the Dutch Target Group Policy. This policy is the central element in the current Dutch system of industrial environmental regulation. Through negotiations with sectors of industry, the Ministry of the Environment, and regional and local governments, agreements are sought concerning the contribution of specific industrial sectors to the goals laid out in the National Environmental Policy Plan. Because some of the agreements are demanding, sectors can opt out if their technologies do not develop at a pace that enables them to reach the agreed-upon goals. The analysis by Hofman and Schrama shows that the effectiveness of the target group policy so far seems satisfactory, as interim goals have been met in most sectors. However, thus far the policy seems to promote mainly innovations of an incremental nature and wider adoption of best available technologies (BAT). It remains to be seen whether the R&D components of this approach along with private sector technology investments will result in the fundamental technology innovation and diffusion necessary to meet the program's longer range targets for deep emission reductions.

In chapter 3, Cary Coglianese and Laurie Allen focus on the Common Sense Initiative (CSI), the prominent example in the United States of a sector-based, collaborative approach to environmental policy. Its goals were to develop "cleaner, better, cheaper" solutions to environmental performance by fine-tuning environmental regulation to the specific circumstances of different industrial sectors. Under CSI, six subcommittees (one for each sector) composed of representatives from industry, NGOs, labor unions, and governments were convened and given the charge to make recommendations to an overarching Council. Allen and Coglianese conclude that CSI resulted in "nothing dramatic." Because of its reliance on consensus-based decision making, its strictly voluntary nature, and the lack of legal authority to make changes to existing regulations, the limited number of agreements that were reached tended to shift toward the lowest common denominator of the players involved, focusing more on the goal of making environmental regulation more efficient rather than the goal of increasing environmental protection.

In chapter 4, Helge Jörgens and Per-Olof Busch focus on the German end-of-life vehicles program. This program consisted of a voluntary agreement between sixteen branch organizations in the automotive recycling and supply sector. Although voluntary, the agreement was reached under the threat of regulation. In response to the voluntary agreement, the German government refrained from a comprehensive regulation and instead introduced an ordinance to facilitate the implementation of the agreement. The agreement focused on the design of cars and their components, as well as environmentally sound recycling and disposal of end-of-life vehicles. Jörgens and Busch argue that the agreement has given great leeway to automobile producers, although it introduces strict, detailed, and costly regulations for dismantlers and return stations. This result is due to the greater political power of the oligopolistic automobile industry compared to the heterogeneous and weakly organized dismantling sector. Thus, while the sectorwide agreement has proved to be effective for short-term and urgent problems (toxic waste from car dismantling), Jörgens and Busch raise questions about the effectiveness for the longer term issue of car recycling and the associated reduction of landfill waste. The voluntary agreement was replaced by a law to implement a European directive on ELV. In this law a few of the shortcomings of the voluntary agreement have been mended.

Chapter 5, by Bruce Paton, focuses on Energy Star, a voluntary program aimed at energy-efficiency improvements in products such as computers and washing machines. Energy Star engages industry in negotiated agreements on the level of energy efficiency necessary to gain the use of the Energy STAR label, while simultaneously working with the government and large companies to specify Energy STAR products in their procurement policies. Paton distinguishes between two mechanisms in voluntary programs: (1) converging mechanisms, which focus on changing the behavior of an entire industry simultaneously, and (2) separating mechanisms, which create best practices in leading firms. The chapter examines two of the Energy STAR product-labeling programs—Office Products and Washing Machines. These examples illustrate the differences in design and outcomes between converging and separating mechanisms. Paton concludes that both programs have led to significant energy savings, but through different routes.

Chapter 6, by Vicki Norberg-Bohm and Robert Margolis, focuses on three US Department of Energy (DOE) R&D collaborations in the power sector: the Advanced Turbine Systems (ATS) program, the Photovoltaic Manufacturing Technology (PVMaT) project, and the Thin-Film PV Partnership project. The goals of these programs were to support the development of a next generation of technology, and included support for innovations in technology systems, technology components and manufacturing processes. Each of these programs included collaborative planning processes as well as collaboration in implementing the R&D programs. The collaboration involved multi-stakeholder partnerships, including companies, governments, Congress, universities, national laboratories, and end-users. The participants from the private sector and the government officials involved in these collaborations viewed them as highly successful, and a good model for future government R&D programs. Norberg-Bohm and Margolis argue that from the standpoint of the environment, R&D collaborations will be most effective if they engage industry and other stakeholders in a planning process that leads to the establishment of "stretch" goals, meaning technological goals with environmental benefits that are beyond what is required by regulation and what the private sector would pursue on its own.

In chapter 7, Ulrik Jørgensen examines cleaner technology programs in Denmark. The Danish government support for cleaner technology innovation and demonstrations was initiated in 1987 with the Development Program for Cleaner Technology. The program offered grants to support identification, development, demonstration, and full-scale implementation of cleaner technologies. Support for this approach continued through a series action plans throughout the 1990s. Jørgenson reports that the Danish collaborative approach to technology innovation and diffusion has had a measurable and significant impact on the availability of cleaner technologies. However, the diffusion and implementation of these technologies, while significant, has not reached its full potential.

The basic difficulty is that local regulators, who have significant discretion in the Danish environmental policy system, have either not learned sufficiently from the clean technology programs or have not forced the implementation of the available clean technologies during their permit negotiations with industry.

Part II: Facility Level Approaches

This part of the book turns to examining approaches that were directed at individual facilities and firms. The first two chapters in this section examine programs that aim to establish EMS within firms: the Dutch Program on Environmental Management and the adoption of the European EMAS Regulation (Eco-Management and Audit Scheme) in the United Kingdom. The next two chapters focus on US programs, Project XL and StarTrack that aim to elicit superior environmental performance in exchange for benefits such as regulatory flexibility and recognition. The final two chapters focus on information disclosure policies, the Toxics Release Inventory in the United States and the Norwegian Accounting Act.

Chapter 8, by Theo de Bruijn and Kris Lulofs, examines the voluntary policy program on environmental management systems (EMS) in the Netherlands. The objective of this program was to have companies introduce an EMS by 1995. Although the ultimate goal of the program was to improve the environmental performance of companies, its main objectives were to generate mutual trust for government-industry collaboration, to enhance capacity building within industry, and to involve third parties in promoting environmental protection. Instead of dealing with SMEs directly, the Dutch government facilitated the formation of networks in which intermediary organizations acted as agents for change. De Bruijn and Lulofs report that this program design proved quite effective. Other factors leading to success in this program included the way in which it was embedded in the wider policy approach of the Netherlands (which includes more coercive approaches), the fit with the general mediating policy style of the Netherlands, and the high level of public concern over environmental matters during the early 1990s.

In chapter 9, Andrew Gouldson focuses on the implementation of the Eco-Management and Audit Scheme (EMAS) in the United Kingdom.

EMAS is a regulation of the European Union (EU), which aims to build capacity for environmental protection and increase the environmental protection activity within firms through the adoption and third-party verification of environmental management systems. EMAS is a voluntary scheme; companies may choose to register their sites if they comply with the requirements of the scheme. Gouldson reports on the implementation and the effects of the regulation in the UK. He suggests that EMAS did raise awareness of the need to innovate and help develop capacities for technological change. However, most of the innovations within EMAS registered companies were low-tech and organizational changes, leading to incremental innovation. While the capacity for more radical change exists, the external incentives or imperatives that could lead to the actual utilization of the capacity are lacking. Gouldson concludes that the drivers for change, including the decision to adopt EMAS, have to come from the wider range of incentives for improving environmental performance.

In chapter 10, Jennifer Nash explores the impact of StarTrack, a program that used the adoption of EMS as part of a tiered system of environmental regulation. The basic premise behind StarTrack, and tiered systems of environmental regulation more generally, is that the environmental authorities offer participating firms regulatory relief and public recognition in exchange for superior environmental performance. For firms to enter the StarTrack program, they had to have a history of compliance and pollution prevention, an EMS or a commitment to adopt an EMS, and a commitment to continued improvement in environmental performance. In exchange, EPA managers promised to forgo inspections, offer penalty amnesty, provide faster permitting, and publicly recognize StarTrack facilities as environmental leaders. Nash concludes that while the idea of tiered environmental regulation is relatively simple, implementation has raised a complex set of issues. One of the main problems was defining superior environmental performance. A second problem is that the benefits EPA provided were less than the agency had promised and not very meaningful to firms. Furthermore the transaction costs were very high, with the program increasing rather than decreasing agency oversight. Nash concludes that unless these issues can be resolved, tiered environmental programs are unlikely to result in tangible benefits.

Chapter 11, by Alfred Marcus, Donald Geffen, and Kenneth Sexton, examines Project XL, a facility-based program aimed at developing bold alternatives to the current approach to standard setting, permitting, implementation and enforcement. Under Project XL, industry and government agencies can petition for regulatory flexibility at a specific facility in exchange for producing an overall increase in environmental quality. Project XL was organized as a stakeholder process that required the participation not only of government and private firms but also community groups. Marcus et al. analyze four cases in which firms tried to negotiate XL agreements, three of which succeeded. In general, Project XL resulted in far fewer agreements than had been expected by the EPA. The chapter examines both the substantive and process issues that created barriers to negotiating Project XL agreements. A central goal of Project XL was to get superior environmental performance. However, it proved very hard to define this concept in practice. In terms of process, Marcus et al. identify the issue of lack of legal authority for changes in permitting practice, and examined how negotiating strategies affected the ability to finalize an agreement. They conclude that Project XL has failed to live up to the expectation that it was a pathway to significant changes in US environmental policy.

Chapter 12, by Mary Graham and Catherine Miller, evaluates the impact of the US Toxics Release Inventory (TRI). The TRI was created in 1986 as part of the Emergency Planning and Community Right-to-Know Act. After several amendments during the 1990s, the TRI now requires most medium and large-scale manufacturing firms to provide facility level data on releases of 602 chemicals to all media (air, water, and land), as well as on-site and off-site storage, treatment, disposal, recycling and energy recovery. It also requires firms to report qualitatively on source reduction activities and to provide a production index, so that changes in releases and transfers can be related to changes in production. The TRI is heralded as a major success, and an important contributor to a 45 percent reduction in releases of listed chemicals between 1988 and 1998. As Graham and Miller discuss, the TRI, however, cannot be given credit for this entire decline. A variety of regulations enacted since 1986, as well as other factors, have influenced firms' decisions to reduce toxic emissions. Furthermore, although examples of preventative action do exist, relatively few facilities cut releases by reducing waste at the source; rather recycling increased substantially. Also, releases declined at a much more rapid rate in early years, raising questions about the long-term impact of this policy approach. Despite these caveats the TRI was clearly path breaking legislation, which has contributed to toxic emission reductions and provided lessons for future information disclosure policies.

Chapter 13, by Audun Ruud, examines a newer effort at information disclosure, the Norwegian Accounting Act (NAA) of 1998. This act requires the boards of directors of all commercial firms subject to external auditing requirements to disclose environmental data on activities that may cause "a not insignificant impact on the external environment." The NAA goes beyond the requirements for reporting on plant specific pollution control, which forms the basis for the Norwegian Pollutant Release and Transfer Register, by requiring firms to report on the life cycle environmental impacts of their products. After its first years of implementation, few firms were fully complying with the act. As outlined by Ruud, the issues here are twofold: to get firms to report on environmental impacts beyond the emissions data they are already reporting, and to get firms to take a life cycle approach, reporting on the environmental impacts of inputs and product use, as well as future plans. This is admittedly a tall order, and one made more complicated by the fact that although the act requires this information in the annual report, it does not require the third party auditor to validate the environmental portion of the report. Thus the engagement of a third party as educator, implementer, and enforcer is not part of this legislation, leaving environmental reporting requirements still quite distinct from financial reporting requirements.

Conclusions

In the final chapter of this book we take stock of the experiences to date with voluntary, collaborative, and information-based programs, assessing whether the pathways for effectiveness identified in this introductory chapter have become a reality, or whether the potential limitations have instead impeded progress toward an environmentally sustainable industrial transformation. From the evidence in the cases presented in this book, we conclude that synergistically employing multiple approaches will provide the best opportunity to guide industry on the path toward sustainable development. Voluntary, collaborative, and information programs can play a useful role in such a comprehensive strategy, but only if they are carefully designed to fit with and complement the other elements of a nation's environmental policy system. In the end the real question therefore is not whether the new approaches should be used, but rather *how* they should be used. Regardless of goals there will remain a role for direct regulations and market-based approaches as part of an overall strategy—they will be needed to create sufficient pressures to push industry along the path toward sustainability.

Notes

1. Critiques of the existing regulatory system, along with prescriptions for improvement came from a variety of sources. In the United States, the high-profile multi-stakeholder processes included the President's Council on Sustainable Development (PCSD 1996) and the Enterprise for the Environment (Ruckelshaus 1998). In Europe the Fifth Environmental Action Plan made a plea for a considerable change of environmental policies both in terms of objectiveness and strategy (CEC 1993; see also Kronsell 1997).

2. See Fiorino 1999 for a probing discussion of the legal and governance issues related to new approaches for engaging the private sector. Many have written about the changing role of business, and the need for private sector leadership to achieve the goals of sustainable development (e.g., Fischer and Schot 1993; Leveque 1996; Roome and Cahill 2001). Influential writings from the private sector include DeSimone and Popoff (2000), Schmidheiny (1992), and Von Weizsacker et al. (1997).

3. Many authors stress the need for changes at a level beyond the individual firm and the involvement of firms in different stages of the product chain; see, for instance, Hart (1995), Schot et al. (1997), and Roome and Cahill (2001).

4. There are many writings in which governments, industry, and NGOs emphasize the need for industry to be engaged; see, for instance, Bendell (2000), CEC (1993), Millais (1994), PCSD (1996), and WBCSD (2001). See also notes 1 and 2.

5. In most step-by-step guides that are available for (environmental) projects, collecting information is the first step for taking action; see, for instance, Sheldon

and Yoxon (1999), Young (1998), Welford and Gouldson (1993). The same idea also stands at the basis of pollution prevention approaches in which a thorough analysis of material flows enables the formulation of preventive measures; see, for instance, Dieleman and Hoo (1993) and USEPA (1989).

6. This summary of the characteristics of regulation that stimulate innovation come from a number of studies, including Ashford (1993), Ashford and Heaton (1983), and Porter and Van der Linde (1995).

7. A broad set of literature supports this claim. See, for instance, Haigh and Irwin (1990), Kemp et al. (1994), Wallace (1995), Gouldson and Murphy (1998), Norberg-Bohm (1999).

8. For a review of the debate on the availability of win–win opportunities, see Norberg-Bohm (2001).

References

Argote, L., and D. Epple. 1990. Learning curves in manufacturing. *Science* 247: 920–24.

Ashford, N. 1993. Understanding technological responses of industrial firms to environmental problems: Implication for government. In K. Fischer and J. Schot, eds., *Policy. Environmental Strategies for Industry*. Washington, DC: Island Press.

Ashford, N. A., and G. R. Heaton. 1983. Regulation and technological innovation in the chemical industry. *Law and Temporary Problems* 46(3): 109–57.

Baker, S., M. Kousis, D. Richardson, and S. Young. 1997. *The Politics of Sustainable Development*. London: Routledge.

Bendell, J., ed. 2000. *Terms for Endearment: Business, NGO's, and Sustainable Development*. Sheffield: Greenleaf.

Blazeczsak, J., D. Edler, J. Hemmelskamp, and M. Jänicke. 1999. Environmental policy and innovation—An international comparison of policy patterns and innovative impacts. In P. Klemmer, ed., *Innovation and the Environment*. Berlin: Analytica Verlagsgesellschaft, pp. 9–30.

Bressers, J. T. A., and A. Hoogerwerf. 1991. *Beleidsevaluatie*. Alphen aan den Rijn: Samsom H.D. Tjeenk Willink.

Caldart, C. C., and N. A. Ashford. 1999. Negotiation as a means of developing and implementing environmental and occupational health and safety policy. *Harvard Environmental Law Review* 23(1): 141–202.

CAVA. 2001. *The Use of Voluntary Approaches Policy brief*. Paris: Cerna.

CEC. 1993. *Towards Sustainability: A European Community Programme of Policy and Action in relation to the Environment and Sustainable Development*. Luxembourg: CEC.

Coglianese, C. 2001. Is consensus an appropriate basis for regulatory policy? In E. W. Orts and K. Deketelaere, eds., *Environmental Contracts: Comparative*

Approaches to Regulatory Innovation in the United States and Europe. London: Kluwer Law International, pp. 93–113.

Coglianese, C., and J. Nash. 2001. *Regulating from the Inside: Can Environmental Management Systems Achieve Policy Goals?* Washington, DC: Resources for the Future.

De Bruijn, T. 2003. Multi-level governance between the European Union and its member states: The Importance of policy style. In H. T. A. Bressers and W. Rosenbaum, eds., *Achieving Sustainable Development: The Challenge of Governance across Social Scales*. Westport, CT: Praeger.

De Bruijn, T., and V. Norberg-Bohm. 2001. *Voluntary, Collaborative, and Information-Based Policies: Lessons and Next Steps for Environmental and Energy Policy in the United States and Europe*. Cambridge: Harvard University, p. 63.

De Bruijn, T., and A. Tukker, eds. 2002. *Partnership and Leadership: Building Alliances for a Sustainable Future*. Dordrecht: Kluwer.

Delmas, M. A., and A. K. Terlaak. 2001. A framework for analyzing environmental voluntary agreements. *California Management Review* 43(3): 44–63.

DeSimone, L. D., and F. Popoff. 2000. *Eco-Efficiency: The Business Link to Sustainable Development*. Cambridge: MIT Press.

Dieleman, H., and S. D. Hoo. 1993. Toward a Tailor-made Process of Pollution Prevention and Cleaner Production: Results and Implications of PRISMA project. In K. Fischer and J. Schot, eds., *Environmental Strategies for Industry: International Perspectives on Research Needs and Policy Implications*. Washington, DC: Island Press.

Dosi, G. 1982. Technological paradigms and tecnological trajectories: A suggested interpretation of the determinants and directions of technical change. *Research Policy* 11: 147–62.

Dowd, J., K. Friedman, and G. Boyd. 2001. How well do voluntary agreements and programs perform at improving industrial energy efficiency. *Proceedings of the 2001 ACEEE Summer Study on Energy Efficiency in Industry*. Washington, DC: American Council for an Energy-Efficient Economy.

Driessen, P. 1998. Concluding remarks. In P. Glasbergen, ed., *Co-operative Environmental Governance: Public-Private Agreements as a Policy Strategy*. Dordrecht: Kluwer, pp. 251–68.

Fiorino, D. J. 1999. Rethinking environmental regulation: Perspectives on law and governance. *Harvard Environmental Law Review* 23(2): 441–69.

Fischer, K., and J. Schot, eds. 1993. *Environmental Strategies for Industry: International Perspectives on Research Needs and Policy Implications*. Washington, DC: Island Press.

Freeman, J. 1997. Collaborative governance in the adminstrative state. *UCLA Law Review* 45(1): 98B.

Geels, F. 2002. *Understanding the Dynamics of Technological Transitions: A Co-evolutionary and Socio-technical Analysis.* Enschede: Twente University Press.

GEMI. 1999. *Environment: Value to Business.* Washington, DC: Global Environmental Management Initiative.

Gilbert, M. J. 1993. *Achieving Environmental Management Standards: A Step by Step Guide to Meeting BS7750.* London: Pitman.

Gray, B. 1989. *Collaborating: Finding Common Ground for Multiparty Solutions.* San Francisco: Jossey-Bass.

Gunningham, N., and D. Sinclair, eds. 2002. *Leaders & Laggards. Next-Generation Environmental Regulation.* Sheffield: Greenleaf.

Haigh, N., and F. Irwin, eds. 1990. *Integrated Pollution Control in Europe and North America.* London: Conservation Foundation and Institute for European Environmental Policy.

Hajer, M. A. 1995. *The Politics of Environmental Discourse: Ecological Modernization and the Policy Process.* Oxford: Oxford University Press.

Hart, S. L. 1995. A natural resource based view of the firm. *Academy of Management Review* 20(4): 986–1014.

Hartman, C. L., P. S. Hofman, and E. R. Stafford. 1999. Partnerships: A path to sustainability. *Business Strategy and the Environment* 8(5): 255–66.

Hartman, C. L., P. S. Hofman, and E. R. Stafford. 2002. Environmental collaboration: Potential and limits. In T. De Bruijn and A. Tukker, eds., *Partnership and Leadership: Building Alliances for a Sustainable Future.* Dordrecht: Kluwer, pp. 21–40.

Hoffman, A. J. 2000. *Competitive Environmental Strategy.* Washington, DC: Island Press.

IHDP. 1999. *Industrial Transformation Science Plan.* Bonn: International Human Dimensions Programme on Global Environmental Change.

Jänicke, M., J. Blazejczak, D. Edler, and J. Hemmelskamp. 2000. Environmental Policy and Innovation: An International Comparison of Policy Frameworks and Innovation Effects. In J. Hemmelskamp, K. Rennings, and F. Leone, eds., *Innovation-oriented Environmental Regulation: Theoretical Approaches and Empirical Analysis.* Heidelberg: Physica, pp. 125–52.

Jänicke, M., and H. Weidner, eds. 1995. *Successful Environmental Policy. A Critical Evaluation of 24 Cases.* Berlin: WBZ.

Jänicke, M., and H. Weidner. 1995. Successful Environmental Policy: An Introduction. In M. Janicke and H. Weidner, eds., *Successful Environmental Policy.* Berlin: WBZ, pp. 10–26.

Jänicke, M., and H. Weidner. 1996. Summary: Global Environmental Policy Learning. In M. Janicke and H. Weidner, eds., *National Environmental Policies: A Comparative Study of Capacity-Building.* Berlin: Springer Verlag, pp. 299–313.

Kemp, R., J. Schot, and R. Hoogma. 1998. Regime shifts to sustainability through processes of niche formation: The approach of strategic niche management. *Technology Analysis and Strategic Management* 10(2): 175–95.

Knaap, G. J., and T. J. Kim, eds. 1998. *Environmental Program Evaluation: A Primer.* Urbana: University of Illinois Press.

Knill, C., and A. Lenschow. 1998. The impact of British and German administrations on the implementation of EU environmental policy. *Journal of European Public Policy* 5(4): 595–614.

Kronsell, A. 1997. Policy innovation in the garbage can: The EU's Fifth Environmental Action Programme. In D. Liefferink and A. S. Anderson, eds., *The Innovation of EU Environmental Policy.* Copenhagen: Scandinavian University Press, pp. 111–32.

Laswell, H. 1951. The policy orientation. In D. Lerner and H. D. Laswell, eds., *The Policy Sciences.* Stanford: Stanford University Press, pp. 1–15.

Lévêque, F., ed. 1996. *Environmental Policy in Europe: Industry, Competition and the Policy Process.* Brookfield, VT: Elgar.

March, J. G., and J. P. Olsen. 1989. *Rediscovering Institutions: The Organizational Basis of Politics.* New York: Free Press.

Mazmanian, D. A., and M. E. Kraft, eds. 1999. *Toward Sustainable Communities: Transition and Transformations in Environmental Policy.* Cambridge: MIT Press.

Mazmanian, D. A., and P. A. Sabatier, eds. 1981. *Effective Policy Implementation.* Lexington, MA: Lexington Books.

Mazmanian, D. A., and P. A. Sabatier. 1983. *Implementation and Public Policy.* Glenview, IL: Scott, Foresman.

Millais, C. 1994. Greenpeace's new solutions. *Chemistry and Industry* (June 20): 484.

NAPA. 2000. *Transforming Environmental Protection for the 21st Century.* Washington, DC: National Academy of Public Administration.

NAPA. 2001. *Leading Change: Advancing Effective Governance in the 21st Century.* Washington, DC: National Academy of Public Administration.

Neij, L. 1997. Use of experience curves to analyse the prospects for diffusion and adoption of renewable energy technology. *Energy Policy* 23(13): 1099–1107.

Nelson, R., and S. Winter. 1982. *An Evolutionary Theory of Economic Change.* Cambridge: Harvard University Press.

Norberg-Bohm, V. 1999. Creating incentives for environmentally enhancing technological change: Lessons from 30 years of U.S. energy technology policy. *Technological Forecasting and Social Change* 65 (October): 125–48.

Norberg-Bohm, V. 2000. Technology commercialization and environmental regulation: Lessons from the U.S. energy sector. In J. Hemmelskamp, K. Rennings, and F. Leone, eds., *Innovation-Oriented Environmental Regulation: Theoretical Approaches and Empirical Analysis.* Heidelberg: Physica, pp. 193–219.

Norberg-Bohm, V. 2001. *Beyond the Double Dividend: Public and Private Roles in the Supply of and Demand for Environmentally Enhancing Technologies.* Paris: OECD.

OECD. 1999. *Voluntary Approaches for Environmental Policy: An Assessment.* Paris: OECD.

Orts, E. W., and K. Deketelaere, eds. 2001. *Environmental Contracts: Comparative Approaches to Regulatory Innovation in the United States and Europe.* The Hague: Kluwer Law International.

Paton, B. 1999. *Voluntary Environmental Initiatives and Sustainable Industry.* Paper presented at the Greening of Industry Network Conference, Chapel Hill, NC.

Patton, M. Q. 1978. *Utilization-Focused Evaluation.* Beverly Hills: Sage.

PCSD (President's Council on Sustainable Development). 1996. *Sustainable America: A New Consensus for Prosperity, Opportunity, and a Healthy Environment for the Future.* Washington, DC: Government Printing Office, pp. 25–55.

Peters, B. G. 1993. *American Public Policy: Promise and Performance.* Chatham, NJ: Chatham House.

Peters, B. G. 1999. *Institutional Theory in Political Science: The "New Institutionalism."* New York: Pinter.

Poncelet, E. 2002. In Search of "win–win": Multistakeholder environmental partnerships and the pursuit of sustainability. In T. De Bruijn and A. Tukker, eds., *Partnership and Leadership: Building Alliances for a Sustainable Future.* Dordrecht: Kluwer, pp. 41–60.

Porter, M., and C. Van der Linde. 1995. Toward a new conception of the environment-competitiveness relationship. *Journal of Economic Perspectives* 9(4): 97–118.

Pressman, J., and A. Wildavsky. 1973. *Implementation.* Berkeley: University of California Press.

Reinhardt, F. L. 2000. *Down to Earth: Applying Business Principles to Environmental Management.* Boston: Harvard Business School Press.

Risse, T., M. G. Cowles, and J. Caporaso. 2001. Europeanization and domestic change: Introduction. In M. G. Cowles, J. Caporaso, and T. Risse, eds., *Transforming Europe. Europeanization and Domestic Change.* Ithaca: Cornell University Press, pp. 1–20.

Roome, N., E. Cahill, F. Berkhout, and W. Potratz. 2001. *Sustainable Production: Challenges and Objectives for EU Research Policy.* Brussels: European Commission.

Rosenbaum, W. A. 1998. Why Institutions Matter in Program Evaluation: The Case of the EPA's Pollution Prevention Program. In G. J. Knaap and T. J. Kim, eds., *Environmental Program Evaluation: A Primer*. Urbana: University of Illinois Press, pp. 61–85.

Ruckelshaus, W. D. 1998. Stepping Stones. *The Environmental Forum* 34 (March–April).

Schmidheiny, S. 1992. *Changing Course: A Global Business Perspective on Development and the Environment*. Cambridge: MIT Press.

Schot, J., E. Brand, and K. Fischer. 1997. The greening of industry for a sustainable future: Building an international research agenda. *Business Strategy and the Environment* 6(3): 153–62.

Scriven, M. 1967. The methodology of evaluation. In R. W. Tyler, R. M. Gagne, and M. Scriven, eds., *Perspectives on Curriculum Evaluation*. Chicago: Rand McNally.

Scriven, M. 1991. *Evaluation Thesaurus*. Newbury Park: Sage.

Sheldon, C., and M. Yoxon. 1999. *Installing Environmental Management Systems: A Step-by-Step Guide*. London: Earthscan.

Ten Brink, P. 2002. *Voluntary Environmental Agreements: Process, Practice and Future Use*. Sheffield: Greenleaf.

USEPA (United States Environmental Protection Agency). 1989. *Waste Minimization Opportunity Assessment Manual*. Washington, DC: USEPA.

Van Waarden, F. 1995. Persistence of National Policy Styles: A Study of Their Institutional Foundations. In B. Unger and F. Van Waarden, eds., *Convergence or Diversity? Internationalization and Economic Policy Response*. Gateshead: Athenaeum Press.

Von Weizsacker, E. U., A. Lovins, and H. Lovins. 1997. *Factor Four: Doubling Wealth, Halving Resources Use—The New Report to the Club of Rome*. London: Earthscan.

Wallace, D. 1995. *Environmental Policy and Industrial Innovation: Strategies in Europe, the US and Japan*. London: Earthscan.

WBCSD, (World Business Council for Sustainable Development). 2001. Sustainability through the Market: Seven Keys to Success. Geneva: WBCSD.

Welford, R., ed. 1998. *Corporate Environmental Management 1: Systems and Strategies*. London: Earthscan.

Welford, R., and A. Gouldson. 1993. *Environmental Management and Business Strategy*. London: Pitman.

Young, C. W. 1998. Measuring Environmental Performance. In R. Welford, ed., *Corporate Environmental Management I. Systems and Strategies*. London: Earthscan.

I

Industry Sector Approaches

2

Dutch Target Group Policy

Peter S. Hofman and Geerten J. I. Schrama

Our understanding of environmental issues and of ways to address them has increased in the last decades. However, while some problems have become increasingly evident and more pressing, often no ready-made solutions are available. In search for more effective and efficient environmental policies, approaches in the Netherlands and elsewhere have shifted from direct regulation to more flexible and consensus-based styles. Many of these new developments are based on the premise that the transition to sustainability requires a cooperative paradigm, with diverse stakeholders negotiating a shared vision of the future and the coordination of their resources (Hartman et al. 1999). It also indicates that these approaches are more cost-effective as the targeted groups can time the development and introduction of new technologies; no short-term standards are imposed that are likely to be suboptimal in the long term (Ashford 1996; Harrison 1999). The Dutch Target Group Policy, formulated as part of the National Environmental Policy Plan introduced in 1989, is one of the best-known examples of this new approach.

This chapter evaluates the Dutch Target Group Policy for the industrial sector. The authors focus on how its cooperative and long-term orientation has influenced processes of innovation and diffusion. The first section below introduces the framework in which the Target Group Policy was formulated, the National Environmental Policy Plan, and its background and philosophy. The next section focuses on the theoretical basis for the Target Group policy as an effective means to enhance environmental change in industry. In the following section, the discussion turns to examine the Dutch Target Group Policy in greater detail and explain the mechanisms for inducing innovation and its diffusion throughout industries. The next two sections then focus on the effectiveness of the Target Group Policy, first by assessing the realization of the intermediate and long-term environmental targets, and second, by examining the role of the Target Group Policy in inducing both incremental and radical innovation. The chapter concludes with a discussion of the ability and limitations of the Target Group Policy to create conditions for inducing innovation and its diffusion.

Context of the Target Group Policy

During the 1980s several developments contributed to the emergence of the new policy approach in the Netherlands. There was increasing awareness of the seriousness of environmental problems, the difficulty of tackling them by conventional means such as command-and-control policies, and the need for more integrative policy. An assessment by the National Research Institute on Health and the Environment (RIVM) indicated that preserving the Dutch environment would require 70 to 90 percent reductions of all pollution, requiring a structural change in production and consumption patterns (RIVM 1988).

Pieter Winsemius, Minister of the Environment in 1982 to 1986, developed a National Environmental Policy Plan (NEPP), published in 1989, as part of a deliberate attempt to change the philosophy of environmental policy. It proposed that care for the environment should no longer be the exclusive responsibility of government:

> Everyone is supposed to be aware of his/her responsibility with respect to the environment and has to let this influence his/her actions. The large scale on which some environmental problems occur does not detract from this.... Without the dedication of the target groups, environmental policy cannot be intensified, and the pursuit of sustainable development becomes a dead letter. (Tweede Kamer der Staten-Generaal 1989: 13, 31)

The NEPP focus on target groups involved the formulation of national objectives for 2010 for the various environmental themes, and subsequent consultations with the target groups and their representatives on their role in the implementation process.[1] For each of the target groups more specific policies and intermediate goals were formulated.[2] It was expected that, once 2010 targets were realized, the Netherlands would be well on its way toward sustainability.

The first National Environmental Policy Plan was a clear break with prior environmental policy based on a much more adversarial approach. Although the targets set in the National Environmental Policy Plan were very ambitious—including emission reductions of most hazardous substances by 70 to 90 percent by 2010—they were acceptable for the industry target groups because the exact reduction targets were less important than the discretion to plan environmental improvements free of short-term legal requirements (VROM and VNO-NCW 1998). In fact the NEPP offered the prospect that willingness to cooperate would spare the target groups more restrictive government policies (De Jongh 1999: 143).

Policy Approaches to Stimulate Innovation

The observation in the NEPP that structural changes are necessary and that the long-term targets cannot be met by end-of-pipe technologies increased the need for policy designed to induce innovative behavior. The traditional command-and-control approach to environmental policy favored the prescription of specific technical solutions to industry. However, the scale and size of the largest environmental problems and their solutions were uncertain. Some promising solutions were not applicable in the short run, and implementation required the cooperation of various sectors in society, a strategy difficult to implement through command-and-control policy making. The ambitious goal of sustainable development required a new approach to environmental regulation.

Schrama and Van Lierop (1999) have analyzed the various options for policy to stimulate innovations that will encourage desirable and discourage undesirable behavior on the part of target groups. Their review of the literature in the fields of policy science, economics, and management and organization has revealed six major dimensions or "steering parameters" that might be considered "levers for policy makers."[3]

The first of these is the degree of freedom of choice granted to target groups; relevant especially for policies aimed at inducing innovation (Ashford 1993; Norberg-Bohm 1999). Related and overlapping concepts include "flexibility" and "self-regulation." If target groups are unwilling, policy makers may choose to restrict their freedom of choice and impose a particular behavioral option. Under different conditions policy makers may even want to enlarge the freedom of choice for target groups for a number of reasons:

- To increase the support for or the acceptance of the policy goals by offering the opportunity for the target group to match these goals with their own preferences.
- To complement the limited knowledge and expertise of governmental actors with knowledge and expertise of the target group.
- To deal with the fact that certain policy goals, such as more innovation, cannot be imposed on a target group; rather, the role of government is limited to creating conditions enabling the target group to innovate.

An advantage of increased freedom of choice is that it makes regulation much more dynamic. For example, long-term targets are established but the routes through which these targets may be reached are open to discussion and negotiation (Tweede Kamer der Staten-Generaal 1989). Policy makers expected that the goals of NEPP 1 would be met, although it was uncertain exactly how these aims were to be achieved. Target groups were given time and leeway to consider an array of pathways. Such a policy stance recognizes that when the best current technology is insufficient to meet targets, decision-making processes in companies committed to the objectives will actively support innovations that put the goals within reach as they continue to move in a more environment-friendly direction.

The second dimension is the extent to which the approach is based on *collaboration* with target groups. Faced with the complexities of environmental problems and the drastic behavioral changes that target groups must adopt, many authors have argued that a cooperative paradigm is needed. According to Jänicke (1992, 1997) "consensual capacity" is an institutional condition for success in environmental policy. Collaboration is expected to add value, as it implies "a process through which parties who see different aspects of a problem can constructively explore their differences and search for solutions that go beyond their own limited vision of what is possible" (Gray 1989: 5). Collaboration can vary from simple forms of communication between policy makers and target groups to "interactive policy making" at the preliminary stages of the policy process. Consensus building is also an essential feature of the often-praised Dutch "polder-model."[4] The importance of consensus can be drawn from its various roles:

- In increasing public support and acceptance of policy.
- In reducing the uncertainty for target groups regarding the purpose and consequences of policies.
- In exchange of knowledge regarding underlying causes of environmental problems and the possibilities for more sustainable behavior from the target group.

A third dimension is the *stringency* or the pressure that is put on the target group. This term was introduced by Ashford (Ashford 1985; Ashford and Heaton 1983). Stringent behavior lies at the core of Porter's hypothesis: If the pressure is high enough, companies will utilize their innovative capacity to comply with environmental standards (Porter 1991; Porter and Van der Linde 1995). From this line of thought an environmental policy with ambitious goals is indispensable in spurring a further greening of industry. In reaction to critics, Porter and Van der Linde (1995) acknowledge that regulatory pressure alone is not enough for the effective stimulation of environmental innovations. They also point out such elements as the time path, freedom of choice, and reduction of uncertainty.

The fourth dimension is the *time horizon*. Fundamental innovation takes a long time to develop, often more than is recognized or provided for in legislative measures. Short-term requirements can lead to suboptimal outcomes as target groups look for ready solutions to their problems to comply with legal requirements. Longer term requirements can lead to technological developments that foster superior solutions. The policy maker must decide how much time to allow the target group to comply with regulatory demands. The element of uncertainty also comes into play in this context. To create innovative solutions to environmental problems, target groups need not only sufficient time but also certainty about legal standards for the present and future.

The fifth dimension is the *instrumentation of policy*. The choice of policy instruments, as the vehicle through which policy incentives are "dispatched" to the target group, is a major parameter for policy formulation. Although it has some overlap with other dimensions (e.g., freedom of choice) the characterization of the policy incentives also has some unique aspects. Policy makers must decide between the following:

- Three basic stimuli: economic or financial, legal, and social or communicative.
- Options for monitoring, accountability, and enforcement.

Policy incentives are usually linked to specific policy instruments. For complex strategies, such as the voluntary approaches discussed here, a mix of policy instruments may be preferable. A judicious mix enables policy makers to employ several different types of incentives.

The sixth dimension involves choosing the appropriate *addressees of policy*, and how to approach them. While the final target group is given, the principal recipients of the policy incentives must be carefully selected, taking three realities into consideration.

First, while well-organized industrial branches can be targeted directly, some target groups are more difficult to reach. It may be more effective, for example, to address consumers indirectly through producers or retailers.

Second, a target group may not be monolithic; it may be desirable to differentiate within it. For policies directed toward a specific industrial branch it may be necessary to treat "front-runners" or "laggards" differently from the majority of companies.

Third, policy makers can take advantage of the increasing use of so-called network approaches which have become more influential in Dutch environmental policies directed at industry (see chapter 8). Current policy theory now generally assumes that the adoption and development of environment-oriented innovations in companies take place in interaction with various company network partners; production processes and technology are influenced by network characteristics (Rycroft and Kash 1999; Van Dijken et al. 1999). Recent strategies therefore focus not only on individual companies but also on the networks in which these companies operate. In the Netherlands, for example, the introduction of environmental management systems in companies is stimulated through the deliberate use of so-called network steering (De Bruijn and Lulofs 1996, and chapter 8 in this volume). Since the beginning of the 1990s, negotiated agreements between branches and government agencies have depended to a large extent on the role of the branch associations, often a relevant actor in the network of companies.

The following sections introduce the Dutch Target Group Policy for industry and analyze how the dimensions presented above have been articulated in the new policy approach.

Target Group Policy for Industry

After the publication of NEPP1 in 1989, implementation took place through the establishment of negotiated agreements with industrial branches. The so-called covenants were not replacements for environmental laws; rather, they set a planning cycle and framework in which companies were asked to implement the NEPP. The main objectives of the Target Group Policy for industry can be presented as a six-step scheme that is re-iterated generally every four years.

Negotiated Agreements under the Target Group Policy

The first step concerns the formulation of emission reduction targets for the industry as a whole. These targets are derived from the macro policy targets in the 1989 first and the 1993 second NEPP. Other environmental policies that affect industry are also taken into account, such as policies regarding energy saving and climate change.

Industry as a sector is too large for a uniform approach, so in the second step, 14 priority branches of industry were selected. These involve 12,000 companies responsible for over 90 percent of industry-based environmental pollution. Negotiated agreements have been reached for 10 branches (see table 2.1).

As a third step, negotiations—sometimes called "consultations"—were started with each of the 14 branches, represented by their trade associations. The first aim of these negotiations was to establish so-called integral environmental targets (IET). The typical IET contains emission reduction targets for air, water, and soil for specific substances or categories of substances (usually 30 to 40). An IET also details guidelines concerning energy conservation, waste, soil sanitation, external safety, odor, noise, and internal environmental management. Targets at the branch level have been set for 1994–1995, 2000, and 2010, in relation to a base year that varies based on available (emission) data. These targets

Table 2.1
Overview of target-group negotiated agreements for industry as of 2000

Industry	Companies[a]	Type	Agreement
Primary metals industry	38 (39)	Heterogeneous	10-03-1992
Chemical industry	137	Heterogeneous	02-04-1993
Printing industry	3400[b]	Homogeneous	08-04-1993
Dairy industry	133	Heterogeneous	06-07-1994
Textiles processing, carpet and floor coverings industry	46 (75)	Heterogeneous	08-03-1995
Electroplating industry	±17.000[b]	Hybrid	19-04-1995
Oil- and gas-producing industry	9	Homogeneous	02-06-1995
Paper (products) industry	26	Heterogeneous	08-03-1996
Concrete and cement industry	440	Homogeneous	02-09-1998
Rubber and plastics processing industry	117/1.100[c]	Hybrid	22-12-2000
Meat industry	168 (285)	Homogeneous	22-12-2000

Source: Data taken from FO Industrie (www.fo-industrie.nl), last update July 2003.
a. In parentheses are the number of companies in the branch if it exceeds the number of participants.
b. No data are available about the number of participants in the branch.
c. The duty to make Corporate Environmental Plans applies to 117 larger companies, whereas the *Environmental Handbook* applies to about 1.100 SME's.

are supposed to reflect what is considered as a fair share for the specific sector in relation to the targets for industry as a whole. Each individual company is supposed to contribute to the realization of the sector targets.

The outcomes of the negotiations were written into an agreement signed by all parties involved: the branch associations, some of the larger companies, the relevant ministries, and representatives of the regulators (provinces, municipalities, and waterboards). The targets were often considered demanding, but part of the deal was the assurance that during the "contract period" no new legislative demands would be imposed and that the environmental permits of individual participating companies would be adjusted to the content of the agreement.

The fourth step involved publicity regarding the agreement and provision of information. When the agreements were signed the companies of the branch had to be informed and persuaded to join the process. A central role in the implementation process was assigned to a specially established independent agency, "FO Industrie."[5] Its tasks involved gathering and providing information on the Target Group Policy for the parties to the agreement, as well as for a larger audience, and support for individual firms.

In the fifth step, translation of policy to the level of the individual companies, a distinction was made between homogeneous and heterogeneous branches. Homogeneous branches are characterized by relatively small differences in firm size (i.e., mainly SMEs) and production processes, for instance, the printing industry. In these cases uniform implementation plans have been developed and articulated in *Environmental Handbooks*, containing relevant norms and standards, and specification of the state-of-the-art technologies the companies were supposed to apply.

In the case of heterogeneous sectors, each participating company had to develop an individual corporate environmental plan (CEP). The CEP involved an assessment of the present state of a company's environmental performance, company-specific targets, measures to be taken in the next four years, and a preview to the subsequent cycle. The measures in the CEP can be conceived as the company's route toward compliance, and also as a blueprint for going beyond compliance.

Companies are not obliged to perform all stated measures under all circumstances; the measures and the intended reductions in emission levels are not commitments to result, only to make an effort. The negotiated agreements state that the companies will not be held to the performance of all measures in case of unforeseen circumstances concerning: (1) the economic development of the sector, (2) discrepancy between national and EU environmental policy, and (3) the lack of technological means to reduce environmental impacts.

The plans must be submitted for approval to the main regulator (either the municipality or province), who must coordinate with other regulators (often the water board). The minimum requirements for approval are compliance with "prevailing policies" and application of "state-of-the-art technology."

The sixth, and final, step involves implementation within a company's environmental management practices and the adjustment of its environmental licenses. Participating companies have to submit to the main regulator annual progress reports on monitoring and control according to a format specified in the negotiated agreement. As of 1999, the (approximately) 300 largest and most polluting companies have a legal obligation to publish an annual environmental report.

Negotiations were started with the 14 priority branches of industry. Three negotiated agreements were in place on schedule (i.e., end of 1993) for the chemical industry, the primary metals industry, and the printing industry. In environmental terms, these were also the most important industries, producing 60 percent the total industrial environmental impact. In four cases the parties involved finally agreed to refrain from a covenant.[6] Table 2.1 provides an overview of negotiated agreements between government and industry.

Covenants in Relation to Other Policy Instruments

Figure 2.1 provides an overview of the mechanisms for stimulating innovation and diffusion by linkages between various policy instruments.

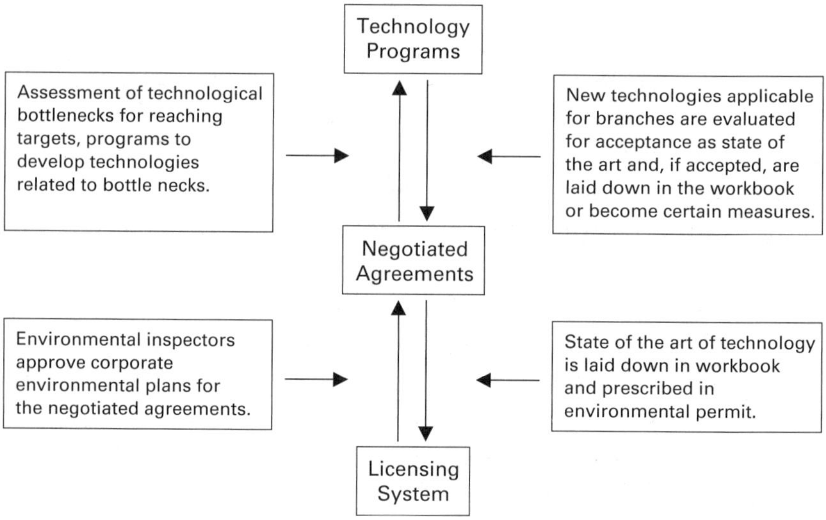

Figure 2.1
Cycle of diffusion and innovation in the Target Group Policy for industry

Part of the Target Group Policy's potential for diffusion and innovation is derived from the relationship of negotiated agreements to technology policy and to direct regulation (environmental permits). The application of state-of-the-art technology is one of the guiding principles of the Target Group Policy. Through the negotiated agreements individual companies are committed to maintain state-of-the-art production processes (insofar as is reasonably achievable). The formulation in the negotiated agreements contains the criterion that the companies can reasonably be assumed to know about the particular technologies. Branch associations play an especially important role in delivering to their members both this message and information regarding what is achievable.

Furthermore companies not bound by the agreement still have to meet the same standards, as the state-of-the-art standard is legally mandated in the comprehensive Environmental Management Act. Through this principle, regulators are legally bound to prescribe measures to companies on the basis of what is reasonably achievable. Companies are required to adapt their processes to the standard prescribed in the workbook or technical assessments of their industrial branch in the process of the negotiated agreement. This way the mechanism of diffusion is kept fueled by the inspectors in the process of giving and renewing permits.

According to Biekart (1994: 31), the legal regulation requires the use of "best technical means," a more severe criterion than "best practical means" which are restricted to proved technology. As such Target Group Policy is equal to the "alara" (as low as reasonably achievable) principle in the Dutch Environmental Management Act, the legal basis of the environmental licenses. Because of the escape clause concerning unforeseen circumstances of an economic or technical nature, the state-of-the-art technology principle can be conceived as "best available technology not entailing excessive costs" (BATNEEC).

The "innovation potential" of the negotiated agreements is related to two mechanisms. First, the targets set for the branches can generally not be met by mere diffusion of the state of the art throughout branches; there are certain bottlenecks for which new or improved technologies need to be developed.[7] These bottlenecks are assessed through detailed branch studies within the agreement which set the directions for technology development.

In exchange for a willingness to commit themselves to specific targets set in the agreements, the Ministry of Environment, through its program "Technology and Environment," provides funds for the development of environment-oriented technologies. The technology program helps the branch and technology developers to identify problems and to develop technologies for the medium and long term. Moreover, when new technologies are developed, and have proved to be applicable to the specific branch through a number of stages, they will eventually be established in the workbooks or as mandated measures. This mechanism ensures that a process of continuous improvement is pursued until the targets set in the negotiated agreement are reached. Apart from the Technology and Environment program aimed at technological bottlenecks, task forces are set up to inform industries about the potential for technological improvement.

In addition other technology programs have been developed, notably by the Ministry of Economic Affairs, to exploit the potential of innovation to improve both economic and environmental conditions. Here the goals formulated in the NEPP guide the direction of innovation. For example, in a program like EET (Economy, Ecology, Technology) the focus is on longer term efficiency gains by a factor of 4 to 10, which requires technological breakthroughs, as well as long-term projects with the cooperation and commitment of different parties as critical conditions. This cooperation should extend from parties involved in fundamental research, such as universities, to partners involved in strategic research, such as research institutes; applied research, such as engineering firms; and finally to companies that market or use the technology. Moreover a specific tax scheme has been developed for new environmentally benign technologies. Companies investing in technologies that qualify for this tax scheme can write off these investments in the year that suits them best, thus offering them considerable tax advantages.

In conclusion, the setup of the negotiated agreements and the linkages to other instruments suggest that mechanisms for continuous technological improvement are in place. This is based on the assumption that technological change is necessary in order to reach the goals set in the negotiated agreements. Such change depends mainly on autonomous developments to which the technology programs are contributing by informing relevant parties about technological bottlenecks, and by providing funds to those willing and able to develop breakthrough technologies.

Technological changes trickle down to individual companies in two stages. First, the new technology must become accepted as state of the art. When this is achieved, it will be incorporated in the workbook for the specific branch. Next, environmental inspectors will use the workbook as their point of reference for framing requirements for an environmental license. In terms of instrumentation this policy system employs steering strategies of a financial, legal, and social nature.

Short-term Effects: Implementation and Compliance

In this section we briefly review the effectiveness of the Target Group Policy for industry in terms of actual environmental improvements. Assessment of this kind can be made in two ways. First, we can look at emission levels in a specific year, and assess environmental improvements against the reference situation. Second, we can analyze the effectiveness of the Target Group Policy in relation to historical trends and ask what would have happened without the negotiated agreement.

Progress Reported in the Sectors

The best available data on the short-term effects are the sector progress or implementation reports. These are consolidated reports drafted by FO Industrie, and based on the annual environmental reports of the individual companies. The larger part of these sector reports describes the achievements in terms of the environmental themes outlined in the negotiated agreement. Numbers of emission reduction targets for specific substances and the extent to which these targets have been achieved are mentioned often. Although the formats used are rather similar, the information provided is not adequate to develop a systematic review for the most important issues.

Highlights of the most recent annual sector reports (2001) concerning the integral environmental targets (IET) 2000 and 2010 for five selected sectors are as follows:

• Primary metals industry. The third planning period (2001–2004) has commenced. Emission reduction targets have been set for 38 substances.

In 2001 the IET 2000 have been achieved for 29 substances, the IET 2010 for 20 substances, while the reduction for 11 substances has been significantly more than the IET 2010. The remaining IET 2010 targets have to be met, but the sector has indicated that extra measures have to be developed, for instance, in the case of NO_x and VOC emissions to the air.[8]

• Chemical industry. The second planning period (1998–2001) has ended. Most emission reduction targets for 2000 have been achieved. With respect to the IET 2010 the sector is on schedule for about 50 percent of the targets concerning emissions to the air and 85 percent of the emissions to water. Special attention is required for NO_x emissions; the development of a NO_x emissions trade system for the sector is being considered.[9]

• Printing industry. The implementation of the measures in the Environmental Handbook has been evaluated in 2001. Overall the results are positive in terms of implementation of environmental measures and achievement of emission reduction target. The emission of hydrocarbons, however, constitute a serious problem for the sector. Emission reduction is far behind schedule, due to lacking implementation of required measures, overestimation of the effects of certain measures, and increases of production volumes.[10]

• Paper (products) industry. All 26 companies in the sector are participating in the process and working on their second Corporate Environmental Plan (2000–2003). Most emission reduction targets for 2000 have been achieved, and most of the IET 2010 are well on schedule. A major problem concerns the emissions related to combustion processes (NO_x and CO_x). A NO_x emissions trade system for the sector is being developed.[11]

• Dairy industry. The sector is in the transition from the second planning period (1999–2002) to the third. Most emission reduction targets for 2000 have been achieved. Major exceptions are NO_x and energy efficiency. For NO_x a new sector approach is being drafted in consultation with the government.[12]

For industry overall, reaching the targets for CO_2 reduction will require intensified efforts. Another assessment study stated that about 80 percent of planned total emission reductions would be realized by 2000. According to Glasbergen (1998: 151) emissions to the air would be reduced more quickly under the negotiated agreements than under a hierarchical regime, while discharge to water would be somewhat higher than otherwise.

Some sector progress reports state that parties to the negotiated agreements are having difficulty meeting their obligations to the reporting system. These groups include companies in the primary metals, chemicals, and dairy industries that have to submit environmental reports and the authorities that must judge them (particularly the primary metals and dairy reports). In several cases, especially those of the chemical and dairy industries, the situation has improved compared to previous years.

General Appreciation of the Results

Government and industry are pleased with the results thus far. Relations between government and industry have improved and industry is actively improving its environmental management and publicly advertising their achievements in this field. Industry also reports that the administrative costs for environmental compliance have been lowered by the covenants. According to Biekart (1998), the chemical companies estimated that the covenant has led to time saving of at least 10 percent in their environmental compliance work. A more recent evaluation concludes that efficiency effects of covenants tend to become more positive in the case of a relatively strong (market) position of the branch, when its environmental image is relatively sensitive to the public, and when it is characterised by fierce competition (De Bruijn et al. 2003: 33). This indicates that the outcome for the chemical industry may be relatively positive compared to other branches.

Initially environmental organizations had strong doubts whether companies, in particular, the large ones, would not simply continue doing as they had under the old regulatory regime, since early evaluations for the primary metals covenant showed that the quality and depth of the Corporate Environmental Plans fell short in some cases (Biekart 1994). Later on, in 1997, it was noted by the environmental organization SNM (Dutch Foundation for Nature and Environment) that considerable progress had been made by the primary metals branch for several environmental themes. The organization concluded that "on the positive side, most goals for air and water pollution while not initially achieved (for the 1995 interim targets) are likely to be reached by the year 2000."[13]

Long-term Effects: Technology and Innovation

This section explores whether negotiated agreements induce technological change. We use data from an evaluation of a technology program set up by the Dutch Ministry of Environment[14] to reduce technological bottlenecks to reaching environmental targets in industry (Arentsen and Hofman 1996). A total of 146 projects have been analysed for the period 1993 to 1995, and the motives for the start-up of these projects have been investigated.[15] Table 2.2 summarizes the outcomes.

Of the 146 projects 67 were motivated by anticipation of legislation. In 10 cases there was specific reference to the negotiated agreement between an industrial branch and the government; interviewees in eight other cases referred to other covenants. Twenty-seven projects took place

Table 2.2
Motives for projects within the Program Environmental Technology, 1993–1995

Environmental program	Number of projects
Total number of projects	146
Motives for projects (more than one answer possible)	
Anticipation of legislation	67 (46%)
Problem of applicant	45 (31%)
Need for technology	76 (52%)
Other motives	29 (20%)
Anticipation of further classified legislation motives (total 67 projects)	
Environmental Act	13
Surface Water Pollution Act	10
Covenant VOC 2000/packaging	8
Negotiated agreements of industrial branches	10
Waste/water policies	11
Other	15

Source: Arentsen and Hofman (1996).

within industrial branches for which negotiated agreements were concluded in the period 1993 to 1995. These branches were the printing industry (agreement in 1993), dairy industry (1994), and textiles industry (1995). For 5 of the 27 projects, respondents referred to the negotiated agreement, while from the printing industry reference was made to 5 projects prompted by the VOC covenant[16] that preceded the negotiated branch agreement. Overall, this indicates that negotiated agreements have some limited influence on innovations in companies of the target group.

To obtain a better sense of the influence of the negotiated agreements, the authors conducted telephone interviews in 1999 with 14 companies in three branches having negotiated agreements: the printing industry, dairy industry, and metal plating and electrical engineering. These companies participated in the same technology program in the period 1995 to 1998. Additional interviews were carried out with six experts from the specific branches and the technology program. Respondents were asked what the influence of the negotiated agreements was on innovation.[17] Table 2.3 summarizes the outcomes.

To summarize, the influence of the negotiated agreements on innovations for the respective branches cannot be clearly assessed. Research results are mixed. However, there are indications that individual companies tend to underestimate the influence of negotiated agreements on innovations, compared to network actors of specific branches. These experts generally acknowledge that negotiated agreements tend to set some degree of direction for what kind of innovation is expected from, or necessary in, specific branches.

Table 2.3
Influence of negotiated agreements on innovative behavior

Branch	Companies	Experts
Average	2	7
Printing industry	1	8
Dairy industry	1	—
Metal industry	4	7

Note: Figures are assessments by respondents on a scale from 0 (no influence) to 10 (high influence).

Apart from the stimulation of new technologies, an important mechanism in the negotiated agreement is that newly developed technologies become state-of-the-art for respective branches. Due to the long-term nature of the development and implementation of new technologies, there is only limited data. For the negotiated agreement of the printing industry, which has been in operation the longest, some data is presented below.

Table 2.4 provides information about the linkage between the workbook of the printing industry and technology projects funded by the Ministry of the Environment. This evidence shows that the state-of-the-art in the respective branches is continuously changing. New technologies that are environmentally superior to predecessors in the workbook of the printing industry are being adopted. The workbook provides the basic requirements under the license system, implying that companies need to adapt their performance to follow the evolution of state-of-the-art technologies. This mechanism clearly has the potential to stimulate the continuous improvement necessary to reach the targets set in the negotiated agreement.

An assessment of the kind of innovations that have been implemented, however, shows that most are of an incremental nature, and do not lead to radical change in the printing industry. The most radical innovation in table 2.4, digital printing, was not accepted for the workbook because it threatens more conventional methods of printing. This is also indicative of barriers for change within negotiated agreements. In the agreements the most powerful companies (and most widely used production technol-

Table 2.4
Linkage of developed technologies in technology program to the workbook of the printing industry

Total projects/technologies	18
Already entered in workbook	2
To be entered in workbook	4
Possibly in the near term to be entered in workbook (additional research needed)	9
Not to be entered in workbook	3

Sources: Memo by VNG to NOVEM and interview with program director NOVEM.

ogies) are strongly represented. Newcomers, or new technologies that may be developed outside the branch, are not part of the negotiated agreement. This implies that the consensual and Target Group approach will generally not be conducive to innovation of a more radical kind. In workshops with actors involved in covenanting processes, designed as part of a recent evaluation of Dutch covenants, this lack of realizing fundamental breakthroughs is confirmed (De Bruijn et al. 2003: 54). The same goes for changes that involve consumers or companies in the economic chain outside the specific industry that makes a product. As these external links are not involved in the negotiated agreement, the development of new production and consumption systems may be well beyond the scope of the negotiated agreements.

Conclusions

In this chapter we analyzed the effects and potential for innovation of the Dutch Target Group Policy for industry. At the beginning of this chapter we introduced several conditions or dimensions of policies that might be relevant to inducing innovation and its diffusion. Table 2.5 summarizes how the Target Group Policy incorporates these dimensions, and how the strategy differs from the traditional command-and-control approach. As is clear from the chart, these two approaches differ in all dimensions

Table 2.5
Differences between command-and-control and Target Group Policy

Dimension	Command and control	Target Group Policy
Degree of freedom of choice	Low	High
Cooperative nature	Low	High
Level of ambition	Variable	Variable
Time horizon	Short term	Medium to long term
Instrumentation	Legal/restrictive/enforceable	Financial, legal, social/ direction setting/more difficult to monitor
Addressees of policy	Generally individual companies/direct approach	Target group/network approach

Table 2.6
Results of Target Group Policy and limitations to its contribution to sustainable development along six dimensions

Dimension	Results	Limitations to contribution to sustainable development
Freedom of choice	Target groups have certain leeway for timing the implementation of measures, but "alara" is the bottom line for individual companies	It is not clear who will take the initiative to develop new technologies; expected market is still the main factor for inducing innovation. Some choices for new technologies and products do not bring sustainable development closer
Consensus seeking or cooperation	Intensive negotiations between target groups and government; regular meetings improve acceptance and basis for environmental policy making	Consultation takes place between existing companies with significant interests in current modes of production and that may impede more radical innovations
Stringency	Long-term targets (2010) are ambitious; short-term targets (1995) reflect state-of-the-art technology and tend to "business as usual"	All depends on the realization of necessary innovations. Targets are mainly emission targets and less focused on input/resource use, whereas this is a condition to progress toward sustainable development
Time horizon	Extension of the time horizon, especially for the mid- (2000) and long-term goals (2010), gives direction and some certainty for industries	Effective links with technology policy are needed in long term to develop new technologies and product-consumption linkages
Policy instrumentation	Mix of instruments works to some extent, as most target groups are on schedule for most targets, but the difficulty is how to change to strict enforcement in time when targets are not within reach	Mix of instruments must give enough pressure and scope to drive and facilitate industries on a path to sustainable development. Coordination between various policies in different policy areas is needed

Table 2.6
(continued)

Dimension	Results	Limitations to contribution to sustainable development
Addressees of policy	Associations and active companies are being reached, but it is difficult to reach laggard/defensive companies. Success of network approach depends on degree of organization of target group	Most goals of sustainable development need collaboration between various actors. Relation to consumption and other industrial sectors is limited, while much of the progress might come from changes in these linkages or technological development outside the branch

except level of ambition. The main conclusion is that on almost all dimensions the Target Group approach provides more of the conditions necessary to stimulate innovation.

In table 2.6 we summarize some results of the negotiated agreements in terms of the six dimensions. We also indicate some of the limitations of the new approach for delivering the kind of change that is needed for sustainable development. The success of the negotiated agreement correlates positively with the degree of organization of the target group, also confirmed in a recent evaluation of Dutch covenants (De Bruijn et al. 2003). This implies that if the target group is differentiated and not well organized, it will progress less easily through the negotiation process. For a relatively new sector, like ICT, the target group approach may be less suitable. This is particularly relevant because the ICT sector is responsible for a considerable increase in, for example, CO_2 emissions due to its fast growing consumption of electricity. The success of the target group approach therefore also depends on the setting and sector to which it is applied.

Within the agreement the target group can time the development and implementation of measures. However, it is still unclear whether this will result in more radical changes in the longer term. The consensual nature of negotiated agreements and the network of actors currently involved in them may promote innovations of an incremental rather than a radical nature because newcomers, or new technologies developed outside the branch, are not part of the negotiated agreement. On the positive side, the process of overcoming bottlenecks to significant improvement and of searching for new technological options leads actors to seek ideas from other actors and discover potential for collaboration. More research is needed on whether the consensual and target group approach is less conducive to innovation of a more radical kind.

For sustainable development the linkage between and within chains of production and consumption is crucial and needs to be reconsidered. A new generation of negotiated agreements that capture these dimensions may be able to further facilitate the contribution of industry to sustainable development. Some of the options available for integrating these considerations in the agreements include the rephrasing of the long-term targets.

In the original agreements these were captured in terms of emission targets (70 to 90 percent) or management tools (introduction of environmental management systems), but not in terms of product features. The advancement of green products may become another important target. Aspects of this idea are well within the scope of feasibility, as evidenced by the concept of green electricity in the Netherlands. Interesting enough, an environmental agreement between the electricity distribution sector and government also played a role in advancing this concept, together with a number of institutional and technological changes (Hofman 2002). Target Group Policy clearly has high potential to contribute to a larger constellation of policies that can facilitate progress toward more sustainable production and consumption patterns.

Notes

1. This took place through ten sessions with stakeholders from the various target groups where ideas were solicited on how to achieve the ambitious goals from the NEPP (De Jongh 1999: 101).

2. The target groups distinguished within the NEPP were the following: industry, agriculture, traffic and transport, construction, gas and electricity sector, drinking water companies, consumers and retail trade, refineries, and waste treatment/disposal companies.

3. Similar schemes have been presented by other authors, for instance, Norberg-Bohm (1999, 2000) focuses on favorable conditions for target groups to inno-

vate, while here the focus is on the effectiveness of policy incentives aimed at policy benign behavior.

4. The polder-model has historically evolved as a process of negotiations between employers organizations, unions, and government agencies over labor issues (wages, conditions). This model is increasingly used in environmental policy making in negotiations among target groups (e.g., industries, energy producers), environmental NGOs, and government agencies, and it has been labeled the "green polder-model."

5. FO Industrie is funded by the Ministry of Environment for facilitating the implementation of the Target Group Policy for industry.

6. This concerned wood preservation, brick and tile, leather, and other mineral products industries.

7. There is some discussion whether this holds for all branches in the negotiated agreements. Representatives from environmental NGOs argue that the chemical industry will be able to reach the targets without development and implementation of new technologies. But agreements for metal-plating and printing industries are expected to be more technology forcing if targets are to be met, especially for reduction of emissions of heavy metals to water and VOCs to air respectively. For the primary metals industry an analysis of technical and economic bottlenecks has revealed that remaining bottlenecks to reach targets for 2000 where predominantly of an economic nature and related to SO_2 and NO_x emissions.

8. See Overleggroep Basismetaalindustrie, Uitvoering intentieverklaring Basismetaalindustrie. Jaarrapportage 2001. Den Haag, November 18, 2002 (www.fo-industrie.nl, accessed on July 11, 2003).

9. See Overleggroep Chemische Industrie, Uitvoering intentieverklaring Chemische Industrie. Jaarrapportage 2001. Den Haag, December 11, 2002 (www.fo-industrie.nl, accessed on July 11, 2003).

10. See Evaluatie MBO Grafische Industrie en Verpakkings-drukkerijen 2000. Den Haag, December 17, 2002 (www.fo-industrie.nl, accessed on July 11, 2003).

11. See Overleggroep Papier- en Kartonindustrie, Uitvoering intentieverklaring Papier- en Kartonindustrie. Jaarrapportage 2001. Den Haag, February 24, 2003 (www.fo-industrie.nl, accessed on July 11, 2003).

12. See Overleggroep Zuivelindustrie, Uitvoering intentie-verklaring Zuivelindustrie. Jaarrapportage 2001. Den Haag, November 18, 2002 (www.fo-industrie.nl, accessed on July 11, 2003).

13. Representative of Stichting Natuur en Milieu (SNM) quoted in de Jongh (1999: 164).

14. Novem is the executive agency for this technology program.

15. This was done by an analysis of the project proposals submitted by the companies to the program committee.

16. This covenant covers the emissions of volatile organic compounds.

17. For companies, this was the influence on the specific technology projects; for experts, it was the general assumed influence of the negotiated agreements on innovative behavior in companies.

References

Arentsen, M. J., and P. S. Hofman. 1996. *Technologie, Schone motor van de economie?* (Technology, Clean Engine of the Economy?) Publicatiereeks milieustrategie 1996/16. Den Haag: Ministerie van VROM.

Ashford, N. A. 1993. Understanding technological responses of industrial firms to environmental problems: Implication for government policy. In K. Fischer K., J. Schot, eds., *Environmental Strategies for Industry*. Washington, DC: Island Press.

Ashford, N. A. 1996. *The influence of information-based initiatives and negotiated environmental agreements on technological change*. Paper for the international conference on Economics and Law of Voluntary Approaches, Venice.

Ashford, N. A., C. Ayers, and R. F. Stone. 1985. Using regulation to change the market for innovation. *Harvard Environmental Law Review* 9(2): 419–66.

Ashford, N. A., and G. R. Heaton. 1983. Regulation and technological innovation in the chemical industry. *Law and Temporary Problems* (Duke University School of Law) 46(3): 109–57.

Biekart, J. W. 1994. *De basismetaalindustrie en het doelgroepenbeleid industrie. Analyse van proces en resultaten op weg naar 2000*. Utrecht: Stichting Natuur en Milieu.

Biekart, J. W. 1998. Negotiated agreements in EU environmental policy. In J. Golub, ed., *New Instruments for Environmental Policy in the EU*. London: Routlege, pp. 165–90.

Bruijn, T. J. N. M. de, and K. R .D. Lulofs. 1996. *Bevordering van milieumanagement in organisaties. Een kwantitatief onderzoek naar beleidsvoering met doelbewust gebruik van beleidsnetwerken*. Ph.D. thesis. Enschede: Twente University Press.

Bruijn, T. de, H. Bressers, K. Lulofs, and A. van der Veer. 2003. *Evaluatie Milieuconvenanten*. CSTM-SR nr 2003/01. Enschede.

Dijken, K. van, Van Dijken, K. Y. Prince, T. J. Wolters, M. Frey, G. Mussati, P. Kalff, O. Hansen, S. Kerndrup, B. Søndergård, E. Lopes Rodrigues, and S. Meredith. 1999. *Adoption of Environmental Innovations: The Dynamics of Innovation as Interplay between Business Competence, Environmental Orientation and Network Involvement*. Dordrecht: Kluwer.

Glasbergen, P. 1998. Partnerships as a learning process, environmental covenants in the Netherlands. In P. Glasbergen, ed., *Co-operative Environmental Governance, Public-Private Agreements as a Policy Strategy*. Dordrecht: Kluwer, pp. 133–56.

Gray, B. 1989. *Collaborating: Finding Common Ground for Multiparty Solutions*. San Francisco: Jossey-Bass.

Harrison, K. 1999. Talking with the donkey: Cooperative approaches to environmental protection. *Journal of Industrial Ecology* 2(3): 51–72.

Hartman, C. L., P. S. Hofman, and E. R. Stafford. 1999. Partnerships: A Path to Sustainability. *Business Strategy and the Environment* 8(4): 255–66.

Hofman, P. S. 2002. Becoming a first mover in green electricity supply: Corporate change driven by liberalisation and climate change. *Greener Management International* 39: 99–108.

Jänicke, M. 1992. Conditions for environmental policy success: An international comparison. *Environmentalist* 12: 47–58.

Jänicke, M. 1997. The political system's capacity for environmental policy. In M. Jänicke and H. Weidner, eds., *National Environmental Policies, A Comparative Study of Capacity Building*. Berlin: Springer.

Jongh, P. E. de, with S. Captain. 1999. *Our Common Journey: A Pioneering Approach to Cooperative Environmental Management*. London: Zed Books.

Norberg-Bohm, V. 1999. Stimulating "green" technological innovation: An analysis of alternative policy mechanisms. *Policy Sciences* 32(1): 13–38.

Norberg-Bohm, V. 2000. Technology commercialization and environmental regulation. Lessons from the U.S. energy sector. In J. Hemmelskamp, K. Rennings, and F. Leone, eds., *Innovation-Oriented Environmental Regulation*. Berlin: Physica.

Porter, M. E. 1991. *America's green strategy*. Scientific American (April), p. 96.

Porter, M. E., and C. van der Linde. 1995. Green and competitive: Ending the stalemate. *Harvard Business Review* 73(5): 120–34.

RIVM (Rijksinstituut voor Volksgezondheid en Milieuhygiëne). 1988. *Zorgen voor morgen. Nationale milieuverkenning 1985–2010*. Alphen aan den Rijn: Samsom H.D. Tjeenk Willink.

Rycroft, R. W., and D. E. Kash. 1999. *The Complexity Challenge: Technological Innovation for the 21st Century*. London: Pinter.

Schrama, G. J. I., and W. van Lierop. 1999. *Nieuwe sturingsmodaliteit in het milieubeleid* (New steering modes in environmental policy). Report for Ministerie van Volkshuisvesting, Ruimtelijke Ordening en Milieubeheer (Dutch Ministry of the Environment). Den Haag.

Tweede Kamer der Staten-Generaal. 1989. *Nationaal Milieubeleidsplan 1990–1994. Kiezen of Verliezen*. Tweede Kamer, vergaderjaar 1988–1989, 21 137, nrs. 1–2. 's-Gravenhage: Sdu Uitgevers.

VROM and VNO-NCW. 1998. *De Stille Revolutie, Industrie en Overheid werken samen aan een beter milieu* (The Silent Revolution, Industry and Government jointly work for a better environment). Publication by industrial and government actors involved in the implementation of the Target Group Policy for industry.

3

Building Sector-Based Consensus: A Review of the US EPA's Common Sense Initiative

Cary Coglianese and Laurie K. Allen

From 1994 to 1998 the US Environmental Protection Agency (EPA) conducted what then-Administrator Carol Browner called a "bold experiment" in regulatory reinvention. The agency brought together representatives from six industrial sectors and sought to forge a consensus within each sector on innovations in environmental management and policy. In this chapter we examine the impact of EPA's experiment with this sector-focused, consensus-based approach to determine how well the EPA's Common Sense Initiative (CSI) achieved the agency's goals of improving technological innovation and environmental results.

The chapter begins by examining the structure and goals of CSI and then considers its relatively modest accomplishments to date. The chapter concludes with a discussion of a key factor that explains the Initiative's failure to achieve its most ambitious goals: EPA's reliance on consensus as a decision rule. By expecting CSI's advising bodies to achieve consensus before EPA would take action, the agency constrained its ability to spur sector-based technological change and achieve significant environmental improvements. Although consensus-based processes have been touted as innovative and promising strategies for regulators to pursue in environmental policy, often in the absence of clear legislative mandates, EPA's limited success with CSI illustrates some of the shortcomings of policy making by consensus and suggests the need for clear goals and legislative authorization in order to make significant regulatory change in the United States.

The Common Sense Initiative

EPA developed the Common Sense Initiative in the 1990s as part of the early Clinton administration's platform to "reinvent government" (Norberg-Bohm 1999). The agency has long faced criticism for the way it deals with environmental problems, including repeated claims that its regulations are burdensome, ineffective, and inefficient. EPA's regulatory activities, grounded principally in environmental legislation adopted in the 1970s, have been characterized as "command-and-control" strategies that are targeted at individual pollutants in individual media, such as air, land, and water. While the agency's past efforts have succeeded in reducing some environmental problems, they have been criticized for being too costly or for failing to achieve other goals, such as creating incentives for continuous environmental improvements or solving cross-media environmental problems. The agency's traditional approach has also frequently been criticized as being too adversarial, leading EPA to miss opportunities for purported win–win policy changes that could improve the environment at lower costs.

EPA launched CSI as the agency's "flagship program" to overcome these limitations attributed to traditional environmental regulation (Browner 1998). From July 1994 through December 1998 the agency attempted to use CSI to develop "cleaner, cheaper, and smarter" approaches to environmental pollution control. By bringing together industry, environmental groups, and other interested parties within each sector, the agency sought agreement on new and better ways of defining and achieving environmental performance goals.

How well did this bold experiment work? The agency has claimed that CSI represented "an innovative approach" and "a pathfinding forum," but the tangible results have been quite modest. CSI clearly had an "ambitious vision" (Fiorino 1996) and made a major commitment of agency resources, staff time, and support from agency leadership. However, nearly four years after the four-year Initiative came to an end, only about five of the approximately 30 subcommittee recommendations (amounting to about 45 distinct projects) that emerged from the CSI process have resulted in actual revisions to EPA regulations. Moreover relatively few of the project's accomplishments, according to the agency's own reports, have produced technological innovations, pollution prevention, or resulted in any other significant policy change. The majority of projects resulted only in the production of educational material or the collection of information.

Goals of the Common Sense Initiative

CSI was the "centerpiece" of the agency's reinvention efforts (GAO 1997). In announcing CSI, Administrator Browner heralded it as "probably the biggest new direction in environmental protection since the founding of the EPA" (Lee 1994). The agency predicted that CSI would "result in significant improvements to current regulations, as well as proposals for Congress to consider" (EPA 1994). Lofty and revolutionary, the overarching goals of CSI were to make dramatic changes that would result in "cleaner, cheaper, smarter" solutions to environmental problems. Upon launching the program, Administrator Browner promised bold action: "I don't think anyone in this country, whether environmental leader or corporate CEO, believes incremental steps will achieve the kind of future we all want" (EPA 1994).

The original advisory committee charter for CSI listed six objectives, or what EPA termed "programmatic elements," of the project (EPA 1996a):

1. Regulation. EPA aimed to review existing regulations for improvement opportunities, better environmental protection, and lower compliance costs.
2. Pollution prevention. EPA sought to integrate pollution prevention into standard business practice within each of the sectors.
3. Reporting and record keeping. EPA hoped to make record keeping easier for industry and more available to public.
4. Compliance and enforcement. EPA wanted to encourage companies to exceed minimal requirements, while enhancing enforcement against intransigent violators.
5. Permitting. EPA aimed to improve its permitting procedures by eliminating duplication and inconsistencies and enhancing public participation.
6. Environmental technology. EPA aimed to provide industry with incentives for adopting innovative technologies to reduce pollution and lower costs.

EPA renewed the original two-year charter of the CSI in 1996, and shortly afterward the CSI Council identified two additional goals for the Initiative:

7. Community involvement. EPA would promote community involvement in environmental management and policy making.
8. Emerging issues. EPA would identify future issues of concern within each sector (Kerr et al. 1999).

Launched in the same year that Philip Howard published his bestselling *The Death of Common Sense* (1994), CSI promised to cut through the senselessness of regulatory red tape.[1] Administrator Carol Browner claimed that CSI would lead to a "fundamentally different system" (EPA 1998), one that would make more sense both in terms of achieving environmental protection and reducing the cost and inconsistency associated with the existing system of environmental control. By focusing on specific industrial sectors, EPA sought to identify specific instances in which the existing regulatory approach hampered efforts to achieve sensible environmental improvements. The involvement of industry, government, and nongovernmental organizations in consensus-based deliberations was designed to take advantage of the insights of those working within each sector and lend legitimacy to the resulting recommendations and projects.

The Structure of CSI

CSI had a two-level structure, a Council and specialized subcommittees. The members of both groups were appointed by the EPA administrator for one-year renewable terms. These members included representatives from industry, national and local environmental organizations, environmental justice and community groups, labor unions, state, local, and federal governments. The Council consisted of approximately 30 members whose responsibility was to evaluate subcommittee project proposals and decide whether any recommended projects within each sector should be reported to the EPA administrator. The EPA administrator would consider those recommendations and if accepted would see that they were shepherded through the agency's normal procedures for implementation.

	1994	1995	1996	1997	1998
Metal refinishing	CSI	----------	----------	----------	Dec
Printing	CSI	----------	----------	----------	Dec
Iron and steel	CSI	----------	----------	----------	May
Petroleum refining	CSI	----------	----------	----------	Dec
Automobile manufacturing	CSI	----------	----------	Mar	
Computers and electronics	CSI	----------	----------	----------	Dec

Figure 3.1
Time line of CSI subcommittees

Six subcommittees reported to the Council, one for each industrial sector included in the project: (1) automobile manufacturing, (2) computers and electronics, (3) iron and steel, (4) metal finishing, (5) petroleum refining, and (6) printing. These subcommittees had the flexibility to do research, propose pilot projects, conduct preliminary information gathering, and recommend demonstration projects for consideration by the CSI Council. As figure 3.1 shows, all but one of the CSI subcommittees for these sectors met for the entire length of the Initiative, from 1994 to 1998. The subcommittee for one sector, metal finishing, actually was an extension of earlier efforts to reform regulation in this sector through EPA's Sustainable Industries Program.

Funded by various EPA program offices, CSI was established as an advisory committee in October 1994.[2] Meetings were conducted under standard advisory committee rules that require advance public notice of meetings, public access to meetings, the keeping of meeting minutes, and the opportunity for public comment. EPA retained the ultimate discretion and responsibility for implementing any CSI-recommended regulatory projects through its normal program offices and within the confines of existing law.

In addition to the six CSI Charter goals, CSI's operating framework consisted of consensus as the decision rule to be used by the Initiative's advisory committee (Norberg-Bohm 1999). Participants interpreted consensus to mean strict unanimity; this gave each player effective veto power (EPA 1997b). Several years into the process, EPA recognized the difficulties posed by strict unanimity as a decision-making principle. In

1996 the agency responded to these difficulties by modestly redefining consensus as something that would be reached "when all Council members at the table can accept or support a particular position, even though the position may not be their first choice" (EPA 1996b). Even under this definition, consensus as a decision rule demanded that all the interests represented within CSI needed to reach agreement on new proposals or projects.

In 1997, midway through the Initiative, two reviews of CSI identified a number of problems associated with EPA's use of consensus as a decision rule. The first review was conducted by the Scientific Consulting Group (SCG) at EPA's request; the second was undertaken by the US General Accounting Office (GAO). In response to these reviews, EPA again re-evaluated its operating definition of consensus.

In a white paper on its consensus–decision-making principles (EPA 1997a), EPA argued that consensus was intended to bring out participants' underlying interests and stimulate creative problem solving. EPA urged the CSI Council, as well as CSI subcommittees and workgroups, to "try to reach full agreement on as many substantive and procedural issues as possible," but allowed the Council to evaluate different member opinions that were submitted separately. The agency also allowed subcommittees and workgroups to determine if a project could go forward without a consensus (EPA 1997a). Some subcommittees, such as the one for the automobile sector, required strict unanimity, while others, such as the metal finishing subcommittee, followed a more informal interpretation (Kerr et al. 1999).

What Did CSI Accomplish?

Each of the six sector subcommittees charted its own course toward a "cleaner, better, cheaper" regulatory system. EPA established broad goals along a number of dimensions of policy concern, and each committee autonomously tried to achieve consensus on more specific performance measures and ways of achieving CSI's goals. The results, perhaps not surprisingly, varied across the several sectors, with over 40 diverse projects emerging from all the sectors taken together. These projects ranged from the development of compliance manuals for small firms to

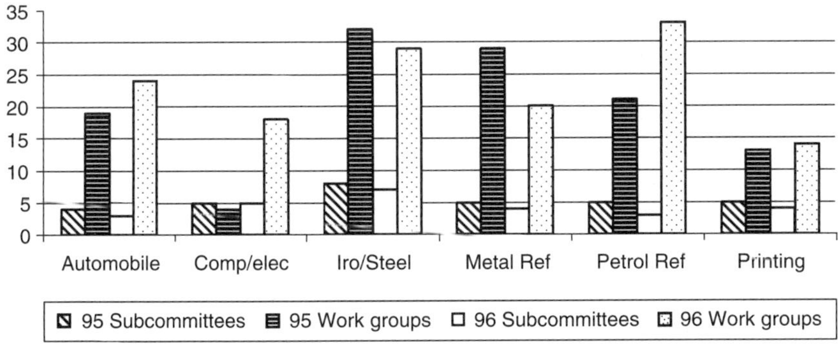

Figure 3.2
Meetings of subcommittees and work groups in 1995 and 1996

the development of voluntary environmental performance targets. This section first reviews the scope of the efforts within each subcommittee, and then provides an overall account of the accomplishments of CSI.

CSI Subcommittee Results

The sectors chosen for CSI varied. Some sectors consisted of a few large firms (e.g., automobile manufacturing), while others included a large number of small firms (e.g., the metal finishing and printing). Some sectors represented long-standing industrial activities (e.g., petroleum refining and iron and steel), while one sector reflected the industrial activity of the new information-based economy (computers and electronics). Figure 3.2 shows the number of meetings held for each CSI sector held during two years of the Initiative.

Metal Finishing The metal finishing industry provides parts that are used in almost every manufacturing process. It is composed of more than 3,000 small job shops and small businesses with limited capital and personnel, and with industry operations that affect the environment across air, land, and water. In addition the sector includes more than 8,000 metal finishing operations that are part of larger manufacturing firms.

Metal finishing has sometimes been characterized as the most successful of CSI's sectors. Unlike the other CSI sectors, the metal finishing industry began working closely with EPA in 1990, developing projects through the Sustainable Industries Initiative (Kerr et al. 1999). The technical studies conducted during this earlier project, as well as the experience in collaborating together, seem to have enabled the metal finishing sector to work more quickly and effectively in the Common Sense Initiative.

The metal finishing subcommittee pursued 13 projects that addressed each of the dimensions of CSI's main goals. Projects were designed to provide incentives for firms to go beyond compliance, but attention was also given to dealing with shops that are habitually out of compliance. Recognizing that in many cases no alternatives exist to substances used in metal finishing, one project focused on reducing waste through improved operations or recycling techniques. The subcommittee developed strategies for testing innovative technologies, such as composite mesh pads and chemical fume suppressants to reduce chromium pollution. Other projects included information and outreach activities, the creation of a regulatory team that addressed permitting and compliance issues, and an incentives program that would reward top performing companies with pollution prevention assistance and enforcement relief.

The Strategic Goals Program (SGP) that emerged from the metal finishing subcommittee in 1998 is cited as one of the most successful CSI projects (Kerr et al. 1999; EPA 1999). This voluntary program set clear national performance targets for facilities that, if met, would eventually result in performance by participating firms that exceeded compliance. Participating firms were to commit voluntarily to reducing the amount of metal disposed of as waste by 98 percent, the amount of water used by 50 percent, and the amount of energy used by 25 percent, all compared to a 1992 baseline. In addition firms were expected to commit to making a 90 percent reduction in emissions of organic chemicals and a 50 percent reduction in the land disposal of hazardous sludges, again compared with 1992 levels.

The overall goal of the SGP was to have 80 percent of the metal finishing facilities in the country achieve these reductions by 2002. By the end of the program, about 510 firms, or only about 15 percent of the independent shops in the industry, had agreed to participate in this program (EPA 2003a), although even fewer have remained active.[3] Nevertheless,

EPA claims that the program has resulted in significant reductions in pollution levels compared with levels reported by SGP firms for 1992, including a reduction of about two billion gallons of wastewater, five million pounds of organic chemical emissions, and 250,000 pounds of metals released into water (EPA 2003a).

While these absolute numbers sound impressive, in relative terms the overall environmental improvements attributable to SGP were probably smaller. According to the latest SGP progress report available on the metal finishing industry's Web site, SGP firms nearly achieved the program's goals for their reductions in organics emissions, but they only came about halfway to achieving other of the program's goals (SGP 2001a). Of the approximately 300 SGP companies included in the progress report, about 130 were reported to be making no progress at all on reducing energy use or reducing the generation and shipment of sludge. Moreover other data available on the industry's Web site indicate that when overall environmental impacts are normalized by sales, participating firms showed little progress since 1998 in all areas, and they even increased their impacts modestly in terms of energy use and shipments of sludge (SGP 2001b, c). The most progress reported by the participating firms occurred prior to the launch of SGP in 1998. At least with the data that the industry has made publicly available, it is difficult to conclude that the Strategic Goals Program has led to substantial environmental improvements, whatever else it may have achieved.

Printing The printing industry consists of more than 70,000 small business print shops diffused throughout the country and employing different printing processes. CSI's printing sector subcommittee pursued two projects during its four-year life span. The first project provided educational outreach to printers in New York City, informing them of pollution prevention measures they could use in their businesses and developing a technical assistance directory for printers in the city. The second project aimed to develop a more flexible, integrated system of issuing environmental permits, with incentives built into the permitting process to encourage firms to achieve a high level of environmental performance. This alternative permitting process has so far been piloted in only a few states.

Iron and Steel The iron and steel industry consists of more than 1,000 facilities making and processing steel, with firms concentrated primarily in the Great Lakes region. This subcommittee addressed six of CSI's goals and pursued 12 projects. One of the subcommittee's projects involved the creation of a Web site designed to make iron and steel firms aware of technologies to improve environmental performance. Another involved the convening of a workshop on ways to reduce spent pickle liquor wastes. Several other projects sought improvements in permitting and reporting requirements, resulting in contributions to the EPA's Permit Reform Action Plan, which EPA approved in early 1999. The iron and steel subcommittee recommendations were included in the "Cross Media Tasks" matrix (EPA 1999).

The sector's biggest project, brownfields redevelopment, was designed to stimulate communities to bring polluted sites back into productive use, but it organized only two pilot projects, one in Alabama and one in Indiana. Only one project resulted in a regulation change, and this was a minor rule revision related to monitoring furnace pressure. Although this sector's subcommittee held over 100 meetings over the course of four years, its resulting projects were quite limited in scope and impact.

Petroleum Refining The petroleum refining industry consists of over 160 operating large and small petroleum refineries, concentrated principally along the Gulf Coast and in heavy industrialized areas on the east and west coasts. The petroleum refining subcommittee's goals were broad and ambitious, focusing on regulation, permits, compliance, reporting, pollution prevention, and environmental technology. Yet, considering the large number of meetings of this subcommittee and its work groups, progress by this committee over the four years turned out to be rather limited.

The petroleum-refining sector subcommittee worked on only three projects. One project studied the air pollution reporting requirements imposed on the sector. This project involved analysis of the impacts on a single refiner in Texas. Contrary to expectations, this study found, according to EPA (1999), that "the reporting requirements did not contain as much duplication as originally anticipated." Nevertheless, the subcommittee still worked to develop an alternative refinery air reporting system. Another project aimed to reduce the pollution associated with leaks from refinery equipment and resulted in the planning of a voluntary program to encourage firms to prioritize their monitoring of equipment for leaks. This led to a third project that developed laser leak-detection technology that, according to EPA's latest reports, is still being tested.

Computers and Electronics The computers and electronics industry is a comparatively "clean" industry when measured against traditional manufacturing. Nevertheless, it does have some significant environmental and occupational health and safety impacts. The subcommittee worked on 12 projects to develop improved strategies for reporting and public access to environmental information, overcome barriers to pollution prevention, encourage product stewardship and recycling, create alternative strategies for environmental management, and coordinate between environmental and workplace health policies.

The one project directed at environmental technology was structured to promote the creation of zero wastewater discharge systems, but was hindered by lack of clarity about how these systems could be incorporated into the current legal framework. Direct environmental results can probably be attained from the subcommittee's cathode ray tube (CRT) recycling and sulfuric acid recycling projects, but preexisting regulations required recycled CRT glass and used acid to be treated as hazardous waste. The subcommittee's work did result in rule revisions to RCRA provisions for hazardous waste for CRT recycling to address this problem. The remaining projects from this sector addressed reporting, information availability, and record keeping.

Automobile Manufacturing The CSI subcommittee for automobile manufacturing set out initially to investigate how communities and the auto industry can interact better, look for alternatives to the existing regulatory system, and identify ways that industry could gain flexibility while reducing cost and maintaining environmental standards. Notwithstanding these goals, the automobile sector is typically considered CSI's biggest failure. This sector had a large number of meetings, worked on three projects, but achieved the least results of any sector, only addressing at best two of the eight CSI program goals.

The projects completed by the automotive subcommittee were little more than inventories and reports, database development, and a meager rule change recommendation. The proposed rule change addressed the mass-per-unit-area approach for total vehicle coating, basically seeking to provide information in a way that is both more understandable and consistent with international regulations. The automobile sector subcommittee concluded its efforts in 1997, a year before the other CSI subcommittees.

Summarizing Overall Results

As this review of CSI subcommittees suggests, the Initiative involved many meetings that resulted in a number of projects. In most sectors, however, these projects failed to achieve results commensurate with EPA's ambitious goals. The Initiative has been reviewed twice by outside consultants commissioned by EPA. As noted above, the Scientific Consulting Group conducted a two-year review (Todd 1997) while CSI was still in progress. The US Government Accounting Office completed a review of CSI (GAO 1997) while the Initiative was in progress. After CSI had ended, the firm of Kerr, Greiner, Andersen, and April, Inc. conducted a review (Kerr et al. 1999).

Taken together, these reviews suggest that CSI was generally tall on ambition but short on meaningful and measurable accomplishment. One subcommittee—metal finishing—composed of smaller businesses made some progress on most of the CSI goals, but most of the other subcommittees achieved much less. As one commentator observed about CSI and other EPA attempts at innovation, "despite the labor and resource-intensive nature of reinvention activity over the past several years, the tangible results of these experimental efforts have been widely viewed as disappointing" (Case 2001).

As table 3.1 shows, the number and magnitude of projects coming out of four years of subcommittee work were rather small, considering the time and resources devoted by the individuals involved in the subcommittees. According to an estimate given to us by an EPA staff member involved in CSI, the agency devoted approximately 50 to 60 FTEs to CSI, while participants from outside the agency undoubtedly devoted still more time and resources. Table 3.1 indicates that most of CSI's

Table 3.1
Summary of CSI projects by CSI goal

Sector projects	Regulation	Pollution prevention	Records/ reports	Compliance/ enforce-ment	Permits	Environ-mental tech-nology	Involving commu-nities	Future issues	Total
Auto	2	1	0	0	0	0	0	0	3
Computers/ electronics	5	3	3	0	0	0	1	0	12
Iron/steel	1	0	1	4	2	1	3	0	12
Metal finishing	1	7	1	1	0	1	1	1	13
Petrol	0	1	1	0	0	0	1	0	3
Printing	0	2	0	0	0	0	0	0	2
Total	9	14	6	5	2	2	6	1	45
Percentage of projects	20.0	31.1	13.3	11.1	4.4	4.4	13.3	2.2	

Source: EPA (1999).

projects were primarily intended to address the goals of regulatory reform and pollution prevention (51 percent combined), but this does not mean that CSI actually achieved this level of results.[4] Kerr et al. (1999) reported that only five projects were complete at the end of CSI and found only eight projects that, if implemented, would be expected to have some level of direct environmental effects.

In order to make our own assessment of CSI, we coded the 45 CSI projects according to the modalities by which the subcommittee worked to achieve the project goals. These modalities, or project types, included (1) education (e.g., directories, reports, Web sites), (2) research and information collection (e.g., databases, technical analyses, reports for data collection), (3) recommended policy change (e.g., regulations, new permit process), (4) development of new technology, (5) voluntary industry action, and (6) discussion/airing of views.

As table 3.2 shows, 47 percent of the projects consisted only of information and data collection, and an additional 24 percent resulted in reports; few of these reports were used to implement any innovative programs with direct environmental results.[5] Nearly three-quarters of CSI's projects aimed only to conduct research or provide some form of educational outreach. Moreover no more than about one-tenth of the total projects were completed by the end of CSI.

In addition only two CSI projects promoted innovations in environmental technology, neither of which was completed by the end of CSI. By that time the handful of subcommittee recommendations endorsed by the council and submitted to EPA had resulted in the agency taking

Table 3.2
Project modalities

	Percentage of total projects
Research/information collection	47
Education	24
Recommended policy change	20
Discussion/airing of views	4
Development of new technology	2
Voluntary industry action	2

steps to propose or issue only four new rules or revisions to existing regulations: (1) amendments to permit requirements for the iron and steel sector, (2) a rule making on a mass-per-unit-area approach for automobile coating, (3) a rule change to extend the accumulation requirement for metal finishing waste to promote on-site recovery, and (4) a proposed rule making to streamline requirements for CRT recycling.

Post-CSI Sector Projects

Although CSI came to an end in 1998, three out of six CSI subcommittees (metal finishing, printing, and petroleum refining) were incorporated for a time as part of a new sector committee under the agency's larger National Advisory Committee on Environmental Policy (NACEPT). Meeting for the first time in April 1999, NACEPT's sectors committee sought to provide EPA a way to continue to receive stakeholder input toward a sector-based approach to environmental problems and incorporate the information into EPA's core functions (EPA 2000).

In addition, before disbanding, the CSI Council prepared a "Sector Action Plan" that would serve as a basis for further work. The 2000 Sector Action Plan proposed continuing projects in six categories that were similar to CSI goals: (1) permitting, (2) enforcement and compliance assurance, (3) rule making, (4) solving regional problems, (5) building voluntary partnerships with the private sector to improve environmental performance, and (6) research and applications of science. Table 3.3 provides a summary of projects in the FY 2000 Sector Action Plan for CSI projects that continued (EPA 2000). None of these projects were new, but some had moved from the discussion stage to pilot projects. One additional regulatory proposal, on zero wastewater discharge systems for the computer and electronics sector, was initiated in 2000. As of two years after the end of CSI, as many as 30 uncompleted CSI projects were apparently still in progress through the NACEPT Sector Program or through internal EPA development (table 3.3).[6]

During the Bush administration, the EPA continued to pursue sector-focused efforts. The agency's goals shifted away from the kind of dramatic changes that CSI had been intended to achieve toward a more limited set of objectives. Proposed in September 2002 and launched in May 2003, EPA's new Sector Strategies Program (SSP) aimed to

Table 3.3
Post-CSI projects—Sector Action Plan 2000

Sector Action Plan program goal	Auto-mobile	Computers and electronics	Iron and steel	Metal finishing	Petro-leum	Printing	Total
Permit	0	0	2	0	0	1	3
Rule making	1	3	1	1	0	0	6
Enforcement and compliance assurance	0	1	2	3	1	0	7
Solving regional problems	0	0	1	0	0	0	1
Building voluntary partnerships	1	2	2	3	0	0	8
Research and scientific applications (pollution prevention, technology)	0	0	0	4	1	0	5
Total	2	6	8	11	2	1	30

Source: EPA (2000).

(1) increase the use of environmental management systems, (2) reduce "government-imposed" barriers to environmental improvement, and (3) measure performance outcomes (EPA 2003b). As of mid-2003, SSP involved a total of twelve sectors, including such industries as agribusiness, higher education, and seaports (EPA 2003c). The only CSI sectors to be associated with SSP were the metal finishing and iron and steel manufacturing sectors.

Assessing the CSI Experience

Administrator Browner envisioned CSI as a means for creating a "fundamentally different system ... [through] a pathfinding forum for breaking through some of the biggest constraints associated with the current environmental regulatory system—the use of single media approach to environmental protection and the adversarial relationships that have built up among stakeholders" (EPA 1998; 2001). As an innovative approach to dealing with the complexities of environmental regulation and management, CSI's sector-based, consensus-driven concept may have held some intuitive appeal, but the changes it produced have been far from significant (Kerr et al. 1999).

Measuring Success

Although EPA described the Common Sense Initiative as an experiment, it was not established in a way that would enable the agency to assess the impact of the program with rigor. Evaluating the impact of the Initiative calls for more than simply listing the various projects of each sector subcommittee. To assess its impact on environmental policy, it should be appropriately compared with other efforts.

Some CSI projects presumably affected the behavior of industrial actors and perhaps even resulted in cost savings and environmental improvements. The key question for evaluation is whether CSI stimulated changes that were more significant than would have or could have been achieved without it. In other words, what difference did CSI make? The answer to this question depends in part on the counterfactual, or what would have happened in the absence of the program (Coglianese 2002). If one were to assume that nothing at all would have been accomplished in the absence of CSI's consensus-based, sectoral approach, then its impact would presumably be viewed as positive. Such an assumption would, however, be unrealistic for two reasons.

First, EPA had pursued other, less prominent sector-based initiatives in the past, such as the Design for the Environment project, Sustainable Industries project, and the Cluster Program. These other initiatives had goals similar to CSI's and they too resulted in some limited projects. These earlier efforts tended to involve only the affected industry and EPA, and seem to have encountered some resistance from environmental groups. Had EPA continued these programs for four additional years, perhaps expanding them to other sectors, it might have achieved the same kind of results that followed CSI. In fact the one CSI sector that appeared to others to have accomplished the most, metal finishing, had been one of three sectors included in EPA's Sustainable Industries Project. It is reasonable to consider how much of this sector's productivity should be attributed to CSI and how much to the head start it made during the Sustainable Industries Project. The Kerr Report (1999) indicates that many participants in the metal finishing subcommittee credit the Sustainable Industries Project for providing the foundation on which the CSI subcommittee achieved its successes. With a comparable amount of effort over four additional years, the metal finishing group within the Sustainable Industries project might well have achieved as much as, if not perhaps more than, the CSI metal finishing subcommittee accomplished.

Second, not only might EPA have achieved similar accomplishments had it continued its pre-CSI sector-based programs, but it also might have brought about more environmental improvement had it devoted a comparable amount of staff effort to exercising its traditional regulatory authority. CSI demanded many hours of EPA staff time in managing the subcommittees and working on projects. Compared with the nearly 300 final regulations the EPA issues each year, CSI stimulated only four proposed regulatory changes over the course of four years' time—one-half of 1 percent of all the rules issued during that time. This is not a significant accomplishment. EPA may well have generated more environmental improvement by devoting the same resources it invested in CSI to the development of even a single regulation, if that rule forced firms to develop promising new technologies or achieve substantial new milestones in environmental protection.

EPA did not develop a way to measure the environmental impacts associated with the Common Sense Initiative that could be used to compare its outcomes with those of other regulatory efforts. It would be quite difficult to develop such measures, if only because most of the CSI projects were educational or research efforts only indirectly connected with environmental improvement (table 3.2). However, even in the absence of such measures, the counterfactual scenario—what would have happened in the absence of CSI—was probably not an empty set. In other words, had EPA not pursued CSI, the agency would undoubtedly have taken other actions to seek environmental improvement. One cannot reasonably conclude that CSI had an impact merely because it successfully completed some projects.

Even if CSI had an impact that went beyond what alternative courses of action would have achieved, the objective of CSI was not simply to have such a greater impact. Instead, CSI was conceived as a groundbreaking program, one that sought to overcome the limitations of the current media-specific system of environmental regulation and to chart a new course for the future (EPA 1998). The hope among EPA staff was to craft new approaches that would achieve integrated environmental management and better environmental results, all at a lower cost. When evaluated against these original aspirations that the program would bring about fundamental change, CSI clearly was not a successful initiative. The Kerr (1999) report indicates that few participants "felt that they had succeeded in addressing issues of the scope they had anticipated at the outset of CSI. CSI made very little progress in addressing broad regulatory changes."

For example, the flagship program of the metal finishing subcommittee was its Strategic Goals Program (SGP), hailed by some as the most significant of all the CSI projects (Kerr et al. 1999). Even if SGP achieved some results that might not otherwise have occurred, its overall impact falls far short of achieving the kind of regulatory transformation EPA anticipated at the outset of CSI. Moreover, while participating companies and their improvements in environmental performance can be applauded, some in the metal finishing sector probably viewed the SGP as a means of staving off impending regulation (Kerr et al. 1999).[7] If the limited voluntary efforts associated with SGP are compared with the likely impact of a new environmental regulation that would have covered all 3,000 firms, instead of just the small fraction who participated in SGP, the environmental impacts of SGP can hardly seem all that significant.

The most cited accomplishment of CSI overall has been its positive impact on relationships between government, industry, environmental groups, and the other organizations involved in CSI discussions (Davies and Mazurek 1996; Todd 1997; Kerr 1999). While this may well be a noteworthy accomplishment, it could be considered little more than a post hoc justification for four intensive years of meetings. In themselves, improved organizational relationships do not translate into direct effects in terms of improvements in environmental regulation, economic efficiency, or improved environmental conditions, the original goals of the Initiative. Moreover almost any intensive group process involving several years of deliberation could claim some credit for helping people learn more about each other and about how to work with each other. In some cases CSI clearly did not even achieve this goal. The automobile sector had a history of adversarial relationships with stakeholders and the CSI consensus process did little to improve them.

Lessons of CSI

CSI faced fundamental limitations on the degree of change it could produce, limitations that resulted, on the one hand, from the parameters specified in current environmental legislation and, on the other, from the kind of consensus-based, "multi-stakeholder" process EPA employed in an effort to overcome problems in the existing system. The key lesson to be gleaned from the Common Sense Initiative is that consensus-based processes are ineffective means of overcoming perceived limitations in existing statutory law. Overcoming any such limitations will require more than just consensus-building in the administrative process. They will require statutory change, however difficult this may be to secure from the legislature.

CSI was designed to overcome the media-specific biases in existing law that inhibit firms from managing their overall environmental impacts in an integrated fashion. The current system of environmental regulation in the United States, as noted earlier, is usually criticized for failing to confront trade-offs and spillover environmental effects across air, water, and groundwater (Aspen Institute 1996; Ruckelshaus and Hausker 1998; Davies and Mazurek 1996; Esty and Chertow 1997). The EPA hoped that CSI would provide more focused improvements to environmental policy based on the specific needs of the industrial sectors included in the Initiative, rather than appearing to take the "one size fits all" approach for which EPA has often been criticized (Fiorino 1996).

Unfortunately, because the problems CSI was designed to address were embedded in underlying environmental statutes, EPA came up against some significant limitations in what it could accomplish through administrative actions. EPA cannot change these statutes; in fact they are written to constrain and direct the kinds of actions EPA takes. Statutes not only direct EPA's priorities (Landy, Roberts, and Thomas 1994) but also stipulate numerous judicially enforceable deadlines that the agency must meet and often specify the regulatory strategies EPA must use in great detail.

Existing environmental statutes limit EPA's opportunities to waive statutory requirements, and they provide no catchall provision granting EPA authority to develop alternative regulatory systems. Moreover most environmental statutes authorize citizen suits against firms that fail to comply with the letter of the law, meaning that even if EPA did grant firms exemptions or waivers from existing requirements these same firms might still be pursued in court by environmental organizations.

Faced with these kinds of legal limitations, EPA tried to use sector-based consensus building to develop innovative new approaches through the Common Sense Initiative. By forging agreement across varied interest groups, the agency hoped to create a degree of legitimacy around the projects that emerged from CSI. Legitimacy, after all, is sometimes thought to be enhanced by consensus processes (Freeman and Langbein 2001). EPA officials hoped that consensus would ensure that the innovations adopted by CSI would unify political support, thus increasing the likelihood that a wide range of actors would view its outcomes as sensible and implement them without much challenge.

Although consensus-building has long been viewed as having this kind of strategic advantage, in practice consensus-building faces significant limitations (Coglianese 2001a). In fact CSI's relatively tepid outcomes can be largely attributed to the limitations of consensus as a strategy for making policy decisions. There was simply no way that EPA could have accomplished through consensus-building anything remotely as dramatic as what it originally set out to accomplish.

Consensus-building relies on agreement, usually unanimous agreement, among participants in a policy-making process. As defined in CSI's original statement of operating principles, consensus would be "reached when all Council members at the table can accept or support a particular position, even though the position may not be their first choice" (EPA 1996a). By February 1997 an evaluation showed that subcommittees were often implementing the consensus standard in a way that required unanimity and that, as a result, CSI participants expressed dissatisfaction with the resulting delays (Todd 1997). As Davies and Mazurek (1996) reported toward the midpoint of CSI, "trying to persuade all parties to come to an agreement has proven so problematic that the agency has considered moving away from a strict interpretation of the term consensus." While EPA continued to support consensus as the preferred approach throughout the duration of CSI, urging subcommittees to "attempt to reach full agreement on as many substantive and procedural issues as possible," the agency did subsequently relax its decision rule to permit projects to go forward even in the absence of full agreement (EPA 1997a).

As is evident in international relations where consensus operates as a decision-making norm, building consensus is not easy. It is difficult to find any broad group of individuals with divergent interests who can come to agreement on major policy issues, especially when they arrive at the table with different opinions, assumptions, and value commitments. For this reason the process of building consensus over policy innovations can take more time and demand more resources than processes not based on consensus (Coglianese 1997; 2001b). In addition the outcomes of consensus processes tend to focus on the most tractable, and often least important, policy problems; to rely on agreement over imprecise or general principles rather than on concrete operational results; and to reflect what amounts to a lowest common denominator of all the participating interests (Coglianese 2001a).

The results of EPA's Common Sense Initiative reflect these limitations. For example, CSI resulted in a number of narrow, relatively tractable projects, instead of the ambitious redesign of the regulatory system announced by EPA at the outset of the Initiative or as reflected in EPA's goals for the Initiative. The Kerr Report (1999) indicates that as time went on, participants in CSI reduced their expectations about what CSI could achieve, choosing to work on areas in which agreement was possible, such as training manuals, case studies, and public outreach, and not necessarily on those areas most in need of fixing (table 3.2). These informational and administrative projects had narrow impacts that failed to generate the kind of conflicts that might have arisen over more ambitious policy efforts.

Nothing dramatic resulted from CSI because the agency chose to pursue its agenda through consensus. In the case of environmental protection, which requires firms to internalize the social costs of their activities, industry players from each sector are unlikely to agree to changes that would require costly new technologies, at least not without some impending threat of government regulation (Caldart and Ashford 1999). After all, innovation usually comes with risks—both to the firm and to the government. As a result it should not be surprising that CSI achieved very little in the way of promoting new environmental technologies.

Furthermore, because the EPA needed to seek consensus in the absence of clear regulatory authority, most of the projects that emerged from CSI were strictly voluntary. The metal finishing sector's Strategic Goals Program, for example, was designed to encourage firms to make environmental improvements that exceeded existing regulatory requirements. SGP may well have led some firms to achieve environmental improvements that they would not otherwise have achieved, but it seems likely that the SGP was more attractive to firms that were already committed to maintaining a solid environmental record and perhaps had already achieved results that went beyond compliance with existing requirements. Any policy program that relies on attracting volunteers runs the risk of attracting mainly the kind of participants that the program least needs to address (Coglianese and Nash 2001).

Conclusion

CSI's focus on fine-tuning environmental regulation to the specific circumstances of different industrial sectors may hold the potential for developing more sensible and effective methods of environmental protection. However, CSI's objectives were pursued under the constraints of consensus as a decision rule, and consequently the Initiative struggled to achieve much of significance. CSI's ultimate failure to transform the existing regulatory system, or even to result in significant innovations, stems from the limitations of consensus-building, especially in the face of statutory constraints on regulatory change.

The chief lesson to be learned from CSI appears to be that fundamental change in a regulatory system that is governed by a highly detailed set of statutes will come about neither without changing those statutes nor through consensus. As Vicki Norberg-Bohm (1998) observed about similar kinds of initiatives aimed at so-called green design and manufacturing, "expecting these initiatives alone to lead to extensive private sector efforts in environmentally conscious design and manufacturing is truly asking the tail to wag the dog. Legislation which provides stronger incentives over an appropriate time frame will be needed." Deliberative, sector-based efforts such as CSI may well serve a useful purpose of generating some new ideas, making incremental changes, or providing feedback to those involved in the regular policy process, but we should not expect that consensus-building will provide the route to a fundamentally "cleaner, cheaper, and smarter" regulatory system.

Acknowledgments

The authors acknowledge helpful comments from Darryl Banks, Theo de Bruijn, Daniel Fiorino, Shelley Metzenbaum, Vicki Norberg-Bohm, and Christian Richter. The views expressed in this chapter are those of the authors and not necessarily those of the commentators or of the institutions with which the authors are affiliated.

Notes

1. When originally conceived, CSI was called the "Green Sectors Project" but was renamed to fit the government reform rhetoric of common sense. For a further example of this rhetoric, see Gore (1995).

2. See Federal Advisory Committee Act (FACA), Pub. L. 92-463, Oct. 6, 1972; 86 Stat 770 as amended by Pub. L. 94-409, Sec 5(c) Sept. 13, 1976.

3. Although more than 500 firms were "part of" SGP, apparently about 115 of these firms never submitted any data on their environmental progress (http://www.strategicgoals.org/reports2/review.cfm?state=all&requesttimeout=200, accessed September 2, 2003). Strikingly, the National Metal Finishing Strategic Goals Program's Web site provides facility reports for the year 2002 from only about 130 facilities (http://www.strategicgoals.org/02cards/card.cfm, accessed September 2, 2003).

4. Some projects addressed more than one goal, but adding the secondary project goals does not dramatically affect the distribution of projects reflected in table 3.1. With secondary goals added, the distribution of projects by goal is as follows: 19 percent regulation, 22 percent pollution prevention, 8 percent recordkeeping/reporting, 11 percent compliance and enforcement, 8 percent permits, 12 percent environmental technology, 14 percent involving communities, and 5 percent future issues.

5. As with the CSI project goals, adding in the secondary modalities or project types does not appreciably affect the distribution of projects among the different modalities (46 percent research/information, 30 percent education, 15 percent policy change, 4 percent discussion, 3 percent new technology, and 1 percent voluntary industry action).

6. The projects in the three subcommittee columns that are not continuing as FACA committees under NACEPT (automobiles, computers, and electronics, and iron and steel) were in progress at the end of CSI, though formal meetings no longer continued to occur.

7. Metal finishing firms were facing brownfields issues and effluent guidelines that, in combination with the general need for flexibility for small business, could have helped provide the impetus for industry involvement in SGP.

References

The Aspen Institute. 1996. *The Alternative Path: A Cleaner, Cheaper Way to Protect and Enhance the Environment.* Aspen, CO: The Aspen Institute.

Browner, C. 1998. Remarks Prepared for Delivery at the American Electroplaters and Surface Finishers Society Conference on Pollution Prevention and Control, Orlando, FL. http://yosemite.epa.gov/administrator/speeches.nsf/0/3af66b32ccd4ff698525659f006c24fc?OpenDocument.

Caldart, C., and N. Ashford. 1999. Negotiation as a means of developing and implementing environmental and occupational safety and health policy. *Harvard Environmental Law Review* 23(1): 141–202.

Case, D. W. 2001. The EPA's environmental stewardship initiative: Attempting to revitalize a floundering regulatory reform agenda. *Emory Law Journal* 50(1): 1–100.

Cavanaugh, S. M., R. W. Hahn, and R. N. Stavins. 2001. National environmental policy during the Clinton years. Regulatory Policy Program Working Paper RPP-2001-10. Center for Business and Government, John F. Kennedy School of Government, Harvard University, Cambridge.

Coglianese, C. 1997. Assessing consensus: The promise and performance of negotiated rulemaking. *Duke Law Journal* 46: 1255.

Coglianese, C. 2001a. Is consensus an appropriate basis for regulatory policy? In E. Orts and K. Deketelaere, eds., *Environmental Contracts: Comparative Approaches to Regulatory Innovation in the United States and Europe*. London: Kluwer Law International.

Coglianese, C. 2001b. Assessing the advocacy of negotiated rulemaking. *New York University Environmental Law Journal* 9: 386.

Coglianese, C. 2002. Empirical analysis and administrative law. *University of Illinois Law Review* 2002: 1111.

Coglianese, C., and J. Nash. 2001. Environmental management systems and the new policy agenda. In C. Coglianese and J. Nash, eds., *Regulating from the Inside: Can Environmental Management Systems Achieve Policy Goals?* Washington, DC: Resources for the Future Press.

Davies, T., and J. Mazurek. 1996. *Industry Incentives for Environmental Improvement: Evaluation of US Federal Initiatives*. Washington, DC: Global Environmental Management Initiative.

Davies, T., and J. Mazurek. 1998. *Pollution Control in the United States: Evaluating the System*. Washington, DC: Resources for the Future.

EPA. 1994. Browner names six industries in plan to improve environmental protection. *Environmental News Release*, July 20.

EPA. 1996a. *Operating Principles of the Environmental Protection Agency Common Sense Initiative Council*. June.

EPA. 1996b. *Federal Advisory Committee Act Charter—Common Sense Initiative*. Filed with Congress, October 17.

EPA. 1997a. *Consensus Decision-Making Principles and Applications in the EPA Common Sense Initiative*. White paper, October 1.

EPA. 1997b. *Review of the Common Sense Initiative: Two Year Review*.

EPA. 1998. *The Common Sense Initiative: Lessons Learned*. EPA 100-R-98-011.

EPA. 1999. *Status of the Implementation of the Common Sense Initiative Recommendations*. Prepared by the US EPA Office of Policy and Reinvention, Updated October 1999, including individual Subcommittee Fact Sheets.

EPA. 2000. *EPA Sector Program Plan 2001–2005*. US EPA Office of Policy, Economics, and Innovation. November 9.

EPA. 2001. *What Is CSI Trying to Achieve? Can We Create an Environmental Regulatory System More Attuned to the Needs of Specific Industry Sectors and Stakeholder Groups?*

EPA. 2003a. Metal Finishing. http://www.epa.gov/sectors/metalfinishing/index.html#links.

EPA. 2003b. Sector Strategies Program (homepage). http://www.epa.gov/sectors/.

EPA. 2003c. Sector Strategies Program: Announcement of program launch. *Federal Register* 68: 23, 303.

Esty, D., and M. Chertow. 1997. *Thinking Ecologically: The Next Generation of Environmental Policy*. New Haven: Yale University Press.

Fiorino, D. J. 1996. Toward a new system of environmental regulation: The case for an industry sector approach. *Environmental Law* 26: 457–88.

Freeman, J., and L. Langbein. 2001. Regulatory negotiation and the legitimacy benefit. *New York University Environmental Law Journal* 9: 60.

GAO. 1997. *Regulatory Reinvention: EPA's Common Sense Initiative Needs an Improved Operating Framework and Progress Measures*. GAO/RCED-97-164.

Gore, A. 1995. *Common Sense Government: Works Better and Costs Less*. New York: Random House.

Howard, P. K. 1994. *The Death of Common Sense: How Law Is Suffocating America*. New York: Random House.

Kerr, Greiner, Anderson, and April, Inc. 1999. Analysis and evaluation of the EPA Common Sense Initiative. Prepared for USEPA.

Landy, M., M. Roberts, and S. Thomas. 1994. *The Environmental Protection Agency: Asking the Wrong Questions*. Oxford: Oxford University Press.

Lee, G. 1994. EPA Chief Plans Major "New Direction." *Washington Post*, July 20, p. A9.

Norberg-Bohm, V. 1998. Reforms in the US environmental policy system: First steps toward stimulating green design and manufacturing. Draft paper, September 28.

Norberg-Bohm, V. 1999. Stimulating "green" technological innovation: An analysis of alternative policy mechanisms. *Policy Sciences* 32: 13–38.

Ruckelshaus, W. D., and K. Hausker. 1998. *The Environmental Protection System in Transition: Toward a More Desirable Future*. Washington, DC: Center for Strategic and International Studies.

Strategic Goals Program. 2001a. Summary of Progress Report. http://www.strategicgoals.org/reports2/review2a.cfm?state=all.

Strategic Goals Program. 2001b. Energy Use Normalized by Sales. http://www.strategicgoals.org/reports2/t7.cfm?state=all&requesttimeout=300.

Strategic Goals Program. 2001c. Hazardous Sludge Shipped to Landfills. http://www.strategicgoals.org/reports2/t6.cfm?state=all&requesttimeout=300.

Todd, J. A. 1997. *The Common Sense Initiative*. Prepared for the USEPA by The Scientific Consulting Group, Inc., Gaithersburg, MD.

Yosie, T. F., and T. D. Herbst. 1998. *Using Stakeholder Processes in Environmental Decisionmaking: An Evaluation of Lessons Learned, Key Issues, and Future Challenges*. Washington, DC: National Academy of Public Administration.

4

Voluntary Approaches in Waste Management: The Case of the German ELV Program

Helge Jörgens and Per-Olof Busch

Traditionally German environmental policy has been characterized as bureaucratic, highly legalistic, inflexible, and based on "conventional attitudes toward regulation," (Jänicke and Weidner 1997: 140; see also Weidner 1995: 67). While the early phases of German waste policy seem to confirm these judgments, cooperative instruments of indirect regulation have gradually found their way into German waste policy since the mid-1980s (Jörgens and Jörgensen 2000). The most prominent cooperative measure in the area of waste policy has been the German Packaging Ordinance of 1991.

In its coalition paper of 1994, the German government laid down a clear preference for voluntary solutions in environmental policy in general and especially in the area of waste policy (SRU 1996: 97, 199). The social-democratic and green coalition government, which came into power in October 1998, has maintained this preference for voluntary solutions.

Until the end of the 1990s, and contrary to what has long been practiced in other European countries like Denmark or the Netherlands, voluntary agreements in Germany have not been formally signed by the government and therefore did not entail any legal obligation for the state. In practice, however, environmental agreements have always been the result of thorough informal discussions and negotiations between industry and public authorities.

Against this background, this chapter analyzes the negotiation and implementation of the German voluntary agreement and complementary ordinance regulating the management of the end-of-life vehicle (ELV). The disposal of ELVs became an important environmental issue in

Germany in the late 1980s. The first national goals in this area, set in 1990, formed the starting point of an intensive debate on regulatory measures for the reorganization of the national system of ELV management. In February 1996, 16 branch organizations of the automotive, recycling, and supply sector submitted the *Voluntary Pledge Regarding the Environmentally Sound Management of End-of-Life Vehicles*, which was informally accepted by the environment ministry. As a reaction to the voluntary pledge, the German government refrained from a comprehensive regulation, but presented a complementary ordinance that established the legal framework for the functioning of the voluntary solution. Both the voluntary agreement and the parallel ELV ordinance came into force in April 1998.

German efforts to regulate the environmentally sound recycling and disposal of end-of-life vehicles were accompanied by the formulation of a corresponding directive of the European Union. An EC directive on end-of-life-vehicles was adopted in October 2000 (directive 2000/53/EC) and was subsequently—in compliance with the directive's requirements—transposed into national law by the German government in May 2002 by means of a new ordinance on ELVs. The new law replaces the voluntary agreement on ELVs and the previous ordinance.

This chapter starts out by briefly outlining a theoretical framework for analyzing the German voluntary program on ELV management. The next section describes the general context in which the voluntary agreement and the corresponding ordinance evolved. The subsequent section focuses on the negotiations leading up to the voluntary agreement as well as on its contents. In the final section, the German program on end-of-life vehicles will be evaluated against the theoretical background developed at the beginning of the chapter.

A Framework for Analysis

Voluntary or negotiated agreements can be defined as agreements between polluters—in most cases industry—and a public authority that result in targets and methods for reaching targets that are neither the result of a purely industrial commitment nor of a purely governmental or administrative proposal (De Clercq and Suck 2002: 10). Since the mid-

1980s the use of negotiated agreements in the environmental field has rapidly increased throughout the whole of Europe. In areas such as waste management and climate protection voluntary agreements have become an important complement to the use of legal or economic instruments (ELNI 1998: 75; Börkey et al. 1999: 30f).

A number of real or perceived advantages of negotiated agreements over legal or market-based instruments have been identified by scholars as well as by practitioners, such as their flexibility, their ability to trigger learning processes, their potential for collaboration, and the encouragement of first movers (see chapter 1 in this volume). Furthermore, in some cases, industrial target groups and public authorities may choose voluntary negotiations instead of legislative action in order to exclude other actors such as parliament, state and local governments, or environmental NGOs from the decision making process (see De Clercq and Suck 2002: 12–13; Wicke 2001; SRU 1998: 132f; Matthijs et al. 1999: 16f; Aggeri 1999).

The actual effectiveness of voluntary agreements compared to other types of instruments, however, depends on a number of additional factors. In a recent comparative study on the preconditions under which voluntary agreements can be successfully employed in the field of environmental protection, De Clercq and Suck (2002) distinguish between (1) factors concerning a specific voluntary agreement and (2) the general political and economic context in which a voluntary agreement is being designed and implemented.[1]

The first relate to the specification of a voluntary agreement, namely its consistency (or "fit") with the underlying environmental problem and policy objectives. Here there are crucial preconditions for the successful application of this type of environmental policy instrument (De Clercq and Suck 2002: 50 f):

• The formulation of stringent, clearly defined and quantified targets, including intermediate targets (milestones) as well as operational targets (measures), which represent a real improvement when compared to a "business-as-usual" scenario.

• The allocation of specific responsibilities to different groups within the broad spectrum of actors (burden-sharing scheme).

• The inclusion of adequate monitoring mechanisms in order to identify any need for corrective action as well as the collection, dissemination, and disclosure of information in order to foster learning processes; ideally these tasks would be carried out by an independent body.

• The formulation of additional guarantees or sanctions regarding the achievement of the targets of the voluntary agreement, which could range for example from simple fines to expulsion from the agreement and/or retrospective taxation.

With regard to the general political and economic context the following factors have proved to have a significant impact on the performance of a voluntary agreement (De Clercq and Suck 2002; De Clercq et al. 2002; Ameels et al. 2000):

• A strong tradition of consensus seeking and joint problem solving in environmental policy, enhancing a voluntary agreement's fit with the existing national environmental policy system (policy hypothesis).

• The general readiness of policy makers to use alternative policy instruments as a "stick" in case a voluntary agreement fails (instrumental hypothesis).

• An environmental problem that is caused by a homogeneous industry sector consisting of a small number of players or being represented by a powerful industry association that can speak for all its members (sectoral hypothesis).

In evaluating the strengths and weaknesses of the German program on ELV management we will look both at the specific characteristics and the wider political and economic context of the program.

Context of the German ELV Program

The Environmental Problem

Environmental problems caused by the disposal of end-of-life vehicles (ELVs) include the following (SRU 1991; Benzler and Löbbe 1995: 4–8; Schenk 1998: 219; Zoboli 1999: 2–4; SRU 2000: 376):

• Direct pollution during the dismantling of ELVs (e.g., soil or groundwater contamination caused by the leaking of operating fluids).

- Generation of shredder waste (automotive shredder residue, or ASR), which may be contaminated with heavy metals, polyvinyl chloride (PVC), oil, and fuel, and which to a large extent traditionally has been landfilled together with normal household waste.
- The irregular ("wild") disposal of ELVs.

Traditionally the system for recycling ELVs has been oriented toward the recovery of steel and metals, a process that, early on, has made the automobile one of the most extensively recycled long-lasting consumer items (Schenk 1998: 1, 95). Long before the coming into force of the voluntary agreement approximately 75 percent in weight of end-of-life vehicles had been reused or recycled (Institut der Wirtschaft 1998). The remaining 25 percent, however, were mostly shredder residue (ASR), intended for disposal and constituting a serious environmental problem, at least when landfilled.[2]

Besides the generation of ASR, the inordinate dismantling of ELVs as well as the uncontrolled disposal of old cars, estimated very roughly at 100,000 vehicles annually, often results in the additional environmental problem of leakage of oils and other liquids.

Actors in the Field of ELV Management

A number of actors with diverging interests and often contradictory incentive structures to reduce harmful behavior have a stake in this issue area. The most important are automobile producers, car dealers, repair stations, dismantlers, scrap processors, and the last owners of ELVs.

Automobile Producers, Importers, and Suppliers The automobile industry is one of the key economic sectors in Germany. It has an oligopolistic structure with a small number of producers operating on a large scale and international or global scope (Zoboli 1999). The automotive producers' decisions during the production of automobiles—such as design, material composition, and durability of new cars—significantly affect the environmental impact at the later stage of ELV management (Lucas 2000: 13). However, until recently environmental considerations with regard to the recycling and disposal of ELVs have played only a very minor role in the strategic decisions of the automotive industry. This is mainly due to a general lack of economic incentives for automobile producers to improve the recyclability of their products and reduce their negative environmental impact at the stage of disposal (Schenk 1998: 18).

In addition to the automobile producers, suppliers, which are mostly medium-size enterprises, play an important role as they produce about 77 percent of the components used in the production of automobiles. To some extent they are responsible for the design and development of new components, but their influence is limited by the requirements set by automobile producers.

Automobile Dealers and Repair Stations As dealers of spare parts, repair stations buy or take back used components and sell used or refurbished parts to car owners. Both dealers and repair stations can buy or take back ELVs either for reconstruction or to recover spare parts. When purchasing a new or used car, customers can usually trade in their old cars. Automobile dealers and repair stations then either resell the cars or pass them on for disposal.

Both automobile dealers and repair stations have only limited interest in improving the system for ELV management. Although, economically, automobile dealers may be interested in trading in old cars as a marketing mechanism, they are generally opposed to holding ELVs on their premises as this can create a negative image (Schenk 1998: 24–25). Furthermore repair stations have little interest in an easier dismantling of cars, because this can lead to reduced repair costs and, therefore, threaten one of their major sources of income.

Dismantlers Contrary to the homogeneous automotive industry, the dismantling sector is characterized by its polipolistic structure. Traditionally most dismantlers have been small businesses. Dismantling is the first phase in the process of automobile recycling. Theoretically, at this first stage of automobile recycling, substantive increases in recycling rates could quickly be achieved that in turn would reduce environmental pressures at later stages of this process (Zoboli 1999: 13). However, in the absence of legal requirements, there would be hardly any incentives for dismantlers to improve their practices. For economic reasons, dismantlers only recover those materials that are suitable for reuse, recycling, or sale. There is no positive economic incentive for them to recover other materials such as operating fluids or plastic components. This situation is aggravated by a weight-based price calculation of shredder companies to whom the dismantled car wrecks are sold (Schenk 1998: 224–25).

Due to their weak organization (at least three branch organizations claim to represent dismantlers at the national level), the small number of employees, their negative public image, and their weak lobbying capacity, the specific interests of dismantlers have generally been weakly considered in the political process (Schenk 1998: 42–43).

Scrap Processors (Shredders) and Scrap Dealers Scrap processing (or shredding) is the second phase in the process of automobile recycling. End-of-life vehicles are the most important source of income for shredders. After reducing the car wrecks to small pieces the metals are extracted and the remaining shredder residue is landfilled.

Like dismantlers, shredder companies have little incentive to reduce the environmental impacts of their operations. On the one hand, they have very limited influence on the composition of the car wrecks that are delivered to them and thus on the degree of contamination of the shredder residue. On the other hand, the possibility of depositing their residues in cheap landfills has kept shredder companies from searching for other, more environmentally sound, ways of disposal.

Consumers Consumers play a dual role in the process of ELV management. As customers they may or may not consider ecological aspects in their purchasing decision. As final owners they decide where to leave their end-of-life vehicles. In general, this decision is influenced not only by factors like the disposal charge to be paid to dismantlers or return stations and the costs of transportation but also by the time and effort needed to dispose of an old car. If the last owner judges the costs of regular disposal in Germany to be too high, he might consider selling the car to an exporter or disposing of it illegally (Schenk 1998: 28, 77).

In sum, the disposal of ELVs results in many environmental problems. Solving these problems involves many different actors with strongly diverging positions of power. Given this context, one can expect complex governance patterns.

The German End-of-Life Vehicles Program

Negotiations

The issue of end-of-life vehicles was first placed on the German political agenda in the 1980s as a result of two major developments. At the international level, the inclusion of ELVs in the group of priority waste streams by the European Commission was a major driving force (Zoboli 1999: 11; UBA 1993: 240–41). At the national level, ELV policy was initiated when prognoses indicated rapidly increasing waste volumes combined with an increasing scarcity of disposal facilities. For these reasons waste management became a priority area within environmental policy (Jörgens and Jörgensen 1998). In the second half of the 1980s and early 1990s these developments resulted in the preparation or adoption of a number of product-related ordinances in areas such as packaging waste, construction rubble, or beverage containers (UBA 1989: 189).

A first proposal for a set of national objectives in the area of ELV management was formulated by the Federal Ministry for the Environment (BMU) in August 1990. Following an extensive debate with the automobile industry and the presentation of a counterproposal by the Association of the German Automotive Industry (VDA), the Ministry for the Environment presented in August 1992 a revised proposal for an ELV ordinance. Most notably, this proposal foresaw the cost-free take back and environmentally sound recycling of end-of-life vehicles by automobile producers or car dealers (Brockmann et al. 2000: 97f). Other important elements of this draft ordinance and subsequent modified proposals included quantified recycling targets for a number of different materials, labeling obligations for plastics, and provisions for the extraction and separate disposal of certain hazardous components.

The idea of a cost-free take back of scrap cars followed the original concept of extended producer responsibility (see OECD 1998) and was intended to set an economic incentive for automobile producers to improve the recyclability of their products. However, the automotive industry—backed by the Ministry of Economic Affairs—strongly opposed this plan throughout the whole negotiation process. They advocated a stronger focus on the possible contributions of other actors in this issue area, namely dismantlers and shredders and argued that this should be accomplished through the introduction of stricter environmental standards and the establishment of a monitoring system for dismantlers and shredders based on certification by independent experts.

In February 1996, after lengthy informal negotiations, the German government finally accepted a *Voluntary Pledge Regarding the Environmentally Sound Management of End-of-Life Vehicles (Passenger Cars)* which was jointly presented by 16 branch organizations of the automotive, recycling, and supply sector. In order to create a legal framework for this voluntary pledge, government was asked to pass a supplementary ordinance, which was adopted by the Ministry for the Environment in November 1996 and approved by parliament in June 1997.

While the voluntary pledge formulated a general commitment for car manufacturers to increase the recyclability of their products and set quantified goals for the recycling of materials from end-of-life vehicles, the complementary ordinance specified minimal technical requirements for the operations of return stations, dismantlers, and shredders. Both the voluntary pledge and the parallel ELV ordinance came into force in April 1998, almost ten years after the issue had first been placed on the political agenda.

Contents and Voluntary Pledge

The main goals of the voluntary pledge on the environmentally sound management of ELV were as follows:

- Improvement of the recyclability of cars and their components.
- Environmentally sound treatment of ELVs, especially during the removal of operating fluids and the dismantling of vehicles.
- Development, setup, and optimization of closed material cycles and facilities for recovery, especially for shredder residue in order to ease the strain on landfill capacities and natural resources.

These goals were to be reached as follows:

- By setting up a nationwide infrastructure for take back and recycling of ELVs.
- By reducing the amount of shredder residue for disposal from 25 percent by weight to a maximum of 15 percent by the year 2002, and a maximum of five percent by 2015.

- By manufacturers taking back any ELV of their brand at market conditions; only cars registered after April 1998 and that are not older than 12 years must be taken back free of charge.

On the crucial issue of producer responsibility, therefore, the automotive industry was successful in avoiding a general cost-free take back.

The responsibilities for reaching the goals of the voluntary agreement were divided among the different actors along the ELV management chain. Automobile producers and suppliers were required to continually optimize the recyclability of their products and—in the case of automobile producers—to set up a nationwide system of return points and dismantlers. Operators of return stations, dismantlers, and shredders had to implement the environmental standards for their operations laid down in the ELV ordinance in order to receive certification (see below). Finally, a car owner was obliged to return an old car to a certified return point or dismantler in order to receive a proof of recycling, which in turn was necessary in order to deregister the vehicle.

Implementation of the self-commitment was to be monitored by a newly created committee—the ARGE-Altauto—appointed by the Association of the German Automotive Industry. Its main tasks were to coordinate the fulfillment of the voluntary agreement and verify the level of progress achieved by preparing a monitoring report to be delivered to the ministry of the environment and the ministry of economic affairs every two years. Additionally ARGE-Altauto was to provide information to all actors involved in ELV management as well as to the interested public and provide a forum for debate.

Ordinance

A legal framework for ELV management was established through the ELV ordinance of July 4, 1997. The law regulates the process of ELV management and the relationships among ELV owners, operators of return stations, dismantlers, and shredders. It aims to guarantee the effective implementation of the self-commitment and to ensure the competitive structure of the ELV market (Bundesregierung 1996). It does so as follows:

- By introducing a legal obligation for final owners to return ELVs to a certified return station or dismantler.

• By requiring a proof of disposal issued by a certified dismantler in order for the last owner to deregister his car.

• By requiring dismantlers to extract spare parts and materials amounting to at least 15 percent by weight from the wreck by the year 2002 and ensure their reuse or recycling. By 2015 the generation of waste for disposal from all stages of ELV management is to be further reduced to a maximum of 5 percent.

• By introducing a system of independent certification for return stations, dismantlers, and shredders.

Summing up the negotiation process and the resulting ELV program, it can be concluded that while the ministry for the environment succeeded in placing the issue on the political agenda and eventually in finding a political solution, it was less successful in pushing through its regulatory concept as presented in the ministry's first draft ordinances. This was mainly due to the unwillingness of the automotive industry to carry the costs of ELV management and to accept detailed recycling quotas for different materials which would limit its choice of possible strategies for ELV management.

Within the cabinet, opposition by the Ministry for Economic Affairs (BMWi) had blocked the enactment of an ELV ordinance that would not be in the interests of the automotive industry. Several draft ordinances were criticized and eventually vetoed by the BMWi, which favored a voluntary solution and aimed at reducing the economic and competitive impacts resulting from regulations in the area of the ELV management.

Recent Developments

The voluntary program on ELV management was in force for less than three years. The adoption of a European Union directive on end-of-life-vehicles (directive 2000/53/EC) in October 2000 came despite the strong resistance of the German and European automobile industry (Wurzel 2000). The European directive obliged manufacturers to take back old cars free of charge, and it also introduced a far-reaching ban on the use of certain heavy metals in the construction of new cars (SRU 2002: 400), so the automobile industry gradually ended its efforts to implement the VA. In May 2002, a new legal act on end-of-life vehicles transposing the European directive into German law officially substituted the previous combination of a voluntary self-commitment and a supplementary ordinance.

Implementation and Results

Because of the short period during which the voluntary ELV program was in actual operation (from April 1998 to early 2001), a conclusive assessment of its environmental effectiveness cannot be easily made. Therefore this section offers predominantly a preliminary evaluation of its initial performance and environmental impact based mainly on the first and only official monitoring report published in spring 2000 (ARGE-Altauto 2000).

ELV Management Processes

According to the first monitoring report, the goal of setting up a nationwide infrastructure for takeback, recycling, and disposal of end-of-life vehicles consisting of independently certified companies had clearly been reached (ARGE-Altauto 2000: 20–24; see also SRU 2000: 376). The evaluation reported a network of approximately 15,000 return stations, 1,400 dismantlers, and 57 shredder facilities (16 of them in neighboring European countries) that were certified and controlled by approximately 90 appointed experts (ARGE-Altauto 2000: 20–25). In addition a nationwide infrastructure for takeback and recycling of old spare parts from vehicle repairs was established (ARGE-Altauto 2000: 25–27; Lucas 2000).

The successful creation of a nationwide infrastructure for the take back of end-of-life vehicles led to a significant qualitative improvement at all stages of ELV management. According to estimates by ARGE-Altauto, return stations, dismantlers, and shredders have invested a total of €256 million since 1997 in an effort to comply with strengthened environmental standards (ARGE-Altauto 2000: 28). A great number of the facilities that were functioning prior to the enactment of the ELV ordinance did not apply for or receive certification under the new regulations. Consequently the number of businesses operating in the sector decreased considerably. The number of dismantlers underwent an especially marked decline of at least 50 percent from an estimated 3,000 to 5,000 facilities to approximately 1,400 certified businesses in the year 2000 (ARGE-Altauto 2000).

Despite these apparent improvements some loopholes and implementation deficits remained. First, evaluations show that still a significant number of certified operators did not fully meet the environmental standards. Second, the independent experts responsible for certification were in many cases underqualified and public authorities had practically no means to penalize incorrect certification (ARGE-Altauto 2000: 55). Furthermore certification practices varied considerably among the different federal states (Bundesregierung 1999). According to an inquiry carried out at 600 certified dismantlers in September 1998, public authorities at the local or state level often continued to tolerate dismantlers failing to fulfill the required environmental standards (Abfallwirtschaftlicher Informationsdienst 1998: 20).

Implementation of the requirement to obtain a proof of disposal as a necessary precondition for deregistering an old car was also seriously flawed. Based on a representative inquiry, the ARGE-Altauto estimated that for approximately three million cars deregistered in the year 1999, only around 450,000 proofs of disposal and 185,000 declarations of whereabouts were submitted to the local authorities. For nearly 77 percent of all deregistered cars, the necessary documentation was not submitted (ARGE-Altauto 2000: 32–33). The authorities responsible regularly allowed car owners to deregister cars without presenting a proof of disposal. Often the only sanction they imposed was an additional processing fee of five euros. Due to scarce administrative capacities and a general lack of incentive to carry the costs of implementing this federal law, most local authorities refrained from further inquiries into the whereabouts of ELVs that had been deregistered without a proper proof of disposal (SRU 2002: 399–400).

Car owners also could opt for a temporary deregistration instead of definitely deregistering their vehicles. For temporary deregistration, a car owner theoretically was obliged to hand in a proof of disposal issued by a certified dismantler or an alternative declaration of whereabouts. In practice, local authorities regularly failed to see that this was actually done (ARGE-Altauto 2000: 51–52; Bundesregierung 1999).

This implementation gap exemplifies a more general problem that voluntary solutions are regularly confronted with: the increased complexity of implementation processes due to a growing number of participants at different levels of decision making. Despite basically being a voluntary program, ultimate responsibility for controlling its implementation, especially of the provisions contained in the ELV ordinance, rested with the public authorities—in the German federal system, the local authorities. These, however, had little incentive to enforce the ELV regulations. On the one hand, the ELV program had been negotiated exclusively between the federal government and industry. The federal states (*Länder*) and local authorities, which normally have a strong institutionalized influence on the outcomes of the legislative process, had been largely excluded from the informal negotiations leading to the VA. As a result their support of the program was rather low. On the other hand, while the local authorities were partly responsible for implementing the program, they were not granted the additional financial or staff resources necessary for effective implementation. Due to the complexity of the implementation process and the large number of private and public participants, it had been impossible to anticipate these problems or to find a quick solution. As a result the new ELV ordinance of May 2002, which was adopted to implement the European Union ELV-directive of October 2000, explicitly no longer foresees any sanctions in case an automobile is deregistered without a proof of proper disposal (SRU 2002: 403).

Recycling-Oriented Car Development and Recovery of ASR

Progress in the recycling-oriented construction of new automobiles is hard to measure due to the lack of clear and quantified goals and concrete measures in the voluntary agreement (Lohse and Sander 2000: 3–5). In general, the automotive industry claimed in its first monitoring report that criteria for recycling had been integrated into the process of developing new automobiles, that the number of materials used in the automobile construction had been reduced, that the amount of recyclable materials in cars had been increased, and that disassembly of old automobiles had been facilitated (ARGE-Altauto 2000: 10–13). Additionally ARGE-Altauto declared that draining of operating fluids was eased, plastic components were marked to indicate their material composition, automobile producers started supplying disassembly information to the recycling industry, and recycled plastic materials were being used in the production of new automobiles.

However, apart from singular examples given in the monitoring report, no general figures are available to confirm these improvements. One of the more prominent examples given in the report is the International Dismantling Information System (IDIS), which was developed jointly by 20 European automobile producers and which contains dismantling information on 364 car types and approximately 20,000 parts. In 1999 IDIS was distributed as a free CD-ROM to more than 2,500 European dismantlers (ARGE-Altauto 2000: 17–19). Other initiatives mentioned in the report had been initiated long before the voluntary agreement was signed, mainly as a reaction to early signals at the German and European level indicating an increasing regulatory activity in the area of ELV management.[3]

While the environmental impact of these voluntary measures cannot be adequately assessed due to the vagueness of the targets and the lack of comprehensive information (Lohse and Sander 2000: 3–5), it is clear that the ambitious goal of reducing shredder residue from end-of-life vehicles to merely 5 percent of the average car weight by 2015 can only be reached if the recyclability of new automobiles is significantly increased.

Analysis

Because of the long time frame of the ELV program's goals and industry's early withdrawal from its implementation, a critical assessment of the program should focus on the design and quality (i.e., the specifications) of the voluntary agreement. In doing so, we will draw on our analytical framework presented above. The specific characteristics of the ELV program to be discussed in this section include the quality of the targets and measures, the provisions for burden-sharing among the participants of the program, the monitoring mechanisms, and the sanctions to be imposed in case of failure. In our conclusions we look at the influence of the wider political and economic context.

Relevance of Targets and Measures

At first glance the goal of setting up a nationwide infrastructure for return and recycling of ELVs did not seem to be very ambitious, as a large number of recycling facilities already existed prior to the development of the VA (Rennings et al. 1996: 249). However, if one takes into account the obligatory environmental standards for dismantlers, shredders, and return stations that were laid down in the ELV ordinance, this goal may be considered an ambitious attempt to reduce direct pollution occurring during the recycling process and thus overcome the previously prevailing implementation deficit in this field.

For the recycling quotas laid down in the VA and their expected effect of reducing the amount of shredder waste for disposal, a twofold picture emerges. The short-term goal of a 15 percent reduction by the year 2002 could have been reached probably relatively easily by merely extracting the operating fluids, tires, and spare parts prior to the shredding phase (SRU 1998: 200; SRU 2000: 377). In fact, as has been shown in the previous section, the first and only monitoring report confirmed that the 15 percent goal was well within reach (ARGE-Altauto 2000).[4] However, the medium-term goal of reducing waste for disposal from ELV management to a maximum of 5 percent by the year 2015 cannot be reached by improving the operations of dismantlers and shredders alone. If one takes into account the increasing use of synthetic materials in the construction of new cars, this goal appears to be especially ambitious.

While the third main goal in the VA, the recycling-oriented design of new automobiles, in principle could have provided an effective and sustainable way of reducing the environmental impacts of ELV management, its formulation was rather vague and unspecified ("continuously optimize the recycling oriented construction of vehicles," "continuously improve the recyclability"). Quantified targets and concrete time frames for reaching this goal were completely missing. Clearly, the voluntary agreement put no direct pressure on the automotive industry to increase the recyclability of their products. Consequently only few and rather unsystematic efforts by automobile producers have been observed so far. The most direct incentive for automobile producers to develop more recycling-oriented cars would have been the imposition of a general obligation for automobile producers to take back all cars of their brand regardless of their age or condition (Rennings et al. 1997: 26; SRU 1998: 200–201).

Overall, it can be concluded that the targets and measures formulated in the voluntary agreement and the ELV ordinance were not entirely adequate to the problem at hand. Clear targets had been adopted for the operations of recycling facilities, but targets for the development and construction of new automobiles were kept very vague.

Especially the introduction of a general cost-free take back of ELV regardless of age and condition would have created a stronger incentive for automobile producers to improve the recyclability of their products. Finally, the option of reducing the environmental impact of shredder residue by systematically treating it as hazardous waste and thus applying stricter environmental standards for its disposal has not been pursued (SRU 2000: 377).

Burden Sharing

The voluntary agreement and the accompanying ordinance clearly allocated specific responsibilities to different groups along the chain of ELV management. However, this "burden sharing" was highly imbalanced and clearly reflected the relative strength, homogeneity, and bargaining power of the different actor groups. While automobile producers had been able to fight off far-reaching measures such as commitments to take back end-of-life vehicles free of charge or to significantly increase the recyclability of new cars, operators of return stations, dismantling facilities, and shredders had to incur considerable expenses in order to comply with the far-reaching environmental standards laid down in the ELV ordinance.

Thus the responsibility for reaching the intermediate target of reducing the amount of nonrecyclable shredder residue to 85 percent by the year 2002 had almost entirely been allocated with the politically and economically weakest actors in the ELV chain. Although the introduction of higher environmental standards for operators of recycling facilities was a necessary measure with an expected positive environmental impact, it was not sufficient to reach the far-reaching target of reducing the amount of shredder residue to 5 percent by the year 2015. To reach this ambitious goal, a greater allocation of responsibilities to the producers of automobiles and their suppliers would have been necessary (Lohse and Sander 2000: 3–5).

Monitoring and Information Dissemination

While the voluntary agreement provided for regular monitoring of its performance, the monitoring task was not performed—as would have been desirable—by an independent body, but by a body established and financed by the automobile industry. According to the voluntary agreement, monitoring reports had to be prepared every two years under the direction of the ARGE-Altauto. The first—and only—monitoring report was submitted in March 2000 to the federal government (ARGE-Altauto 2000).

As early as 1999 a study commissioned by the Federal Environment Agency had pointed out deficits in the essential preconditions for an effective monitoring of the voluntary agreement and the ELV ordinance (Institut für Ökologie und Politik 1999). The study pointed out difficulties in assessing the number of end-of-life vehicles recycled by certified operators, determining whether recycling quotas were actually being met, and measuring improvements in the recycling-oriented design of new vehicles. A follow-up evaluation of the first monitoring report confirmed these concerns (Lohse and Sander 2000) and the report itself acknowledges a severe lack of reliable data (ARGE-Altauto 2000).

While the reliability of the data provided in the monitoring report is to some extent open to discussion, the ARGE-Altauto had been very active in fostering the exchange of information between the different actors in the ELV chain by initiating workshops, conferences, meetings, and other internal events. Furthermore the ARGE-Altauto provided comprehensive information about progress in the implementation of the voluntary agreement such as the official monitoring report, regularly updated lists of certified dismantlers, and shredder companies, third-party evaluations of the VA, and official documents regarding different aspects of ELV management on its Internet Web site.

Additional Sanctions

While the voluntary agreement itself did not contain sanctions against the automobile industry should it fail to reach its targets, recycling facilities had to meet the requirements for independent certification introduced by the ELV ordinance. Shredders and dismantlers who did not comply with the recycling targets and environmental standards laid down in the ordinance ran the risk of losing this certification. Similar to the unequal allocation of responsibilities between the different actor groups, the additional sanctions provided for in the German ELV program mainly affected the actors operating in the recycling sector.

Conclusions

The assessment of the German ELV program shows two contradictory trends. On the one hand, the environmental standards for the recycling sector and the long-term recycling goal clearly went beyond any business-as-usual scenario. On the other hand, the German ELV program failed to meet important preconditions for a successful application of voluntary agreements that were outlined in our framework of analysis.

First of all, the ELV program suffered significant shortcomings in its design. In terms of stringency and clarity of goals, the program lacked well-defined operational goals and measures for improving the recyclability of newly constructed automobiles. Important alternative measures that could have changed the competitive environment more effectively—such as the cost-free take back by car manufacturers—were rejected during the negotiation process. These shortcomings most likely prevented first mover strategies by targeted businesses, especially by car manufacturers. Furthermore the burden that had to be shouldered to reduce harmful impacts of ELV management was unequally shared among the actors involved. In other words, the voluntary agreement and its complementary ordinance did not exert similar pressure on all targeted actors and as a consequence new relationships between all targeted actors with the potential to foster learning processes and to generate innovative solutions had not been created to a sufficient degree. It is therefore very unlikely that the program would have enabled Germany to reach the 95 percent recycling goal for the year 2015.

The inadequate specification of the voluntary agreement and the complementary ordinance can be explained by a number of factors, that are closely related to the three hypotheses outlined in our framework of analysis. Above all, the relatively homogeneous producing sector with common interests and a powerful industry association had been able to avoid a more balanced allocation of responsibilities (sectoral hypothesis). By contrast, the heterogeneous and weakly organized recycling sector had not been able to obstruct or influence the formulation of the actual measures and to successfully demand an even burden-sharing. Furthermore the characteristics of the industry sector, namely the distance in time (and possibly space) between car producers (responsible for the design of new automobiles) and the final owners of ELV, may explain the difficulties in setting up an effective and coherent voluntary agreement. This separation, for example, practically excludes recycling characteristics from being an important element in consumers' purchase decisions.

In addition the instrumental hypothesis partially explains the uneven distribution of responsibilities among actors. The diverging views between the federal Ministry for the Environment and the Ministry of Economic Affairs impeded the consensual passing of an ambitious ELV ordinance that would have brought about a different distribution of responsibilities and costs for ELV management. The resulting lack of a credible threat of a stricter regulatory response should negotiations for the voluntary agreement fail, eased the car manufacturers obstruction of meaningful obligations for them.

Finally, with regard to the policy hypothesis the relatively short tradition of consensual policy making based on voluntary negotiations in German waste policy adds a further explanation. Partially due to the inexperience with voluntary agreements, the industry had been reluctant to make far-reaching concessions.

While the German government has recognized the weaknesses of the German ELV program (Bundesregierung 1999), in the shadow of the parallel development of an EU directive it has refrained from trying to amend the ELV ordinance and from renegotiating the voluntary agreement with the automobile industry. The European directive on end-of-life vehicles (2000/53/EC) was adopted in October 2000. Like its German antecessor, the directive aims to reduce waste from ELV disposal to only 5 percent of the average car's weight by the year 2015. An intermediate goal of 15 percent of waste for disposal is set for 2006. Besides these overall recycling goals, the directive contains measures that significantly exceed those of the German program. From 2007 on it obliges car manufacturers to take back end-of-life vehicles of their brand from the last owner free of charge and regardless of the car's age. Furthermore the directive prohibits the use of heavy metals in the construction of new automobiles except for a limited number of exceptions.

In order to transpose the EC directive, the German parliament approved a new ordinance on end-of-life vehicles in May 2002. With this, the German legislator has finally opted for a full allocation of the costs for ELV recycling to the automobile producers. Furthermore the new law strengthens the requirements for the authorization of independent experts and expands the obligations of return stations, dismantlers, and shredders to pass on relevant data and information to the responsible authorities. As a consequence of the insufficient handling of the required proof of disposal on the side of the local authorities, the new law no longer foresees any sanctions in case an automobile is deregistered without a proof of proper disposal (SRU 2002: 403).

These recent developments clearly show that while the German voluntary agreement was—at least to some extent—vaguely worded, it has triggered a learning process within government and led to significant improvements in the course of the necessary transposition of the European regulation into national law.

Acknowledgments

The authors would like to thank Franck Aggeri, Bart Ameels, Theo de Bruijn, Marc De Clercq, and Vicki Norberg-Bohm for their valuable comments on earlier versions of this chapter.

Notes

1. The research project "NEAPOL—Negotiated Environmental Agreements: Policy Lessons to Be Learned from a Comparative Case Study," of which our study on the German ELV program is a part (Jörgens and Busch 2002), was coordinated by Marc De Clercq at the Centre for Environmental Economics and Environmental Management of the University of Ghent; it was funded by the Environment and Climate Programme of the European Commission (ENV4-

CT97-0560). Based on the evaluation of twelve voluntary agreements in six European countries the project analyzed the specific impact of the political and economic context of a voluntary agreement on its performance (De Clercq 2002).

2. There is some disagreement on whether end-of-life vehicles are partly responsible for PCB contamination of shredder waste or whether PCB contamination is caused exclusively by other shredder inputs such as electronic appliances (see Schenk 1998: 222f).

3. For a detailed account of recycling initiatives by German producers, see Zoboli et al. (2000: II-22–II-37).

4. With the adoption of the EC directive 2000/53/EC on end-of-life vehicles and the German automobile industry's subsequent decision to halt implementation of the VA, no further monitoring efforts were made. Although the EC directive also sets an intermediary goal for reducing waste from ELV to 15 percent of a car's weight, this reduction is due only by the year 2006 and official data on its implementation is not available yet (SRU 2002: 398–404).

References

Abfallwirtschaftlicher Informationsdienst. 1998. No. 5 (September 16, 1998). Berlin: Rhombus Verlag.

Aggeri, F. 1999. Negotiated Agreements and Innovation: A Knowledge-Based Perspective on Environmental Policies. In CEEM, ed., *Negotiated Environmental Agreements: Policy Lessons to be Learned from a Comparative Case Study. Theoretical Report*. Gent: CEEM, pp. 73–95.

Ameels, B., A. Suck, and M. De Clercq. 2000. *The Diffusion of Voluntary Agreements in the European Union: Critical Conditions for Success*. Paper presented at the international workshop on Diffusion of Environmental Policy Innovations. Berlin, December 8–9.

ARGE-Altauto. 2000. *1. Monitoringbericht gemäß Punkt 3.6 der Freiwilligen Selbstverpflichtung zur umweltgerechten Entsorgung von Altfahrzeugen (PKW) im Rahmen des Kreislaufwirtschafts-/Abfallgesetzes. Der Bundesregierung vorgelegt am 31 März 2000. Berichtszeitraum: 1 April 1998 bis 31 März 2000*. Frankfurt: ARGE-Altauto.

Benzler, G., and K. Löbbe. 1995. Rücknahme von Altautos—Eine kritische Würdigung der Konzepte. RWI-Papiere 40. Essen: RWI.

Börkey, P., and F. Lévêque. 1999. *Voluntary Approaches for Environmental Policy: An Assessment*. Paris: OECD.

Brockmann, K. L., S. Deimann, and F. Wallau with B. Dette. 2000. *Endbericht zum UBA-Forschungsvorhaben "Evaluierung von Finanzierungsmodellen zur Durchführung der kostenlosen Rückgabe von Altautos."* Mannheim, Darmstadt, and Bonn: ZEW, IFM, and Öko-Institut.

Bundesregierung. 1996. *Verordnung der Bundesregierung über die Entsorgung von Altautos und die Anpassung straßenverkehrsrechtlicher Vorschriften.* 07.11.1996, 13/5998. Bonn.

Bundesregierung. 1999. *Unterrichtung durch die Bundesregierung. Entsorgung von Altautos*, 14.06.1999. Bonn.

Busch, P.-O., and H. Jörgens. 2002. Self-commitment on the collection and recovery of spent batteries and the reduction of mercury content in batteries. In M. De Clercq, ed., *Negotiating Environmental Agreements in Europe: Critical Factors for Success.* Cheltenham, Northampton: Elgar, pp. 67–86.

De Clercq, M., ed. 2002. *Negotiating Environmental Agreements in Europe: Critical Factors for Success.* Cheltenham, Northampton: Elgar.

De Clercq, M., and A. Suck. 2002. Theoretical reflections on the proliferation of negotiated agreements. In M. De Clercq, ed., *Negotiating Environmental Agreements in Europe: Critical Factors for Success.* Cheltenham, Northampton: Elgar, pp. 9–64.

De Clercq, M., A. Suck, B. Ameels, and R. Salmons. 2002. Comparative Evaluation of the Case Studies. In M. De Clercq, ed., *Negotiating Environmental Agreements in Europe: Critical Factors for Success.* Cheltenham, Northampton: Edward Elgar, pp. 337–87.

ELNI (Environmental Law Network International). 1998. *Environmental Agreements: The Role and Effect of Environmental Agreements in Environmental Policies.* London: Cameron May.

Institut der Wirtschaft. 1998. Automobiler Stoffkreislauf. *IW-Umwelt-Service* No. 1. http://www.iwkoeln.de/Umwelt/U1-98-4.htm.

Institut für Ökologie und Politik. 1999. General requirements for monitoring the recycling of long-lived, technically complex products with an in-depth-analysis of end-of-life vehicles. Summary. http://www.Oekopol.de/Archiv/Stoffstrom/Autoeng.htm.

Jänicke, M., and H. Weidner. 1997. Germany. In M. Jänicke and H. Weidner, eds., *National Environmental Policies—A Comparative Study of Capacity-Building.* Berlin: Springer, pp. 133–55.

Jörgens, H., and P.-O. Busch. 2002. The voluntary pledge regarding the environmentally sound management of end-of-life-vehicles. In M. De Clercq, ed., *Negotiated Environmental Agreements in Europe: Critical Factors for Success.* Cheltenham, Northampton: Elgar, pp. 88–113.

Jörgens, H., and K. Jörgensen. 1998. Abfallpolitik in der Bundesrepublik Deutschland. In G. Breit, ed., *Neue Wege in der Umweltpolitik.* Schwalbach: Wochenschau, pp. 40–52.

Jörgens, H., and K. Jörgensen. 2000. Von der Abfallbeseitigung zur Kreislaufwirtschaft; Abfallpolitik in der Bundesrepublik Deutschland. *Geographische Rundschau* 52(6): 4–8.

Lohse, J., and K. Sander. 2000. *Stellungnahme zum ersten Monitoringbericht der ARGE-Altauto*. Hamburg: Ökopol.

Lucas, R. 2001. End-of-life vehicle regulation in Germany and Europe—Problems and perspectives. Wuppertal Papers 113. Wuppertal: Wuppertal Institut für Klima, Umwelt, Energie.

Matthijs, F., S. Baeke, and M. De Clercq. 1999. Literature analysis. In CEEM, ed., *Negotiated Environmental Agreements: Policy Lessons to be Learned from a Comparative Case Study. Theoretical Report*. Gent: CEEM, pp. 7–21.

OECD. 1998. *Extended and Shared Producer Responsibility. Phase 2. Executive Summary*. Paris: OECD.

Rennings, K., K. L. Brockmann, H. Koschel, H. Bergmann, and I. Kühn. 1996. *Nachhaltigkeit, Ordnungspolitik und freiwillige Selbstverpflichtung. Ordnungspolitische Grundregeln für eine Politik der Nachhaltigkeit und das Instrument der freiwilligen Selbstverpflichtung im Umweltschutz*. Heidelberg: Physica.

Rennings, K., K. L. Brockmann, and H. Bergmann. 1997. *Voluntary agreements in Environmental protection—Experiences in Germany and Future Perspectives*. Mannheim: Zentrum für Europäische Wirtschaftsforschung.

Schenk, M. 1998. *Altautomobilrecycling. Technisch-ökonomische Zusammenhänge und wirtschaftspolitische Implikationen*. Wiesbaden: Deutscher Universtitäts Verlag.

Schrader, C. 1998. Die deutsche und die europäische Altauto-Regelung aus ökologischer Sicht. In J. Beudt and S. Gessenich, eds., *Die Altautoverordnung. Branchenwandel durch neue Marktstrukturen. Chancen und Grenzen für die Abfallwirtschaft*. Berlin: Springer, pp. 53–66.

SRU (Rat von Sachverständigen für Umweltfragen). 1991. *Abfallwirtschaft. Sondergutachten September 1990*. Stuttgart: Metzler-Poeschel.

SRU (Rat von Sachverständigen für Umweltfragen). 1996. *Umweltgutachten 1996. Zur Umsetzung einer dauerhaft-umweltgerechten Entwicklung*. Stuttgart: Metzler-Poeschel.

SRU (Rat von Sachverständigen für Umweltfragen). 1998. *Umweltgutachten 1998. Umweltschutz: Erreichtes sichern—neue Wege gehen*. Stuttgart: Metzler-Poeschel.

SRU (Rat von Sachverständigen für Umweltfragen). 2000. *Umweltgutachten 2000. Schritte ins nächste Jahrtausend*. Stuttgart: Metzler-Poeschel.

SRU (Rat von Sachverständigen für Umweltfragen). 2002. *Umweltgutachten 2002. Für eine neue Vorreiterrolle*. Stuttgart: Metzler-Poeschel.

UBA (Umweltbundesamt). 1989. *Jahresbericht 1989*. Berlin: UBA.

UBA (Umweltbundesamt). 1993. *Jahresbericht 1993*. Berlin: UBA.

VDA (Verband der Automobilindustrie e.V.). 2000. Zahlen aus der Automobilwirtschaft (*www.vda.de/zahlen/*).

Weidner, H. 1995. 25 Years of modern environmental policy in Germany. Treading a well-worn path to the top of the international field. WZB Discussion paper FS II 95-301. Berlin: Social Science Research Centre (WZB).

Wurzel, R. 2000. Flying into unexpected turbulences: The German EU presidency in the environmental field. *German Politics* 9(3): 23–42.

Zoboli, R. 1999. Environmental regulation and innovation in the end-of-life-vehicle sector. Paper presented at the International Conference on Innovation-Oriented Environmental Regulation—Theoretical Approaches and Empirical Analysis, Potsdam, May 27–29, 1999.

Zoboli, R., G. Barbiroli, R. Leoncini, M. Mazzanti, and S. Montresor. 2000. *Regulation and Innovation in the Area of End-of-Life Vehicles*. Seville: Institute for Prospective Technological Studies (IPTS).

5

Dynamics of Voluntary Product Labeling Programs: An Energy Star Case Study

Bruce Paton

While many voluntary environmental initiatives have focused on interactions between firms and regulators, voluntary product-labeling programs are designed to alter the relationship between firms and their customers. The mechanics of these programs and the industry dynamics they produce remain largely unexplored. Such programs have powerful potential, providing opportunities for corporations to advertise and profit from their superior environmental performance. By providing information that prices alone cannot, product labeling enables customers to respond to the "green" attributes of a product. The preferences of customers and other stakeholders can influence company management to pursue energy efficiency or other environmentally desirable product attributes or modes of production.

In the United States this type of initiative includes the Energy Star programs administered by the US Environmental Protection Agency (EPA) and the US Department of Energy (DOE). This chapter examines two specific efforts, the Energy Star Office Products program administered by EPA and the Energy Star Clothes Washer program administered by DOE. The Clothes Washer program illustrates the dynamics that occur when a voluntary program heightens competition among manufacturers. The Office Products program shows what happens when a voluntary program provokes a common response among virtually all manufacturers in an industry.

This analysis is based on industry and government publications, interviews with government program officers and contractors, and telephone interviews with company managers. The industry interviews[1] covered firms representing more than 99 percent of the US clothes washer industry (six interviews), and more than 55 percent of the US desktop computer industry (five interviews and one written response to questions).

Both programs have created mechanisms that have begun to fundamentally transform the markets for these two product categories.[2] However, they have driven change in very different ways. This chapter explores the mechanisms underlying these programs and the different industry dynamics they have produced.

I begin this chapter by exploring the logic of voluntary product-labeling programs and introduces the separating and converging mechanisms illustrated in this chapter. Next I introduce the Energy Star program and contrast the two initiatives considered in this chapter. In the following two sections I describe each effort in greater depth. My analysis then focuses on how the separating and converging mechanisms work. In the final section I summarize the insights gained through comparing the two programs.

The Logic of Voluntary Product-Labeling Programs

In many instances firms could improve the energy efficiency of their products voluntarily instead of waiting for mandatory performance standards. However, market responses to the presence of "hidden information"—knowledge available to one party to a transaction but not to another—may prevent voluntary action (see Macho-Stadler and Perez-Castrillo 1997) and thus lead to "adverse selection."

Adverse selection occurs when uncertainty about product characteristics reduces the number of transactions, modifies the terms of transactions, or eliminates them altogether. The archetypal example is the "market for lemons" problem, described by Akerlof (1970), in which used car sellers have private information about the quality of their vehicles, but potential buyers must assess their value using only publicly available information. The possibility that any car offered for sale can be a "lemon" reduces the potential value of all cars and discourages the owners of better cars from offering them for sale. In extreme cases adverse selection can prevent any products from being sold.

Adverse selection has greatly limited the development of energy-efficient products. Howarth and Andersson (1993) observe that growth in the market for energy-efficient products has been inhibited by inefficient transfer of information between producers and consumers. They point out that consumers:

> rely heavily on manufacturer reputation and previous experience in owning and operating equipment—factors which reflect past rather than present equipment performance. (p. 268)

Reliance on old information may perpetuate the use of outdated technologies, even though the net effect is economically inefficient. Howarth and Andersson's model indicates that public policies can offset this informational asymmetry.

Studies on "voluntary overcompliance" have examined how voluntary initiatives might encourage firms to improve the energy efficiency of their products. Arora and Gangopadhyay (1995) demonstrate that under complete information some firms will modify their manufacturing processes or product mix if customers will pay more for greener products. In this situation less environmentally conscious firms will meet legal requirements, while more environmentally conscious firms will voluntarily overcomply.

Kirchoff (1999) demonstrates how firms might benefit from an institution that encourages them to overcomply. Firms can inform customers about an environmentally superior offering by publishing claims about their own product, or adding a label indicating participation in an externally validated program recognizing greener products. Under asymmetric information consumers are uncertain about the validity of firms' product claims, and are therefore less willing to pay a premium. A third-party labeling system can certify producer's claims and deter false claims. Such a mechanism enhances social welfare and economic efficiency by increasing the supply of the green products that customers prefer while providing higher profit level to firms.

The economic literature on signaling and screening describes mechanisms that can deter producers from making invalid claims and allow consumers to make effective choices under asymmetric information (Spence 1974; Stiglitz 1975). Under carefully specified conditions both signaling and screening models lead informed parties to provide signals that allow the uninformed parties to make decisions as if they had detailed knowledge of the informed party's private information.

To use a signaling mechanism, producers invest in communicating the important attributes of their products or services (Spence 1974). In a favorable response, customers will pay more for these products. Stiglitz (1975) introduced the related concept of screening, a model in which buyers create market opportunities that lead producers to tout the desirable characteristics of their products. Voluntary initiatives include a third alternative, closely resembling the screening model, in which a third party, such as a government agency or a nongovernment organization (NGO), encourages producers to signal the superiority of their products to potential customers.

Rothschild and Stiglitz (1976) introduced the concepts of "pooling" and "separating" to address the effectiveness of a screening mechanism. A separating mechanism enables the creator of mechanism to distinguish among categories of participants. For example, a medical insurance policy that offers a trade-off between higher deductibles and lower monthly premiums could be expected to separate relatively healthy patients from patients with a recent history of higher medical bills. A pooling mechanism fails to differentiate among participants or customers. For example, a medical insurance policy that charges all patients the same deductibles and premiums will pool all customers into a single, undifferentiated group.

This chapter introduces the concept of a "converging" mechanism,[3] to supplement the pooling and separating mechanisms identified by Rothschild and Stiglitz. A converging mechanism has the effect of changing all products without differentiating among them, and signaling to the buying public that the *industry* has adopted the desired behavior.[4]

The Energy Star Office Products and Clothes Washer programs illustrate separating and converging mechanisms common in voluntary environmental initiatives. Table 5.1 summarizes the key attributes of these mechanisms. Each approach can reward firms for improving their environmental performance and discourage firms from providing misleading information. Although converging and separating mechanisms work in very different ways, both can lead to substantial market transformations.

The Energy Star Program

The Energy Star initiative is a family of voluntary programs designed to increase energy efficiency and reduce carbon emissions. EPA introduced

Table 5.1
Mechanism types

Type	Key features	Example
Separating	Firms choose whether to participate, or level of participation, and then separate into a small number of types	Energy Star Clothes Washer program
Converging	Firms in a targeted group choose to make desired level of improvements in performance	Energy Star Office Products program

Energy Star in 1993 as a voluntary labeling program to identify and promote energy-efficient products. EPA subsequently partnered with DOE in 1996 to promote the Energy Star label and broaden the range of activities it covered. Over the past few years Energy Star has expanded to cover 31 product categories, including residential and commercial buildings, residential heating and cooling equipment, major appliances, lighting, and consumer electronics. Cumulatively, the Energy Star program has saved an estimated 1,130 petajoules (10^{15} joules) of primary energy and avoided the emission of an estimated 20.7 MtC of carbon (Webber et al. 2000).

The mission of the Energy Star program is to "realize significant reductions in emission and energy consumption by permanently transforming markets for energy-consuming products" (Brown et al. 2000). Energy Star initiatives pursue several interrelated strategies including setting standards for the label; labeling energy-efficient products; providing objective information to consumers; working with national, regional, and local groups to promote energy efficiency; and lowering the costs of owning energy efficient equipment and products through alternative financing (EPA 1997).

The program has achieved considerable success, reducing energy consumption, carbon emissions, and expenditures. In 1999 Americans purchased more than 100 million Energy Star–compliant products, representing approximately a 20 percent market share for the product categories addressed by the program (Brown et al. 2000). In addition the Energy Star label and standards have become a de facto international standard. It has been adopted by Japan, New Zealand, Australia, and recently, the European Union.

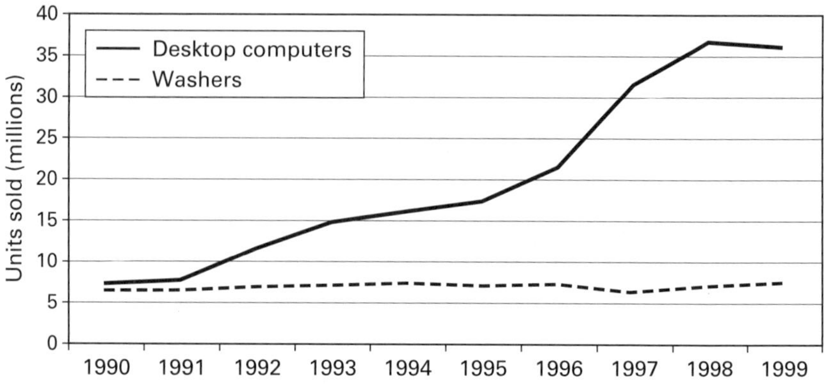

Figure 5.1
Unit shipments of clothes washers and desktop computers, 1990 to 1999. (Sources: Appliance 1999a, b, 2000)

The Energy Star Clothes Washer program and the Energy Star Office Products programs have been successful in two industries with striking differences. The washing machine industry is relatively mature, with very modest growth, while the personal computer industry grew by nearly 500 percent between 1990 and 1999. Figure 5.1 shows unit shipments for these two industries. They also differ in their degrees of market concentration, pace of innovation, and intensity of price competition. Table 5.2 highlights several significant differences between the industries.

The US clothes washer industry has traditionally been quite stable. It is highly concentrated, with the top three companies (Whirlpool, Maytag, and GE) accounting for 89 percent of the US market, and the top five companies accounting for more than 99 percent. Whirlpool has been the market leader for more than 40 years, with 53 percent share of US sales in 1999 (Appliance 2000).

The Energy Star Clothes Washer program has achieved moderate success to date. The majority of washers sold are still not Energy Star models and industry participants believe that Energy Star–compliant clothes washers constitute a small percentage of the existing stock. However, as described below, the entire industry has recently committed to mandatory energy efficiency standards for all clothes washers that exceed the current Energy Star voluntary standards.

Table 5.2
Characteristics of two Energy Star programs

Characteristics	Clothes washers	Office products
Industry organization	Highly concentrated: top 5 firms command 99%+ of US market share	Moderately concentrated: top five firms command 57% of the US market
Basis of competition	Products differentiated on features, quality and price	Leading products very similar in price and features
Price trends	Prices relatively stable	Price/performance ratio continually moving downward; price a dominant factor in customer choice
Brand	Products sold under national brand names	More than 50% of products sold under brand names; many products sold under retailer's private label
Innovation	Moderate innovation; innovations enabled by key suppliers and fundamental innovations relatively infrequent	Constant rapid innovation driven in part by chip manufacturers and software firms; turnover of basic technology every 1–2 years
Regulatory threat	Energy-efficiency standards in place before program began; legislative mandate to review standards every 4 years	No legislative mandate for energy-efficiency standards
Program administration	US Department of Energy	US Environmental Protection Agency
Retailer involvement in program	High	Low

Cumulatively, the Energy Star clothes washers have saved an estimated 31 petajoules of primary energy from 1996 to 2000, and prevented emissions of an estimated 0.0076 MtC of carbon. The program is expected to save 340 petajoules of primary energy from 2001 to 2010, and prevent emissions of an estimated 18 MtC of carbon. These savings have reduced energy expenditures by $220 million from 1996 to 2000, and are expected to reduce energy bills by $16 billion from 2001 to 2010[5] (Webber et al. 2000).

The US office products industry includes manufacturers of personal computers, monitors, printers, multifunction devices, and copiers. This chapter focuses on the desktop computer segment of that industry. This industry is intensely competitive, with more than 50 manufacturers vying for market share. Market leadership is fragmented, and the leading firm has changed several times over the last decade. Competition is characterized by rapid turnover of product models, and constant pressures to reduce prices and increase performance. Although branded products command more than 50 percent of the market, a sizable portion of the market is addressed through "white box" products sold under the names of retail establishments.

The Energy Star Office Products program has achieved an estimated 80 percent market share for computers, 95 percent for monitors, and 99 percent for printers (Brown et al. 2000). Cumulatively the program has saved an estimated 360 petajoules of primary energy from 1993 to 2000, and prevented emissions of an estimated 2.8 MtC of carbon. The program is expected to save 2,200 petajoules of primary energy from 2001 to 2010, and prevent emissions of an estimated 33 MtC of carbon. These savings have reduced energy expenditures by $2.5 million from 1996 to 2000, and are expected to reduce energy bills by $14 billion from 2001 to 2010 (Webber et al. 2000).

These two industries provide striking contrasts in terms of the competitive environment, the pace of technological innovation, and many other variables. The Energy Star programs that address these two industries also differ in very striking ways. The next two sections describe the Energy Star Clothes Washers and Office Products programs in greater depth.

The Energy Star Clothes Washers Program

The Energy Star Clothes Washers program, administered by DOE, illustrates the dynamics that can occur when a voluntary program encourages firms to differentiate their products on a desired dimension. It allows manufacturers to use the Energy Star label on any washers that meet a level of energy efficiency significantly exceeding that required by current standards. The Energy Star program contributed to a technology race among clothes washer manufacturers, leading to product innovations affecting energy efficiency as well as several other elements of product design.

Interviews with industry participants revealed sharply different reactions to the program in terms of customer demand for energy-efficient products and the nature of the opportunity the program presented. Some firms viewed the Energy Star program as an opportunity to act on their commitments to energy efficiency and environmental protection, and in the process call attention to their energy-efficient products. Others viewed their decision to participate as a defensive move in response to threats by DOE to raise standards for washers under the National Appliance Energy Conservation Act of 1987. While all of the participants interviewed now believe that energy efficiency represents a customer preference, they continue to disagree over its importance relative to other needs.

The program initially created a sharp division within the industry because firms differed in their ability to create products that qualified for the label. At the time the program was proposed, none of the market leaders had announced products that could qualify for the Energy Star label. None of the US market leaders had released washers that used the energy-saving horizontal-axis design that is dominant in Europe.

Maytag began development of a horizontal axis washer in response to the 1994 DOE proposal to consider the performance of horizontal-axis washers in setting the next round of energy-efficiency standards. This effort led Maytag to introduce its Neptune washer in 1997. Neptune was the first clothes washer to qualify for the Energy Star label, and sold at nearly double the price of conventional washers. Over the next two years all of the other US market leaders began development of clothes washers that qualified for the label. Over time the agency has expanded the options for complying with program requirements. This has resulted in 64 models representing 20 brands qualifying for the Energy Star label.

The Energy Star Clothes Washers program has contributed to a major burst of technological innovations in the appliance industry. These improvements have allowed clothes washer manufacturers simultaneously to improve product performance, energy efficiency, and water conservation. The innovations have been driven by a combination of consumer demands for new functionality and convenience and regulatory pressure to increase energy efficiency. These improvements have created "innovation offsets"—product or process improvements resulting from environmental-performance improvements—that provide the firm with positive net returns (Porter and van der Linde 1997).

These improvements are made possible by diffusion of existing technologies from Europe as well as innovations in component technologies and the rapid transformation from electromechanical to electronic controls (McHenry and Houston 2000). Electronic controls, originally found only in high-end products, have begun to migrate to mainstream products. At least four technologies—variable speed motors, digital signal processors, microcontrollers with embedded flash memory, and electronic sensors—have enabled clothes washer performance to improve while conserving water and energy. These innovations have allowed American manufacturers to achieve significant energy savings in washers using horizontal-axis technologies imported from Europe, as well as in washers using vertical-axis technologies developed in America.

Efficient horizontal-axis washers require variable speed motors and digital signal processors to control them (Murray 2000). Variable speed motors allow washers to operate at speeds ranging from 35 rpm with intermittent pauses during tumble wash cycles to 1,200 rpm during spin cycles. Increases in spin cycle speeds have allowed washers to reduce the moisture remaining in clothes at the end of wash cycle, allowing for significant reductions in the energy required to dry clothes. This improvement has also allowed manufacturers to meet customer demand for a closer match between the time required to wash a load of clothes and the longer time required to dry it.

The development of microcontrollers with embedded flash memory has created benefits for both manufacturers and consumers. Programmable microcontrollers allow manufacturers to incorporate sophisticated algorithms to control wash cycles. Incorporating flash memory gives manufacturers the opportunity to optimize product performance during product design, and to fine-tune adjustments during production. Flash memory also creates the opportunity to upgrade a product by providing additional algorithms after the washer is installed in the customer's home. This capability can potentially extend product life and provide additional value to customers by allowing them to develop customized washing cycles.

The development of electronic sensors has allowed manufacturers to fine-tune washer performance in several ways (Mnif 2000). Water level sensors allow clothes washers to automatically adjust the volume of water based on the size of the laundry load. Load balance sensors are necessary to reach the higher spin cycle speeds mentioned above. These sensors detect when a load is unbalanced and adjust washer speed in order to rebalance the load before the washer shifts to very high speeds. Pressure sensors let clothes washers fine-tune the amount of water used during a particular cycle. They also allow the sophisticated spray-rinse cycles necessary to make vertical-axis washers energy and water efficient.

The incorporation of these electronic sensors and controls has allowed manufacturers to reduce dramatically the number of mechanical parts in clothes washers, simultaneously reducing manufacturing and repair costs. These savings have helped offset the significant investments required to build new factories and engineer radically new designs for enclosures and mechanical parts. The innovations also help pave the way toward incorporating clothes washers into home networks. The capability to control home appliances from a network is beginning to appear in high-end products, and is expected to become mainstream over the next decade.

Technological innovation appears to have had significant effects on the clothes washer industry. The industry has been mature in the United States for more than a decade, with replacement of worn-out washers creating the bulk of new sales (Appliance 2000). Recently, however, the volume of sales has increased, and buying patterns have begun to shift. One industry executive reported that, "people are finding reasons to upgrade their current appliances. For the first time in the history of the industry, people are buying laundry equipment to replace units that aren't broken" (LaPat 2000). Customers are realizing that more efficient new clothes washers can save them $90 to $100 per year in their energy and water bills. New buying patterns also suggest that customers believe the information about potential savings that have led them to consider operating costs along with initial purchase costs.[6]

The threat of regulation has contributed to industry decision making concerning energy efficiency throughout the life of the Energy Star Clothes Washers program. The combination of a voluntary program and regulatory development has led recently to an historic agreement on requirements for the Energy Star label and efficiency standards among DOE, industry, energy activists, and other stakeholders.

The National Appliance Energy Conservation Act of 1987, which amended the Energy Policy and Conservation Act, imposed standards for clothes washers as part of a larger program of energy conservation requirements for consumer products. Initial levels were relatively modest, requiring only that clothes washers manufactured on or after January 1, 1988, have an unheated rinse option. The 1987 Act required the DOE to decide by January 1, 1990, whether this standard needed to be amended. The rule, issued on May 14, 1991, took effect three years later (56 FR 22279).

In the 1991 rule, DOE spokesmen announced plans to accelerate the second review of energy efficiency standards for clothes washers because the department had become aware, after the rule making was closed, of the horizontal-axis design used in Europe that had not been considered during the rule making. On November 14, 1994, DOE issued an Advance Notice of Proposed Rulemaking (ANPR) that began the second review of energy efficiency standards for clothes washers, dishwashers, and clothes dryers. DOE presented the technologies to be considered and the product classes they planned to analyze along with the analytical framework and models to be used in performing analyses. The Federal Register notice specified DOE's intention to consider horizontal-axis washers as one of the feasible technologies.

In 1996, however, Congress required DOE to revise its standard-setting process to include stakeholder participation. In 1996, DOE published the final rule called "Procedures for Consideration of New or Revised Energy Conservation Standards for Consumer Products," which became known as the "process rule." In 1998, DOE issued a Supplemental ANPR, beginning the first rulemaking under the process rule. The new notice presented the product classes to be analyzed, the analytical framework, and preliminary analyses of life-cycle cost, payback, and national energy savings.

DOE convened a series of meetings between manufacturers, energy groups, and other stakeholders, beginning in 1996 and continuing until May 2000. In May 2000 the working group reached agreement on a proposed standard. DOE responded favorably to the proposal and in October endorsed it with minor modifications in a proposed rule that recognized the joint stakeholders' proposed standards to be "technically feasible and economically justified." The final rule was published in January 2001 and survived a subsequent review by the incoming administration.

The joint stakeholders agreement included four provisions relevant to clothes washers. First, it created new energy standards based on "modified energy factors" (MEFs) to take effect in 2004 and 2007. Second, it set new MEFs for machines to use in qualifying for the Energy Star program. Third, it provided tax credits for the production of energy efficient clothes washers as well as refrigerators and freezers. The tax credit will create two energy-efficiency standards. A firm will receive $50 per unit that reaches the first level and $100 for each unit that reaches the second level, up to $30 million per company per level.

Finally, the new rule included an agreement for firms to disclose voluntarily the water usage factors for each model that meets the Energy Star standards, beginning in 2001. The new Energy Star standard provides an initial 22 percent reduction in energy consumption over the current standard by January 1, 2004, and a 35 percent reduction by January 1, 2007. Table 5.3 summarizes the events leading to this agreement.

Industry participants differ in their assessment of the most recent round of negotiations. While most describe themselves as pleased with

Table 5.3
Legislative and regulatory history

1987	National Appliance Energy Conservation Act of 1987 authorizes prescriptive standards and sets initial standard.
1991	Final rule issued. DOE's announces intention to accelerate review of standards.
1994	Advance Notice of Proposed Rulemaking (ANPR). DOE declares intention to consider horizontal axis washers as a feasible technology.
1996	DOE publishes final "process rule," on a stakeholder participation process for standard setting.
1998	Supplemental ANPR announces rule making under the process rule.
May 2000	Working group reaches agreement on proposed standard.
October 2000	DOE proposes rule, based on working group agreement.
January 2001	Final rule based on working group agreement.

the outcome, others describe the process as a marathon, and expressed concern that combining the negotiations over mandatory and voluntary standards strained the process.

Despite the recent agreement, the combination of energy efficiency standards and the Energy Star label has produced mixed results. Progress in actually reducing energy consumption has been relatively slow. Less efficient machines continue to account for the majority of all units sold. Six years after the initiation of the process, the great majority of washers still are not high-efficiency models. Rather than driving the entire industry to change, the program has created two tiers of energy use in the industry.

This two-tier system has positive and negative consequences. Energy efficient machines occupy the high end of the market and capture a significant price premium. Opportunities to differentiate products based in part on energy efficiency have helped spur a wide range of innovation in an industry that had previously lagged behind other sectors in innovation. Innovations at the higher end, along with publicity about the Energy Star program, have raised consumer awareness and built demand for energy-efficient products. Innovations incorporated initially only in high-end products have begun to affect the design of mainstream clothes washers, and have begun to shift the entire market.

At the time of the 1994 Advance Notice of Proposed Rulemaking, none of the major US manufacturers offered high-efficiency washers with performance comparable to European-made horizontal-axis washers. However, by August 2000, all of the major manufacturers in the US market had high-efficiency models on the market or under development. Maytag, Electrolux, and General Electric all offered horizontal-axis machines, and Goodman Manufacturing (Amana) reportedly had a horizontal-axis machine under development. Whirlpool, the market leader, has introduced an energy-efficient vertical-axis machine.

Although the Clothes Washer program has created a sizable energy-efficient segment, the program has not driven the less efficient models from the market. In the absence of continued pressure from DOE, and requirements built into energy legislation to periodically reassess product standards, the clothes washer market might divide into two permanent tiers.

In summary, the Energy Star Clothes Washer program has stimulated technological innovation that has increased the supply of energy-efficient products. It has allowed manufacturers to differentiate energy efficient washers, allowing customers to identify energy-efficient washers. Use of the label has led some of the participants to increase sales and gain favorable publicity. At least two manufacturers have found that the Energy Star product has improved their product's overall reputation for quality and reliability.

In combination with continuing legislative pressure to upgrade energy-efficiency standards periodically, the Energy Star program appears to have begun a major market transformation. However, by the time the newly proposed standards take effect in 2004, the process will have taken more than a decade. The long product life of clothes washers ensures that a major transformation will take many more years.

The Energy Star Office Products Program[7]

The Energy Star Office Products program illustrates the dynamics that occur when a voluntary program provokes a common response among virtually all manufacturers in an industry. The program allows manufacturers to use the Energy Star label on any office equipment that meets a level of energy efficiency specified by the program.

EPA describes the office products initiative as its flagship because it is the first and largest of the Energy Star programs (Thigpen et al. 1998). EPA initiated it in 1992 to expand markets for energy-efficient goods. The objective was to publicize the cooperative efforts of industry groups, create awards to recognize superior efforts by individual firms, and conduct extensive media campaigns to raise public awareness of the Energy Star "brand." At first the office products segment of the program focused narrowly on computers and computer monitors. In subsequent years it expanded to encompass printers, copiers, and multifunction devices.

This initiative was the first US voluntary program focused on products. It began with a proposal from EPA that defined performance standards for computers and monitors and allowed participating firms to use EPA's Energy Star logo to differentiate program-compliant products. The program was designed "to create a market for energy-efficient desktop computers, by providing a clearer market incentive for manufacturers to improve the efficiency of their products and an effective mechanism for consumers to make informed purchasing decisions" (Thigpen et al. 1998). Nearly 100 percent of the firms in the computer manufacturing industry have signed memoranda of understanding (MOUs) committing them to participate. In setting up the program, the Agency stated that it had attempted to "balance the desire to set challenging specifications that maximize per unit energy savings with the desire to set specifications that allow somewhat less savings, per unit basis, but expand the overall market for energy-efficient products" (EPA 1992).

In effect, the agency chose the highest level of energy efficiency that could still allow broad participation. Whether EPA expected the entire industry to enroll in the program is unclear from both published documents and interviews with agency program officers.[8]

The initial guidelines for Energy Star computers and monitors called for products to enter an energy-efficient "sleep state" when not in use. These specifications were relatively easy for many manufacturers to implement because an inexpensive technology used in laptop computers was readily available from a major supplier. This solution allowed manufacturers to adapt products without major problems in either technical design or organizational coordination.

The fact that firms had failed to implement this technology before the program began reflects the limited prior demand for energy-efficient office equipment. Although interviewees expressed support for efforts to reduce energy consumption, none felt that their customers, other than government purchasing departments and a few large clients, demonstrated any interest in energy savings as a purchasing criterion. A spokesperson for one firm indicated, however, that recent marketing data showed energy efficiency had begun to be an important purchasing criterion for its customers.

The subsequent evolution of Energy Star requirements for computers and monitors illustrates a key limitation of the program's design: EPA's attempts to develop more stringent standards for subsequent rounds have led to lengthy and occasionally acrimonious negotiations with virtually all major office equipment manufacturers. Industry participants commented on the change in the collaborative stance of program managers between the agreement on the initial requirements and the most recent round of negotiations in 1998 and 1999.

Although none of those interviewed indicated that their companies had substantial disagreements with the original program requirements, interviewees from all but one firm—which didn't participate in the most recent negotiations—expressed concern about the increasing complexity and the adversarial character of the most recent negotiations. Some participants thought that EPA had moved away from the cooperative tone that the program had initially attempted to create.

During the negotiations EPA invoked the possibility that the Energy Star memos of understanding might not be renewed if an appropriate agreement could not be reached. The agency also raised the specter of more stringent European regulations to persuade manufacturers to compromise on further reductions in energy consumption. In each round of negotiations the parties have eventually reached an agreement, but not without imposing substantial costs on everyone involved.[9] Some participants felt that the program was likely to continue evolving in the future, but several commented that they might be reaching the point of diminishing returns.

The Energy Star Office Products program is somewhat unusual in that it actively intervenes in the behavior of both manufacturers and customers. EPA approached all the leading producers of personal computers and monitors at virtually the same time. Rapid acceptance by several leading manufacturers raised the stakes for others and led many others to sign on. The initial list of 10 participating companies in September 1992 grew to more than 600, including component suppliers, by November 1998 (EPA 1998).

The program influences the purchasing decisions of large customers by providing them with an easy way to identify energy-efficient equipment. By limiting purchases to products that comply with Energy Star guidelines, procurement managers can assure that their organizations will receive relatively energy-efficient products without having to develop detailed energy use criteria. In 1993, the Energy Star programs received a significant boost from Executive Order 12845, which ordered government procurement offices to purchase Energy Star–compliant products whenever possible. The program has also made substantial efforts to encourage state and local government procurement organizations to specify office equipment bearing the Energy Star imprimatur.

These demand-side interventions have created a sizable market for energy-efficient goods. Concentrated demand has built momentum by guaranteeing manufacturers a level of sales that justify the manufacture of energy-efficient products. In fact with very few exceptions manufacturers can no longer afford to produce noncompliant models.

Although EPA has publicized the advantages of participating in the program, firm participation appears to have been motivated by a general desire to make improvements where the costs of doing so are not substantial and by the fear of lost revenues if the firm does not participate. Interviews with industry representatives confirmed the importance of potential negative consequences in motivating them to participate in the Energy Star program. Five of the six computer manufacturers interviewed said that the management of their companies believed that participating in Energy Star was a requirement if they wanted to continue doing business with units of government. Some also mentioned that their large corporate customers also required it. An interviewee from the sixth manufacturer said that the decision was straightforward for her firm because it had already developed a technology that would meet the requirement.

The program works, in part, because it prevents having participants' prices undercut by nonparticipating manufacturers. Otherwise, in the very competitive markets for office products, any manufacturer that raised prices to pay for energy-efficiency enhancement would be at a competitive disadvantage relative to those who left energy consumption unchanged. Few firms would choose to improve energy efficiency unilaterally, especially considering the low priority that customers have traditionally placed on it.

The threat of regulation was not a factor in the Energy Star Office Products program. EPA does not have legislative authority to develop energy efficiency standards for office products. However, the threat of sanctions was very real. Firms that failed to develop Energy Star–compliant products would be unable to sell to the federal government, and probably could not sell products to large customers and energy conscious consumers. Facing fierce competition, virtually all firms in the industry joined the program.

The Energy Star Office Products program illustrates a hazard inherent in voluntary initiatives that produce a collective response—the difficulty of making standards more stringent than their initial levels. The program focused initially on a readily achievable source of energy savings, energy consumption while the computer or monitor is idle. Recently the program has begun to address the issue of energy consumption when the devices are in use. This would require more extensive technical change, for example, incorporating technology that shifts individual components of a computer such as hard disc drives to energy-saving states when not in use. Proposals to tighten the requirements for program participation met initial resistance because participating firms differed in their ability to implement these additional measures.

In summary, the EPA's Energy Star Office Products initiative quickly enlisted virtually all manufacturers in the industry in a voluntary program to reach a level of performance that was feasible for most, if not all, manufacturers of computers and monitors. Most participants complied, utilizing a previously existing technical solution they had no prior incentive to implement. After calculating the benefits and costs of participating, these firms used the Energy Star label to signal the energy efficiency of their products to customers. Subsequent negotiations to raise the standards for participation required negotiations and compromises among participating firms, and between participating firms and the agency. However, the overall result has been a significant level of energy savings, along with periodic increases in energy efficiency achieved by the entire industry.

Separating and Converging Mechanisms

The Energy Star programs for office products and clothes washers illustrate the dynamics of separating and converging mechanisms in voluntary programs. The Clothes Washer program is an example of a separating mechanism, one that drives individual firms to behave differently from their competitors (Rothschild and Stiglitz 1976). Such a mechanism enables individual firms with superior environmental performance to differentiate themselves from others. Separating mechanisms may focus on the performance of a firm's operations or the environmental impact of its products, or both.

Separating mechanisms in voluntary initiatives include programs managed by nongovernment organizations (NGOs), governments, and standards organizations. For example, product eco-labeling programs such as the Blue Angel and Green Cross labels recognize products with superior environmental performance. The Coalition for Environmentally Responsible Economies (CERES)—an NGO—has created a program that asks firms to commit to a stringent code of practices and provide detailed reports on their accomplishments.[10] Separating mechanisms also include government programs, such as Climate Wise and Green Lights, which give recognition to firms that adopt energy-efficient technologies.

Figure 5.2 illustrates the development of a separating mechanism: The "initiating party" proposes a voluntary program to identify firms that are outstanding in a particular dimension of environmental performance. The initiating party announces the program to the target audience(s) and the general public. Firms then decide whether to participate in the program or not, based in part on their estimate of the response from targeted audiences. Target audiences in turn decide whether to reward firms

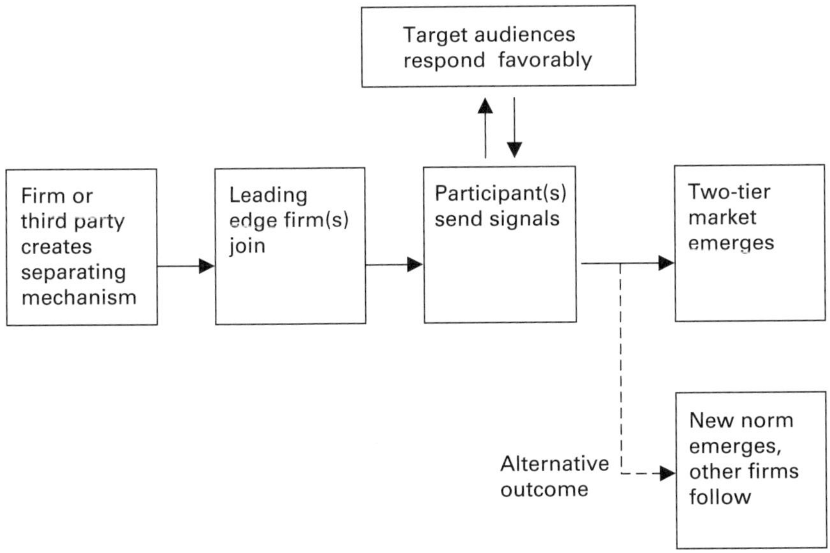

Figure 5.2
Development of a separating mechanism

that choose to participate, to sanction firms choosing not to participate, or to ignore the program altogether. A separating mechanism may create customer or stakeholder expectations that drive trailing firms to follow the leaders in improving environmental performance or energy efficiency.

The Energy Star Office Products program is an example of a converging mechanism, one that leads all targeted firms to adopt a desired behavior. Such approaches tend to impose less ambitious requirements than separating mechanisms, at least initially, because standards must be achievable for all firms in the targeted group. Converging mechanisms produce significant results by stimulating all firms in a given industry to participate. This approach is relatively simple in concept, as shown in figure 5.3. Typically the initiating government group, NGO, or standards organization proposes a voluntary improvement in the environmental performance of the target population. The initiating party contacts the target population—all firms within an industry. All parties then negotiate the requirements and, if agreement can be reached, the entire group commits to them.

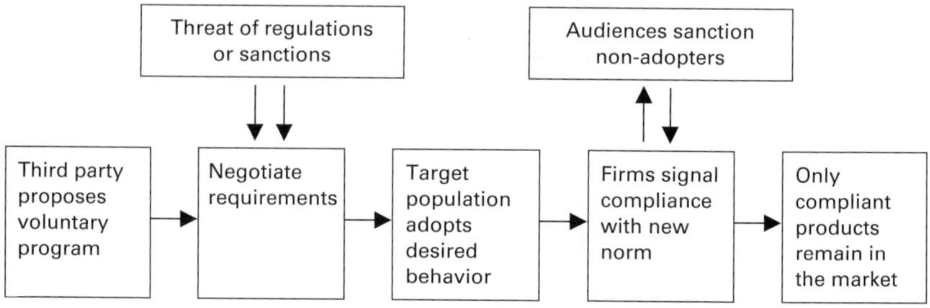

Figure 5.3
Development of a converging mechanism

Participation in a converging mechanism is a mixed blessing for participating firms. No firm suffers a competitive disadvantage from participating, even though participation may raise production costs. On the other hand, no firm gains competitive advantage relative to other participating firms.

In influencing corporate decision making, a converging mechanism relies more heavily on external pressures, such as regulatory threats or changes in customer preferences. The success of a converging mechanism depends more on the possibility of sanctions for nonparticipation than on rewards for participation.

The effectiveness of a converging mechanism depends, in part, on assuring that customers or other stakeholders actually deliver the expected rewards or sanctions. In practice, the most effective sanctions associated with a converging mechanism are often indirect. A company is at risk if it is the only firm or the most visible firm in the target population that does not participate. This threat is difficult to evaluate quantitatively, but can lead to a significant loss of general reputation, potential loss of revenues and market share, and sometimes decreased stock price. In many instances, the nature and likelihood of sanctions may be difficult to assess precisely. This uncertainty concerning sanctions creates the possibility of strategic behaviors by both the initiators and the target population.

The two programs discussed in this chapter may not be entirely representative of their respective mechanism types. The Office Products program developed an industrywide transformation without a significant threat of regulation. The Clothes Washer program made creative use of a very real threat of regulations, and a cooperative rule-making process that has started an evolution toward a converging mechanism. These complexities suggest that converging and separating mechanisms may represent ends of a theoretical spectrum rather than mutually exclusive types.

Conclusion

The Energy Star programs for clothes washers and office products have created incentives for manufacturers to improve the energy efficiency of their products beyond legal requirements. These successes were achieved through two different policy approaches—separating and converging mechanisms—reflecting factors in the nature of and market for the products targeted. As shown in table 5.4, these two mechanisms diverge dramatically in the way that they drive change; in the requirements that they impose on participants; and the circumstances in which they can be employed. Yet both have led to significant energy savings and have instigated major market transformations.

The success of these two programs illustrates the potential power of voluntary programs designed to alter the relationships between manufacturers and customers. But the two programs also illustrate potential limitations of separating and converging mechanisms. The Energy Star Office Products program led to rapid, but limited transformation within the industry. Subsequent improvements in energy efficiency have been difficult to achieve. The Energy Star Clothes Washers program led initially to innovation within the industry, but very little change in energy usage. However, special circumstances have led to a technology race that has propelled a major change in the industry. Here, the separating mechanism along with the market structure has led to a continuous stream of innovations.

These mixed results suggest that the choice between these two mechanisms may not be clear-cut for any particular situation. This chapter does not identify the conditions in which each mechanism is most appropriate. However, a few speculations may be in order.

Separating mechanisms may be most effective in industries where firms already compete through strong product differentiation. In such industries

Table 5.4
Comparison of separating and converging mechanisms

Attributes	Separating mechanism	Converging mechanism
Primary impact on market	Harness demand for greener products	Intervene directly in the supply of all products in the same category, prevent less-energy efficient products from under-cutting prices
Primary "levers"	Reward leaders by increasing sales of their products	Sanction nonparticipants, or impose regulation if agreement isn't reached
Change model	Create incentives for followers to catch up	Move entire group forward in negotiated steps
Requirements for success	Ability for firms to differentiate their products	Compelling reason to cooperate, such as threat of regulation or sanctions
Strengths	Can create competitive advantage for leaders, can provoke competition based on improved environmental characteristics	Can improve entire industry's performance simultaneously
Pitfalls	Can have limited influence on firms that can't qualify	Can produce least common denominator, can be less ambitious than regulations

a voluntary initiative can create incentives to further differentiate products, based on an environmentally desirable characteristic such as energy efficiency. In this way a separating mechanism provides more incentives for continuous innovation in the industry.

Converging mechanisms may be most effective where an existing technology or management strategy has been underutilized. In such industries a voluntary initiative can help overcome the disincentive for any individual firm to undertake energy-efficiency gains or other environmental improvements that would raise costs relative to competitors' products. Converging mechanisms, therefore, are more effective at diffusing innovations that have not been widely adopted within an industry.

Voluntary product labeling initiatives have the potential to become a powerful category of policy instruments by waking the sleeping giant of consumers' preferences for environmentally preferable products. Carefully designed initiatives can harness powerful market forces by overcoming information barriers and allowing customers to vote effectively through their purchasing decisions.

Acknowledgments

The author gratefully acknowledges financial support from the Atmospheric Pollution Prevention Division of the US Environmental Protection Agency, the Switzer Foundation, and the Environmental Studies Department at the University of California, Santa Cruz during the early development of this work. Managers from twelve companies shared valuable insights into their companies' experiences in the Energy Star program. While maintaining their anonymity, I want to acknowledge their critical contributions. This chapter also benefited from insights provided by many individuals in government and academea. These included Richard Howarth, Brent Haddad, Steve DeCanio, Steve Gliessman, Skip Laitner, Jerry Dion, Steven Sylvan, Andrew Fanara, John Rivera, Jeanne Briskin, Jeremy Simons, Eric Dolan, Vicki Norberg-Bohm, Theo de Bruijn, Dan Fiorino, Gabriel Labbate, Joan Brunkard, Jonathan Koomey, and participants at the Association for Public Policy and Management (APPAM) conference in Seattle, Washington, November 2000, the Environmental Studies Department Graduate Seminar at the University of California, Santa Cruz, May 2000, and the Workshop on Voluntary, Collaborative and Information-Based Policies, Cambridge, Massachusetts, May 2001.

Notes

1. To protect their anonymity, the industry interviewees are not cited. All comments attributed to individual firms are taken from published sources.

2. The phrase "market transformation" is widely used in energy-efficiency literature, but does not appear to have a consistent definition. In this chapter the phrase refers to conversion of a large percentage of a market to more energy-efficient products, in response to an intervention.

3. Rothschild and Stiglitz refer to the outcomes as "equilibria". This study refers to the outcomes as "mechanisms" to allow for the possibility of outcomes that

are not stable equilibria. Although application of the term is new in this study, converging mechanisms have been common in voluntary environmental initiatives.

4. I gratefully acknowledge an unidentified participant at the Association for Public Policy and Management (APPAM) conference in Seattle, Washington, November 3, 2000, for calling my attention to this distinction.

5. Cost savings estimates for both industries are expressed in 1998 dollars.

6. Investigation of consumer decision-making processes is outside the scope of the current study. Detailed investigation of the changes in customer response to energy-efficiency information could provide significant insights into the effectiveness of voluntary initiatives such as the Energy Star Washing Machine program.

7. This section builds on the analysis presented in Howarth at al. (2000).

8. Subsequent Energy Star programs for televisions, video-cassette recorders, and stereos attempted explicitly to enlist the participation of their entire industries (Sylvan 1999).

9. Discussion was quite heated at one negotiating session observed by the author. Industry participants indicated during informal discussions that the negotiation process had strained their relations with the EPA program management team.

10. Committing to the CERES principles is both a signal and a formal commitment to a "reporting" mechanism in which participants provided detailed information.

References

Appliance. 2000. 23rd portrait of the U.S. appliance industry. *Appliance* (September): 83–89.

Appliance. 1999a. 22nd portrait of the U.S. appliance industry. *Appliance* (September): 75–81.

Appliance. 1999b. Portrait of the European appliance industry. *Appliance* (November): 61–65.

Appliance. 1999c. Statistical review: A ten year review 1989–1998 of the U.S. Appliance industry. *Appliance* (April): 51–54.

Brown, R., C. Webber, and J. Koomey. 2000. Status and future directions of the Energy Star program. In *ACEEE Summer Study on Energy Efficiency in Buildings*. Washington, DC.

Howarth, R. B., B. M. Haddad, and B. Paton. 2000. The economics of energy efficiency: insights from voluntary participation programs. *Energy Policy* 28(6–7): 477–86.

Kawamoto, K., J. Koomey, B. Norbman, R. Brown, M. Piette, and A. Meier. 2000. Electricity Used by Office Equipment and Network Equipment in the U.S. In *ACEEE Summer Study on Energy Efficiency in Buildings*. Washington, DC.

LaPat, K. 2000. Loads of innovation—53rd annual report on laundry appliances. *Appliance* (September): 38–45.

McHenry, F., and M. Houston. 2000. Applying flash memory in white goods appliances. In *Appliance* (June): 70–73.

Mnif, K. 2000. Watching water-sensors make clothes washers smarter. *Appliance Manufacturer* (February): 52–54.

Murray, K. 2000. DashDSP simplifies washing machine control system development. *Appliance* (May): 88–90.

Perlman, K. 1999. Clothes calls, 52nd annual report. *Appliance* (September): 44–48.

Perlman, K. 1998. Clothes to home. *Appliance* (September): 46–50.

Rothschild, J. E., and M. Stiglitz. 1976. Equilibrium in competitive insurance markets: An essay on the economics of imperfect information. *Quarterly Journal of Economics* 90: 629–49.

Stiglitz, M. 1975. The theory of "screening," education, and the distribution of income. *American Economic Review* 65(3): 283–300.

Sylvan, S. 1999. *Personal communication.* Washington, DC. September.

Thigpen, S., A. Fanara, A. ten Cate, P. Bertoldi, and T. Takigawa. 1998. Market transformation through international cooperation: The Energy Star Office Equipment example. In *ACEEE Summer Study on Energy Efficiency in Buildings*: 5.315–5.326.

US Environmental Protection Agency (EPA). 1992. *The Climate Is Right for Action: Voluntary Programs to Reduce Greenhouse Gas Emissions.* Washington, DC: Office of Air and Radiation.

US Environmental Protection Agency (EPA). 1998a. *Energy Star and Related Programs: 1997 Annual Report.* Washington, DC: Atmospheric Pollution Prevention Division.

US Environmental Protection Agency (EPA). 1998b. *Energy Star Office Products Program.* Washington, DC: Atmospheric Pollution Prevention Division.

Webber, C. A., R. E. Brown, and J. G. Koomey. 2000. Savings estimates for the Energy Star voluntary labeling program. *Energy Policy* 28: 1137–49.

6

Reaching Environmental Goals through R&D Collaboration: Lessons from the US Department of Energy Programs for Gas Turbines and Solar Photovoltaics

Vicki Norberg-Bohm and Robert M. Margolis

Introduction

Over the 1990s the US Department of Energy (DOE) launched several collaborative R&D programs aimed at reaching environmental and competitiveness goals. This emphasis on partnerships and collaboration was part of a larger trend toward collaboration in both environmental policy and technology policy. In environmental policy there has been a growing recognition of the potential for industry to be part of the solution and thus an effort to reshape environmental policy and politics away from adversarial interactions and toward joint problem solving.

In technology policy, although still a contended issue, there is today a growing acceptance that government should support civilian technology innovation that is in the public interest, for example, to improve national economic competitiveness and environmental sustainability (Branscomb and Keller 1998; Stiglitz and Wallsten 2000). For civilian technology, the major challenge for governments is to find ways to encourage private firms to commercialize new technologies that are in the public interest. To accomplish this goal, over the past decade DOE has included collaborative elements in many of its programs. In this chapter we examine three collaborative programs focused on the development of advanced energy technologies that have been organized and funded by DOE: the Advanced Turbine Systems (ATS) program, the Photovoltaic Manufacturing Technology (PVMaT) project, and the Thin-Film PV Partnership project.[1]

In the context of using R&D collaboratives to meet environmental goals, we ask two sets of questions. The first relates to goal setting, focusing on whether collaborative R&D programs provide opportunities for setting "stretch" goals, meaning goals that result in technological innovation that provides significant improvements in environmental performance and that move the technology faster or further than the private sector would have done without the program. In other words, we examine whether these collaborative R&D programs create nonregulatory incentives for improving environmental performance. The second set of questions focuses on the role of collaboration in reaching R&D goals, namely during implementation. We look specifically at the ways in which the collaboration reduced risks or spread risks, thus leading to greater success in technology development. Based on the answers to these two questions, we examine the conditions under which collaborative R&D programs offer a viable model for achieving environmental goals.

This chapter is organized as follows. The next section examines the potential benefits of collaborative R&D, addressing the two questions outlined above. The third section introduces the three US DOE R&D collaboratives evaluated in this chapter. The fourth section focuses on the role of collaborative goal-setting while the fifth looks at the role of R&D collaboration in meeting these goals. The concluding section draws policy lessons from this analysis.

The Benefits of Collaborative R&D

Although collaborative R&D programs have existed for much of this century, there has been a significant growth in government sponsored collaborative R&D over the last two decades. During the 1980s the US Congress passed several laws that paved the way for greater collaboration on civilian technology: the Technology Innovation Act of 1980 (also known as the Stevenson-Wydler Act), the Patent and Trademark Amendments Act of 1980 (also known as the Bayh-Dole Act), the Trademark Clarification Act of 1984, the Federal Technology Transfer Act of 1986, and the National Competitiveness Technology Transfer Act of 1989. Together this set of laws made it possible for small businesses and nonprofit organizations, such as universities, to retain the rights to intellectual property resulting from government-sponsored research. This set of laws also enabled large firms to negotiate to retain the rights to intellectual property generated in cost-shared contracts with the government (Schacht 1999). In the wake of these laws, several government-sponsored R&D programs encouraged or required collaborative efforts.

Collaboration is increasingly being recognized as essential for building the technological and market capabilities needed to successfully commercialize new technologies (Fountain 1998; Powell 1998; Teece 1988). While advocating this position, Jane Fountain argued that, "The federal government should aggressively provide incentives and information to promote the use of networks and consortia in order to connect firms to universities, national laboratories, and state and federal partnership programs" (Fountain 1998: 86–87). Such enthusiasm not withstanding, collaboration has both advantages and costs, with competition remaining an important driver of technology innovation (Mowery 2002). In order to formulate policy recommendations on the use and design of R&D collaboration, there is an urgent need to learn from recent experiences with government-sponsored R&D collaboratives (Mowery 1998).

This research takes on this challenge, with a specific focus on evaluating the potential of using R&D collaboration to meet environmental goals. In this context it examines the value of collaborative mechanisms for both goal setting and for implementing R&D. Government sponsorship of collaborative R&D can provide an opportunity to establish "stretch" goals, the technological goals with environmental benefits that are beyond what is required by regulation and what the private sector would pursue on its own. By developing these environmentally motivated goals within the context of a partnership or collaborative program, the government creates an incentive for all firms in an industry to participate in the program. Firms that do not participate face the risk that their competitors will be ahead on the development of new environmentally sound technologies, technologies that could later provide competitive advantages. Competitive advantage is created when a new technology proves to be win–win, meaning both economically and environmentally beneficial, or when the technology sets the standard for regulations that require better environmental performance, such as stricter emission controls.

In terms of implementation the potential benefits of collaboration include both risk spreading and risk reduction. Risk spreading occurs when firms share the risk of developing new technologies with the government or with each other. Risk reduction occurs when collaborative R&D results in better or faster solutions. Firms are likely to engage in collaborative R&D as long as the total decrease in risk outweighs the loss in benefits from collaboration. Potential losses to individual firms include shared ownership of intellectual property or loss of trade secrets through the collaborative venture.

The most direct way in which government sponsored R&D spreads risk is through cost-sharing, in which government finances a portion of the R&D undertaken by firms. Although cost-sharing does not necessarily involve collaboration in the R&D process, in practice it involves some cooperation in R&D planning between the government and the private sector. As noted above, this cooperative planning can lead to setting "stretch goals," and in so doing can be the most critical avenue for the provision of nonregulatory incentives for improved environmental performance.

Beyond the role of government cost-sharing, R&D collaboration has the potential to spread risks by sharing activities that are outside the scope or capability of an individual firm. Collaborative programs can encourage risk sharing through five potential pathways.[2] First, collaboration may provide firms with access to complementary assets. These assets may include "hardware," meaning laboratory or manufacturing equipment, or "software," meaning human know-how, both codified and tacit.

A second method for risk spreading is to develop shared visions through standards or roadmaps (Mowery 1998). Industry roadmaps can be of particular importance in spreading the risk of developing new technological paradigms (Teece 1988). Government sponsorship can be important to gain the benefits of coordination, as it can guarantee protection from antitrust enforcement that firms might otherwise be subject to for industrywide collaborations (Kelley 1997).[3] While shared visions can spread risk, they also run the risk of creating a lock-in to suboptimal technological trajectories.

A third way to spread risk is by collaborating to gain the benefits of scale economies (Mowery 1998). Scale economies in R&D can be exploited when there are joint problems to be solved within an industry that do not impinge on the competitive advantage of individual firms. This is most likely to occur for precompetitive research. Equally important, when there are scale economies in production, collaborative efforts can create a market that is large enough to reduce the business risk a supplier firm faces in moving a technology through commercialization.

Fourth, reducing duplication can spread R&D costs and the associated risks across multiple firms (Mowery 1998; Stiglitz and Wallsten 2000). From a societal perspective this may lead to economic efficiency, but it is at the risk of limiting the number of technological paths that are being explored.

Finally, collaborative R&D can also spread risk by allowing firms to capture knowledge spillovers (Mowery 1998). In particular, government sponsored R&D projects can reduce business risk if they allow firms to maintain the intellectual property rights from collaborative research done with national labs or universities, or to enter into contracts that share intellectual property rights with other parties on an exclusive basis.

R&D policy has embraced collaboration not only for its potential to spread risks but also for its potential to reduce risks. The normative arguments favoring R&D collaboration suggest that it can reduce risks through three pathways. First, collaboration can enhance and increase the speed of information flows, creating greater access to information by more parties, and providing for the exchange of tacit knowledge (Powell 1990). Second, these enhanced and novel channels of information flow as well as jointly planned and implemented R&D can facilitate technology transfer (Powell 1990). Third, R&D collaborations can increase the technological and organizational capability of the participants. In particular, they offer a better focus for government and university researchers on work that is relevant to industry. The increased information flow, technology transfer, and improved capabilities can lead to creative and novel solutions as well as reducing time and transaction costs.

From the perspective of risk reduction, there are also potential disadvantages to collaboration that must be considered. First, collaboration can create significant transaction costs. These costs must be weighed in considering the benefits of a specific collaboration, and collaborations should be organized to keep transaction costs to a minimum. Second, government-sponsored research that focuses government and university researchers on work relevant to industry raises the concern that near-term technological goals will dominant the research agenda of national laboratories and universities. This can be countered by designing overall government R&D portfolios to distribute research efforts between near-term and long-term technology development.

R&D Collaborations in the Department of Energy

The Advanced Turbine Systems Program

At the beginning of the 1990s a convergence of market potential, technological potential, and public interest made the idea of an advanced gas turbine R&D program attractive to industry and government. Although the market for gas turbines was nearly dead in the United States in the late 1980s and early 1990s, the industry anticipated a strong and growing market by the end of the decade. Due to the lack of current sales, the gas turbine firms were in no position in the early 1990s to invest R&D in a next-generation turbine. Nonetheless, there were promising engineering ideas for further technological development for land-based turbines; some ideas came from advances in jet engine technology, others from R&D directed specifically at combined-cycle gas turbines (CCGT). This opportunity for further technology development for a market that was expected to bloom within the decade coincided with two federal administrations (Presidents George H. Bush and Bill Clinton) that embraced the idea of government support for moving technology toward commercialization. Gas turbines also held the promise of addressing three areas of public interests: reducing CO_2 emissions, reducing emissions of local pollutants, and reducing dependence on foreign oil. In this context government and private sector entrepreneurs gained the support of the relevant congressional committees for a land-based turbine program, which was established through the 1992 EPACT with little fanfare or debate.

The goals of the ATS were to develop ultra-high-efficient, super-clean, and cost-competitive turbines. The ATS was a nine-year research, development, and demonstration program with a total budget of \$826 million of which \$325 million (\$1999) was industry cost share.[4] It involved development of utility-scale turbines (approximately 400 MW and larger); industrial-scale turbines (5–15 MW); and technology base development for materials and manufacturing, combustion, and coal and biomass applications. In this chapter we focus only on the utility-scale turbines, although we recognize that the technology base development was directed toward the needs of both utility and industrial turbines. The ATS goals for utility-scale turbines were efficiency greater than 60 percent, NO_x emissions of less than 10 ppm, and electricity costs 10 percent less than the existing generation of gas turbines. A fourth goal of the program was fuel flexibility, although over time this goal took a backseat to the other three goals. At its inception the ATS program was divided into four phases: (1) systems studies, (2) concept development, (3) technology readiness testing, and (4) precommercial demonstration. In practice, because of budgetary constraints, phases 3 and 4 were combined, and the program did not support full-scale demonstrations.

With the ATS program now officially over, most participants evaluate it as an unqualified success. GE and Siemens-Westinghouse turbines are reaching the ATS goals for efficiency and environmental performance (DOE FETC 1999a,b).[5] General Electric has completed testing and began demonstrations in 2001. Siemens-Westinghouse is in the process of field testing and will conduct demonstrations soon (Layne 2000).

The key system innovation in the ATS turbines is steam cooling, which plays an important role in simultaneously meeting efficiency and emissions goals.[6] The ATS turbines are very high temperature machines with turbine inlet temperatures of 2600°F. New materials were needed to operate at these high temperatures, including thermal barrier coatings and single crystal castings for blades. These technologies had already been deployed in aircraft engines but required significant additional research in order to be applied in large land-based turbines. Innovations were also required to bring NO_x to single digits in these high temperature turbines. Dry combustion systems with lean pre-mix multistage designs were further developed to reach this goal.

The Photovoltaic Manufacturing Technology (PVMaT) Project

Between 1980 and 1990 the US share of global photovoltaic (PV) cell production fell from 76 to 32 percent (OTA 1995: 234). This dramatic decline in market share lead to rising concern within Congress, DOE, the National Renewable Energy Lab (NREL), and the Solar Energy

Industries Association (SEIA) about the long-term competitiveness of the PV industry in the United States. In 1991 in response to the prospect of a vanishing US PV industry, the DOE initiated a new program focused on improving PV manufacturing technology. This initiative, the Photovoltaic Manufacturing Technology (PVMaT) project, was initially envisioned as a five-year industry-government cost-shared project that would help the US PV industry remain competitive.

The PVMaT project's original goal was, "to ensure that the US [PV] industry retains and extends its world leadership role in the manufacture and commercial development of PV components and systems" (Witt et al. 1993). Striving to meet this goal the project has focused on improving manufacturing processes and equipment, accelerating manufacturing cost reductions, improving commercial product performance and reliability, and laying the groundwork for substantial scale-up of US-based PV manufacturing plant capacities (Witt et al. 1998).

The PVMaT project has completed five rounds of solicitations to date. Total project funding through FY1999 was $80.5 million in DOE funding and $59.1 million in industry funding (Witt et al. 1999).

After roughly a decade of operating the PVMaT program is viewed, both within the DOE and by members of the US PV industry, as one of the most successful DOE R&D programs (Herwig 1996; NREL 1999). It helped US PV manufactures continue to reduce costs, expand production and stay competitive in a rapidly growing global PV market during the 1990s. For example, the 12 PVMaT participants with active production lines in 1999 realized significant cost reductions during the 1990s. Their weighted-average cost for manufacturing PV modules declined by 36 percent, from $4.23 per peak watt in 1992 to $2.73 per peak watt in 1999. In addition their manufacturing capacity increased by more than a factor of seven, from 13.6 megawatts in 1992 to 99.3 megawatts in 1999 (Witt et al. 2000: 4).

Thin-Film PV Partnership

The roots of the Thin-Film PV Partnership go back to 1978, when DOE formed the Solar Energy Research Institute (SERI).[7] While SERI's mission was to develop the myriad forms of solar energy—wind, biomass, PV, and solar thermal—from its inception it maintained a heavy emphasis on thin-film PV (Zweibel 1990: 140). In 1978 the research effort began with amorphous silicon (a-Si) materials and devices as part of a Photovoltaic Advance Research and Development program. During the 1980s and early 1990s SERI, and then NREL, carried out a series of three research projects focused on improving a-Si technology. The Thin-Film PV Partnership was organized into a formal project between 1992 and 1994.

The Thin-Film PV Partnership expands on these earlier efforts in two important respects. First, it covers a broader range of thin-film PV technologies including amorphous silicon (a-Si), cadmium telluride (CdTe), and copper indium diselenide (CIS). Second, the project encourages industry–government–university collaboration through its four national teams. Three of the national teams are organized around specific technologies (a-Si, CIS, and CdTe), while a fourth team, the Environmental Safety and Health (ES&H) team, was organized to focus explicitly on the potential environmental impacts of PV. Awardees of contracts under the Thin Film PV Partnership are required to participate on the teams. Each team is a close-knit group typically made up of 5 to 10 researcher from NREL, 15 people from universities, and 5 people from industry. The teams are not a consortium per se, but they do meet regularly (about twice a year) to organize collaborative research activities among the team members. NREL management typically lets the industry team members lead the process, with NREL functioning primarily in a monitoring role.

Total funding under the Thin-Film PV Partnership project between 1994 and 1999 was $102 million in DOE funding and $30 million in industry funding. The project focuses its funding on cost-shared contracts with companies that are working to bring thin-film technologies from the prototype to pilot production phase of development. It refers to these firms as "technology partners." The project also funds NREL, universities and firms to undertake R&D aimed at solving more fundamental problems. It refers to these participants as "R&D partners."

The Thin-Film PV Partnership has contributed to increases in the efficiency of both laboratory cells and large-area modules, and has helped firms move technologies from the lab to pilot production. For example, work on multi-junction a-Si cell structures carried out under the partnership has resulted in a steady increase in the efficiency of a-Si cells: Record laboratory cell efficiencies increased from roughly 9 percent in 1990 to 12 percent in 1999 (NREL 2000a).

The experience with CIS under the Thin-Film PV Partnership has been similar. The CIS team has focused on a range of issues, including improving junctions, developing non-CdS junctions, and improving molybdenum contacts, that have resulted in increased cell and module efficiencies. Record laboratory cell efficiencies increased from roughly 13 percent in 1990 to 19 percent in 1998 (NREL 2000b).

Collaborative Planning and Goal Setting

All three projects included a collaborative planning process. In two of the programs, the ATS and Thin-Film PV Partnership, participation by industry and government in the planning process was important for setting goals that provided nonregulatory incentives for beyond compliance environmental performance. For all three programs, a collaborative planning process was used to set the R&D agenda, helping the government make better decisions about the use of its R&D resources as well as creating legitimacy in the eyes of industry for the government-sponsored R&D program.

Nonregulatory Incentives

DOE program managers worked both formally and informally with turbine manufacturers, end-users, Congress, and other stakeholders to launch the ATS, including setting technological goals and organizing the R&D program. This process provided one input to setting the goals outlined above. A key question for this research is whether these were stretch goals, pushing the industry beyond its current technology trajectory.

Although reconstructing the "counterfactual" history is fraught with difficulties, our conclusion is that the Advanced Turbine Systems project significantly increased the pace of development of the next generation of turbines. As discussed earlier, the weak market in the early 1990s limited private sector investments in R&D. In this context the ATS program was pivotal in persuading firms to invest in new concepts that could push efficiency to 60 percent. Furthermore, having the program in place helped keep the firms committed to this goal through setbacks in the technology development process. In particular, the industry likely would not have pursued steam cooling beyond systems studies, or dropped it prior to the point of commercialization. Steam cooling represented a radical change in the system's configuration, and thus a change from the incremental approach to innovation previously found in the land-based turbine industry. It also proved to be a greater engineering challenge than originally expected.

The ATS has contributed to efficiency gains for current as well as future generations of gas turbines. In terms of the ATS turbines now being demonstrated, and which expect to be widely commercialized at the end of the decade, experts estimate that efficiencies would have likely reached 58 percent rather than 60 percent.

The goal of single digit NO_x (less than 10 ppm) was being driven by regulation and competition, not the ATS program. Thus, the ATS program's contribution was not in setting this performance goal, but in using in-combustion rather than end-of-pipe control. In-combustion control represented a pollution prevention approach, an idea that was just gaining political saliency in the early 1990s when the ATS program started.

The ATS NO_x emissions goal was extensively debated before being set at less than 10 ppm. While EPA was quite impressed and pleased by the possibility of reaching 10 ppm through a pollution prevention approach, the Agency was not able to legally provide any guarantees of future regulatory requirements. The industry had mixed views regarding this goal, with concern not simply about technical feasibility but, more important, about EPA's inability to commit to not making changes to regulatory requirements in the future. For over twenty years NO_x emission requirements had been a "one-way ratchet down."[8] Because emission restrictions are set by the states, and because states that are not in compliance with national air quality standards must set standards at the lowest level technologically achievable (a standard known as LAER, Lowest Achievable Emission Reductions), firms had been competing to put lower emission gas turbines on the market. At the inception of the ATS program, the gas turbine industry was aware of the need to push NO_x to single digit levels. By 1989 the national LAER was 9 ppm, and the California South Coast Air Quality Management District was requiring 9 ppm for siting new gas turbines (Alfonso 2001). Several new gas turbines were capable of meeting this emission target, although these turbines operated at lower temperatures (and efficiencies) than the then highest-efficient turbines. Higher temperature turbines needed to add an end-of-pipe technology to achieve single digit NO_x. Thus the ATS program basically set the goal of reaching the LAER through in-combustion controls. Manufacturers were already looking at methods for in-combustion control, but the higher temperature of the ATS turbines made this even more of a challenge.

In sum, getting to very low NO_x was being forced by local regulations. It thus became a competitiveness issue and all firms were moving in this direction, balancing the not-insignificant cost of end-of-pipe controls against the cost of developing in-combustion controls. The ATS program contributed to the goal of single digit NO_x through in-combustion approaches by forcing all participating companies to work on this technology, and by providing resources to support its development. Given that currently many state environmental agencies are requiring 2 to 3 ppm for sitting new turbines (California Air Resources Board), there is much second guessing about the wisdom of this in-combustion target in the ATS, and renewed discussion about approaches for guaranteeing future emission targets. We will discuss this issue in the concluding section.

The Thin-Film Partnership project also provided nonregulatory incentives, in this case for environmental improvement through the EH&S team. The activities of the ES&H team cut across the thin-film technologies to address the environmental, safety, and health issues associated with the manufacture and disposal of PV cells and modules. The team has focused on a range of issues such as in-plant safety, material toxicology, handling of recycling wastes materials, EPA toxicity tests for both CdTe and CIS modules, and developing methods of removing materials from substrates for reuse (NREL 2000c). The EH&S team was organized during the early 1990s to build on work that had been carried out at Brookhaven National Laboratory (BNL) during the 1980s. BNL, DOE, and NREL managers have provided joint guidance for the team. The

EH&S team has provided an incentive for the PV firms to be more proactive in considering potential environmental impacts and solutions during product and process development.

Setting R&D Agendas

The initial phase of each of the three R&D partnerships was dedicated to setting the R&D agenda. During phase 1 of the ATS, DOE gave firms initial contracts of $100,000 to explore technological paths for reaching the programs goals for energy efficiency, emissions reduction and cost. Another important aspect of the planning for the ATS was a requirement that firms engage in the examination of multiple technological approaches for meeting goals and in more extensive technology testing than firms would have otherwise undertaken. Having these requirements and the funding to implement them resulted in technological knowledge and advances beyond what the firms would have achieved without this program.

In the first phase of the PVMaT project, twenty-two firms were selected through a competitive bidding process to identify key problems that needed to be addressed in PV manufacturing. Each participant was awarded up to $50,000 to conduct a three-month study. These studies—which identified numerous opportunities to improve manufacturing processes, reduce manufacturing costs, increase product performance, and support a scale-up of US-based manufacturing capabilities—were used to define the research agenda for later stages of the project (Witt et al. 1993: 2). Because manufacturing was not a significant focus of the federal PV R&D program during the 1980s, this collaborative planning process was necessary to bring NREL up to speed on the critical manufacturing problems facing the US PV industry, and in so doing helped establish NREL's legitimacy and capability to sponsor PV manufacturing R&D.

Goal setting in the Thin-Film PV Partnership project has been an ongoing process. NREL initiated this process by hosting a team-formation meeting for each team. At these meetings participants defined key goals for the teams to address, and formed sub-teams to work on specific problems. Within each sub-team the members prioritized their common research goals, as well as developing research plans for meeting these goals. In addition each sub-team appointed a leader who was responsible for reporting to a guidance team made up of a group of industry representatives, and DOE and NREL management. This basic structure met the needs of the DOE and NREL project management staff and provided a direct voice from the team to DOE and NREL.

Collaborative Implementation: Spreading and Reducing Risk

Collaboration in implementation contributed to successful technological development by both spreading and reducing risk. This was done through several program elements, including cost-sharing as well as a variety of collaborations among industry, universities, and national laboratories.

Cost-Sharing

Notwithstanding the many collaborative elements in these programs, cost-sharing was the major way in which the programs reduced the risks of the participating firms. For ATS, in the phase 1 systems studies, firms contributed 10 percent. By the later phases of technology development, firms were contributing from 40 to 70 percent. For PVMaT, no cost-sharing was required during the initial problem identification phase, but cost-sharing was built into the proposal evaluation process for subsequent phases in the program. In the later phases of the project a multi-tiered system was set up, with different cost-sharing requirements for large and small firms. The Thin-Film PV Partnership project also developed a multi-tiered cost-sharing approach. In addition to setting higher cost-sharing requirements for large firms than small firms, the Thin-Film Partnership distinguished between technology development and more basic R&D, setting higher rates for the former.

Cost-sharing not only helps to keep firms committed to the project goals, even through setbacks in technological progress, but also helps the firm's internal technology advocates compete for in-house R&D dollars. A key design element was entering these cost-shared contracts with the business units rather than corporate research laboratories, thus providing DOE with a partner that was strongly focused on technology commercialization.

Industry–University and Industry–University–National Laboratory Collaboration

Each of these three programs designed innovative approaches for involving university researchers, with the main goal of focusing the capabilities of universities on industry relevant R&D. Although only a small part of the total budget of these programs (< 10 percent), the collaborations between firms and universities (sometimes including national laboratories) contributed to both risk spreading and risk reduction. In terms of risk spreading, these collaborations focused on longer-term, precompetitive R&D. These collaborations were used to either reduce duplication of efforts or to reduce the risks faced by individual firms. In terms of risk reduction, they not only focused the talent of universities on industry-defined problems, but in so doing increased the technological and organizational capability of universities and the industry as a whole. For some of the firms, the university contribution to creative and innovative problem solving was valuable, resulting in technological developments beyond what in-house efforts alone would have achieved. For others, the university programs' key contribution was not the R&D performed as part of the collaborative, but rather it was the new networks formed through the collaboration that resulted in training future employees, and in follow-on contracts to specific universities to undertake proprietary R&D. In addition to these intellectual complementary assets, in some cases the universities brought complementary physical assets to the collaboration in the form of specialized laboratory and testing facilities. A final role played by industry–university–national laboratory collaboration, in the case of PVMaT, was independent proposal review that provided expertise and credibility to the project.

The ATS program's university–industry consortium, the Advanced Gas Turbine Systems Research (AGTSR), was managed by the South Carolina Institute for Energy Studies at Clemson University. The consortium grew from 53 universities and 6 industry partners at its inception in 1993 (DOE July 1993) to 95 universities and 8 industry partners in 2000. The industry partners included 6 gas turbine and component manufacturers that formed the Industrial Review Board and 2 utilities that had the status of associate members. Each industry partner paid $25,000 annually to belong to the consortium with the bulk of research funding provided by DOE. The Industrial Review Board had full decision-making authority, with oversight by DOE. Every year the Industrial Review Board developed a request for proposals (RFP) based on the research needs identified by the participating companies. This RFP was sent to the participating universities. The companies then ranked responses to the RFP. Based on company rankings, the Industrial Review Board made a decision about which proposals to fund. Although DOE had final decision-making authority, it never reversed the Industrial Review Board's choices. AGTSR was able to provide multiple-year funding contracts, an advantage generally not available to DOE and the national laboratories. In addition to the research, AGTSR had an education program which sponsored undergraduate research fellowships and industrial internships, faculty internships at companies, and short courses. Furthermore it sponsored annual workshops and seminars on key issues including heat transfer, materials, and combustion. The universities retained the intellectual property rights to research sponsored by AGTSR (Fant and Golan 1996; Ali 1997; Padgett 1997).

The PVMaT project encouraged firms to collaborate with universities through generic-teamed research. Under generic-teamed research the PVMaT project allowed firms to waive the cost-sharing requirement for the portion of their contracts used for research at universities. In phase 2 up to \$100K of a contract could be allocated to collaborative research at universities without the cost-sharing requirements. This basic approach has been expanded in subsequent phases of the project. In the most recent solicitation up to 10 percent of a contract's total cost was allowed to go to universities without cost-sharing. In 2000, out of seven contracts, four had "lower-tier" university contracts included in them.

The PVMaT project also used an industry–government–university collaborative process to independently evaluate project proposals. Typically the evaluation panels included one person from NREL and one person from Sandia on a panel of 12 or more people. The panels were set up to have a broad mix of representatives, yet members were required not to have any conflicts of interest. In addition the panels were designed to have a heavy industry representation to ensure their credibility with industry. The panels were not just a rubber stamp, they had real control over the allocation of resources under the project.

Under the Thin-Film PV Partnership the national team process has encouraged firms to collaborate with universities and national labs on fundamental research problems. As discussed previously, the teams include researchers from the national labs, universities and industry. A percentage of each contract awarded under the partnership had to be used in the national team process; the most recent RFP specified 20 percent for the teams. This meant that 20 percent of each contract remained open for negotiation during the team process. The remaining 80 percent could be used for proprietary research. Setting aside a significant share of the R&D funds for the team process has encouraged people to come into the national teams with an open attitude toward collaboration.

In sum, these industry–university collaborations contributed to risk spreading and risk reduction in each of the programs. Although accomplished in varying ways, the key elements for success included incentives for industry to collaborate with universities and industry-defined research agendas. Through these partnerships, universities (and national labs) were able to stay current with industrial trends, interests, and the technical state-of-the-art, making them a greater asset to firms and to the ultimate goal of technology commercialization. These university components were valuable not only for reducing duplication in generic, precompetitive R&D but also for adding some riskier and longer term research into DOE research collaborations that focused on relatively near-term technological goals.

National Laboratory–Industry Collaboration

The national laboratories were not only involved in project planning and management, they also made technological contributions to these R&D programs.

For the ATS program, access to advanced testing facilities at the national laboratories provided significant value to firms. In some instances the national laboratories also provided joint problem solving and information exchange. Firms and universities made use of the NETL gas turbine combustion research facility, which is larger scale and higher pressure than most university laboratories and many industry laboratories. This lab made it possible for small companies and universities with innovative technologies to test their technology. With certified, independent results they were more likely to gain either venture capital or sales of the technology to turbine manufacturers, thus improving their opportunity for commercialization.

In the Thin-Film PV Partnership project, the ES&H team used a collaborative framework to encourage the US PV industry to work proactively with national labs (primarily NREL and Brookhaven National Laboratory) to address the potential ES&H impacts of each of the thin-film technologies. NREL and Brookhaven had worked on these ES&H issues in the 1980s, and thus brought relevant expertise to this mission.

Industry–Industry Collaboration

The ATS program stimulated increased collaboration between firms in its Materials and Manufacturing Program, which was instrumental for the development of single crystal blades and new core materials and processes (Karnitz et al. 1997). As with the ATS program more broadly, the Materials and Manufacturing Program was based on solicitations that required industry cost-sharing. The development of single crystal blades involved the two casting firms, Howmet and PCC, as well as the users of those castings which included all of the turbine manufacturers: GE, Solar Turbines, Westinghouse, Allison, and Pratt & Whitney. Because of the large investment and technical risk involved in scaling up the single crystal castings to the size needed for large land-base turbines, Howmet and PCC needed the entire business base in order to go forward with this technological development; that is, they needed all turbine manufacturers. All the manufacturers were interested in single crystal blades because they are necessary to operate at the higher temperatures needed to reach ATS efficiency goals. In sum, the ATS program, through its material program, provided an avenue for industry–industry collaborations that facilitated economies of scale in research and production, thus reducing the business risks to the innovating supplier firms. These industry–industry collaborations were an outcome of the stretch goals set by the ATS program. In other words, government did not need to promote industry–industry collaborations directly.

In the Thin-Film PV Partnership, industry–industry collaboration occurred through the team process. The teams meet regularly to discuss progress and revise their goals. The systematic approach to collaboration adopted in the teams helped firms avoid duplication and has resulted in faster sample preparation, more rapid sample exchange, better feedback on sample processing and characterization of sample experiments, and improved and broadened feedback on experimental results.

Networking

Each of the three R&D collaborations provided regular forums for all the players in industry, including university researchers, equipment manufacturers, suppliers, end-users, and government personnel. These regular meetings provided the opportunity for information to flow between the various participants.

Conclusion

Industry, university, and government participants view the ATS program, the PVMaT project, and the Thin-Film PV Partnership project as effective in increasing the pace of technological development and as good models for future government R&D programs. Our assessment concurs, as these programs have all been successful in the development of new technologies.[9] However, while successful, they do have limits as models for reaching environmental goals over the long term. This is not a criticism of the programs, per se, as they each recognized the medium-term nature of their goals. Rather, in this conclusion, we use our evaluation of the strengths and limitations of these R&D collaborations as a jumping-off point to consider the potential of collaborative R&D programs to reach environmental goals not only in the near term, but also for the type of transformation needed for long-term sustainability.

From the standpoint of environmental policy, a key element in these R&D programs was collaborative planning, as it provided the opportunity for setting technological goals that reach environmental objectives. In all these programs DOE engaged in a collaborative planning process involving a range of stakeholders. In addition to goal setting for the environment, all three programs used collaborative planning to help the government identify key technological challenges and opportunities, and thus make better decisions about the direction of R&D.

Collaboration took many forms in the implementation of these programs, and contributed to success by both spreading risk and reducing risk, as summarized in Figure 6.1.

In sum, through goal setting and implementation, these collaborations moved industry to develop new technology at a faster pace than it would have otherwise. For both gas turbines and PV, this resulted in having better products available for the rapidly expanding market at the turn of the century. In order to stimulate technology development that stretches industry beyond what it would have done without the program, government needed to provide significant cost-sharing, and industry had to believe that the goals were both reachable and that the resulting technology would be competitive in the market. In this sense these programs operated in tandem with energy and environmental regulation that was creating markets for these technologies, both at home and abroad.

A focus on the role of market opportunity leads to an important cautionary note. The collaborative elements of these programs provided clear benefits, but within a competitive environment. Collaboration alone did not stimulate innovation; competition provided the underlying context in which firms strove for and excelled in innovation. For success, government policy and programs, especially those based on collaboration, must be careful to support and not hinder this competitive environment.[10]

These three cases suggest that in designing R&D collaborations, the uniqueness of each industry sector or subsector must be considered. The role that should be played by each type of organization depends on the technological and organizational capabilities that they can bring to the table, and this may vary considerably from one industry or technology to another. Having said that, some crosscutting institutional lessons suggested by these cases are (1) the private-sector partner should base their program in their business unit, and not in their R&D laboratory, (2) universities should be engaged on industry-defined research agendas, and focused on riskier, longer term and/or precompetitive technologies, (3) the national laboratories should be involved based on the specific complementary physical and/or human assets they can contribute to the collaboration, and (4) industry–industry collaboration will be fostered when the program sets goals that require collaboration for success.

Benefits of R&D collaboration	ATS			PVMaT				Thin film	
	Ind.-gov. (planning)	National lab ind. (technology base)	Ind.-ind. (materials and manufacturing)	Ind.-univ. (AGTSR)	Ind.-gov. (planning)	Ind.-gov.-univ. (evaluation)	Ind.-univ. (generic-teamed)	Ind.-gov.-univ. (planning)	Ind.-gov.-univ. (activities)
Risk spreading									
Access to complementary assets		*		*			*	*	
Coordination through industry standards and roadmaps								*	
Exploitation of scale economies			*						
Reduction of duplication				*				*	
Capture of knowledge spillovers				*					
Risk reduction									
Increased information flows leading to creative solutions	*	*		*	*		*		*
Facilitation of technology transfer leading to reduced transaction cost				*					
Increased technological an organizational capacity				*		*	*		*
Provide nonregulatory incentives									
Stimulate beyond compliance behavior beyond current technological trajection	*								*

Figure 6.1
Comparison of the benefits of different types of collaboration in public-private partnerships

In moving from evaluating the success of these programs to considering them as a model for R&D collaboration for industrial transformation, there are three issues that must be addressed: balancing near-term and long-term goals, integrating evolving environmental science and regulation, and the ownership of intellectual property rights.

The R&D collaboratives evaluated in this chapter had near- to midrange commercialization goals, and thus are not the appropriate mechanism for addressing sustainability over the long run. There is much potential for R&D collaborations to address longer term goals, although this will require systems studies (sometimes known as roadmapping) that look at longer term technological possibilities and then backcast what the current research trajectory should be, perhaps investing in both near-term commercial opportunities and more basic R&D necessary for future breakthroughs. Systems studies also require different mechanisms for effective industry involvement, as the high cost share in the programs we examined are not feasible as most firms have short- to medium-time horizons for their R&D investments. The R&D collaborations evaluated in this chapter were only part of DOE's investment in fossil fuel and solar technologies. Further analysis of DOE's portfolio and approach for longer term R&D planning, as well as examination of innovative programs such as the Industry for the Futures Program, which creates twenty-year roadmaps for specific industries, may hold better lessons for thinking about longer term R&D collaboration.

The second issue is finding ways to integrate evolving environmental science and regulation into technology-driven programs. The importance of this is made clear from the NO_x issue in the ATS program. While industry participants evaluate the ATS program generally as effective and as an excellent model for future government R&D programs, they do have a strong criticism related to the lack of coordination between the environmental goals in the ATS program and the environmental regulatory process. The turbine manufacturers went to great efforts to meet the single digit NO_x goal of the ATS program. However, throughout the decade, the end-of-pipe technology improved in terms of performance and cost and urban ozone, for which NO_x is a key precursor, continued to be a serious environmental problem. By the end of the 1990s, NO_x emission requirements for new gas turbines were 2.5 ppm in many parts of the country (California Air Resources Board). Turbine manufacturers have argued that they have wasted much of their technological efforts in reaching 9 ppm, as quite different technologies will likely be needed to meet 2 or 3 ppm with in-combustion technologies. In other words, if they had been aiming for 2 to 3 ppm from the start, they would have pursued alternative technologies for NO_x control. Although it is not possible or desirable to completely insulate industry from such regulatory changes, there may be mechanisms to lessen the frequency or surprise. This would require the active participation not just at the beginning of a program like ATS, but throughout, from the environmental community. Midstream evaluation of environmental goals is crucial.

A final set of concerns pertains to the ownership of intellectual property. First, some DOE project managers have expressed concern that cost-share programs are allowing firms to retain the rights to generic technologies. To the extent this is happening, it may not properly balance incentives for private sector participation with the public interest in technology diffusion. Second, firms are required to manufacture the technology developments from government-sponsored R&D in the United States. It is worth considering whether this balances the public interest in US competitiveness with our interest in international cooperation for solving global environmental problems.

Notes

1. Our analysis of these programs is based on a combination of archival research and interviews with a wide range of government, industry, and university participants in these three programs. The interviews were carried out between September 1999 and April 2001. In addition to the programs discussed in this chapter, the Program for a New Generation of Vehicles (canceled by the George W. Bush administration for Freedom Car. See: http://www.cartech.doe.gov/freedomcar/index.html) and DOE's Industries for the Future (see http://www.oit.doe.gov/industries.shtml) provide examples of the efforts in the United States to use collaborative R&D programs to promote technological innovation to reach environmental objectives.

2. Mowery (1998) and Powell (1990) both develop taxonomies for the ways in which R&D collaborations can provide risk reduction. Many of their categories bear resemblance to the categories in the taxonomy presented in this chapter. We cite these authors specifically only in cases of direct correspondence. As noted in these citations, Mowery focused more on risk spreading, and Powell focused more on risk reduction.

3. The National Cooperative Research Act of 1984 has made it easier for firms to join in research consortia without fear of antitrust action (Stiglitz and Wallsten 2000: 57).

4. The ATS was set up as an eight-year research, development and demonstration program with total federal support of $700 million. The program was not fully funded, leading to a reworking of phases 3 and 4, as discussed below. The market for gas turbines had picked up by the end of the program, reducing the need for government funding of demonstrations.

5. Competitiveness is more difficult to evaluate at this point, given that the technology is in the demonstration phase.

6. Steam cooling (compared to the previous approach of air cooling) reduces the temperature drop between the combustor and the turbine inlet. This allows for an efficiency gain without further increases in firing temperatures. Because NO_x emissions increase with firing temperatures, this results in improved efficiency without increased emissions.

7. SERI was designated as a national laboratory in 1991 and renamed the National Renewable Energy Laboratory (NREL).

8. Quote from government regulator.

9. A recent NRC report (2001) also concludes that public-private partnerships are an effective model for technology innovation.

10. In an evaluation of the government's role in technology innovation in the information and electronics industries, Mowery (2002) shows that when government policies expand competition, they create a fertile environment for technological innovation. A key finding of the pathbreaking research led by Nelson and Langlois (1982) concluded that government investment in generic R&D is effective only if done in a way that does not pose a threat to the established positions of competitors.

References

Ali, S. 1997. An Industry Perspective on AGTSR Accomplishments. Prepared for ATS Annual Conference.

Branscomb, L. M., and J. H. Keller, eds. 1998. *Investing in Innovation: Creating a Research and Innovation Policy the Works*. Cambridge: MIT Press.

Branscomb, L., and P. Auerswald. 2001. *Taking Technical Risks: How Innovations, Managers and Investors Manage Risk in High-Tech Innovation*. Cambridge: MIT Press.

California Air Resources Board. BACT Clearinghouse Database. http://www.arb.ca.gov/bact.

Department of Energy (DOE). 1999. *The Next Generation of Gas Turbines*. Washington, DC: DOW, OFE, OEERE, and OIT, nd1.

Department of Energy (DOE). *Program Facts: Advanced Turbine Systems Program.* Washington, DC: DOE, FETC and OIT, nd2. http://www.epa.gov/performancetrack/about/about.htm.

Department of Energy (DOE) and Federal Energy Technology Center (FETE). 1999a. *Applying Tomorrow's Technology to Today's Gas Turbines.* Washington, DC: DOE and FETE.

Department of Energy (DOE) and Federal Energy Technology Center (FETE). 1999b. *Advanced Turbine Systems: The GE-II Class.* Washington, DC: DOE and FETE.

Department of Energy (DOE). 1993. *Report to Congress: Comprehensive Program Plan for Advanced Turbine Systems.* Washington, DC: DOE.

Environmental Protection Agency (EPA). 1998. *Reinventing Environment Protection: 1998 Annual Report.* Washington, DC: Office of the Administrator and EPA.

Fant, D. 1997. The AGTSR Industry-University Consortium. Prepared for ATS Annual Conference.

Fant, D., and L. Golan. 1998. AGTSR: A Virtual National Lab. Prepared for ATS Annual Conference.

Farrell, A., and T. J. Keating. 1998. Multi-Jurisdictional air pollution assessment: A comparison of the eastern United States and western Europe. ENRP Discussion Paper E-98-12, Kennedy School of Government, Harvard University.

Fountain, J. 1998. Social capital: A key enabler in innovation. In L. M. Branscomb, and J. H. Keller, eds., *Investing in Innovation: Creating a Research and Innovation Policy the Works.* Cambridge: MIT Press.

Hewig, L. O. 1996. *U.S. Department of Energy (DOE) Assessment of the Photovoltaic (PV) Industry's Needs, Priorities, and Views Regarding the DOE Photovoltaic Program: A Summary of Feedback from Visits to 22 PV Companies.* Washington, DC: US Department of Energy.

Hill, C. 1998. The advanced technology program: Opportunities for enhancement. In L. M. Branscomb, and J. H. Keller, eds., *Investing in Innovation: Creating a Research and Innovation Policy the Works.* Cambridge: MIT Press.

Karnitz, M. A., I. G. Writght, M. K. Ferber, R. S. Holcomb, and M. Rawlings. 1997. ATS materials/manufacturing. Prepared for ATS Annual Conference.

Kelley, M. 1997. From mission to commercial orientation: Perils and possibilities for federal industrial technology policy. *Economic Development Quarterly* 11(4): 313–28.

Layne, A. 2000. Next generation power systems. National Energy Technology Center, Washington, DC.

Marcus, A. 2000. Cooperative regulation: Setbacks and accomplishments of Project XL. Paper prepared for Association for Public Policy Analysis and Management, 22nd Annual Research Conference, Seattle, WA, November 2–4.

Maycock, P. D. 1993. *Photovoltaic News, Annual PV Market Survey*. Warrenton, VA: PV Energy Systems.

Maycock, P. D. 2000. *Photovoltaic News, Annual PV Market Survey*. Warrenton, VA: PV Energy Systems.

Mitchell, R. L., M. Symko-Davies, H. P. Thomas, and C. E. Witt. 1998. *PVMaT 1998 Overview*. Paper presented at NCPV Program Review Meeting, Denver, CO, September 8–11.

Mowery, D. 1998. Collaborative R&D: How Effective Is It? *Issues in Science and Technology* (Fall): 37–44.

Norberg-Bohm, V. 1999. Creating incentives for environmentally enhancing technological change: Lessons from 30 years of U.S. energy technology policy. *Technological Forecasting and Social Change* 65 (October): 125–48.

Norberg-Bohm, V. 1999. Stimulating "green" technological innovation: An analysis of alternative policy mechanisms. *Policy Sciences* 32: 13–38.

NREL. 1996. Solarex: *Cast Polycrystalline Silicon Manufacturing, PVMaT Fact Sheet*. Golden, CO: NREL.

NREL. 1999. A partnership that works. *In NREL PV: Working with Industry*. Golden, CO: NREL, p. 1.

NREL. 2000a. Thin Film Partnership Program: Amorphous Silicon Team Web Page. http://www.nrel.gov/ncpv/aisteam.html.

NREL. 2000b. Thin Film Partnership Program: Copper Indium Diselenide Team Web Page. http://www.nrel.gov/ncpv/cisteam.html.

NREL. 2000c. Thin Film Partnership Program: Environment, Safety, and Health Team Web Page. http://www.nrel.gov/ncpv/eshteam.html.

Nutech. Nutech internet resource for air-pollution control technologies. http://www.nutech.org.

OTA. 1995. *Renewing Our Energy Future*. Washington, DC: Office of Technology Assessment, Congress of the United States.

Padgett, G. 1997. Solar turbines collaborative project with universities. Prepared for ATS Annual Conference.

Powell, W. 1990. Neither market nor hierarchy: Network forms of organization. *Research in Organizational Behavior* 12: 295–336.

Roos, D., F. Field, and J. Neely. 1998. Industry consortia. In L. M. Branscomb and J. H. Keller, eds., *Investing in Innovation: Creating a Research and Innovation Policy the Works*. Cambridge: MIT Press.

Schacht, W. H. 1999. *Technology Transfer: Use of Federally Funded Research and Development*. Washington, DC: Congressional Research Service.

Stiglitz, J. E., and S. J. Wallsten. 2000. Public-private technology partnerships: Promises and pitfalls. In P. V. Rosenau, ed., *Public-Private Policy Partnerships*. Cambridge: MIT Press.

Teece, D. 1988. Profiting from technological innovation: Implications for integration, collaboration, licensing an public policy. In M. Tushman and W. Moore, eds., *Readings in the Management of Innovation.* New York: HarperCollins.

Unger, D., and H. Herzog. 1998. *Comparative Study on Energy R&D Performance: Gas Turbine Case Study.* Cambridge: MIT Press.

Watson, J. W. 1997. Constructing success in the electric power industry: Combined cycle gas turbines and fluidised beds. PhD dissertation. University of Sussex.

Witt, C. E., R. L. Mitchell, and G. D. Mooney. 1993. Overview of the Photovoltaic Manufacturing Technology (PVMaT) Project. Paper presented at National Heat Transfer Conference, Atlanta, GA, August 8–11.

Witt, C. E., R. L. Mitchell, M. Symko-Davies, H. P. Thomas, R. King, and D. S. Ruby. 1999. Current status and future prospects for the PVMaT Project. Paper presented at 11th International Photovoltaic Science and Engineering Conference (PVSEC-11), Sapporo, Japan, September 20–24.

Witt, C. E., R. L. Mitchell, H. Thomas, and L. O. Herwig. 1994. Photovoltaic Manufacturing Technology Project (PVMaT) after three years. Paper presented at the 29th Intersociety Energy Conversion Engineering Conference, Monterey, CA, August 7–11.

Witt, C. E., R. L. Mitchell, H. P. Thomas, M. Symko Davies, R. King, and D. S. Ruby. 1998. Manufacturing improvements in the Photovoltaic Manufacturing Technology (PVMaT) Project. Paper presented at 2nd World Conference and Exhibition on Photovoltaic Solar Energy Conversion, Vienna, July 6–10.

Witt, C. E., T. Surek, R. L. Mitchell, M. Symko-Davies, and H. P. Thomas. 2000. Terrestrial photovoltaics technologies—Recent progress in manufacturing R&D. Paper presented at ASME 34th National Heat Transfer Conference, Pittsburgh, August 20–22.

Zweibel, K. 1990. *Harnessing Solar Power: The Photovoltaics Challenge.* New York: Plenum Press.

Zweibel, K. 1997. *Assessment of R&D Teaming within the Thin-Film PV Partnership.* Golden, CO: National Renewable Energy Laboratory.

7

Cleaner Technology in Denmark: Support Measures and Regulatory Efforts

Ulrik Jørgensen

This chapter explores the impact on industry of a set of Danish programs that provide government grants for cleaner technology development. The analysis of this relationship sheds light on the interaction between voluntary actions and regulatory efforts in environmental protection. More specifically, this case study, covering the period from the late 1980s to 2000, focuses on two related topics:

1. What are the limitations for technology diffusion, and thus the impact, of a rather successful and innovative voluntary support program for the development and deployment of cleaner technologies?
2. What difficulties do regulating authorities face in integrating traditional command-and-control approaches with the results from voluntary innovation programs?

In answering these questions, the chapter explores two issues in the contemporary, often still hostile, discourse on environmental regulation. First, it addresses the need to rebuild institutions and regulatory practices in both government and industry that are capable of supporting the development of the vision and need for "shared responsibility" for further environmental improvements. Second, it points to the necessity of building trust and constructing new ways to integrate concepts of self-control and continuous improvement into traditional standard-based command-and-control regulation.

The first section introduces the concept of regulatory regimes that is used to analyse the difficulties of integrating different regulatory measures in support of the innovation and adoption of cleaner technologies. The next three sections focus on the Danish cleaner technology programs, discussing their results and impact, and the context for the environmental reform in the early 1990s in Denmark. The two concluding sections offer an analysis of the unsuccessful attempt for integration of the results in the existing regulatory regime, and suggest some measures that would help overcome regime-based resistance to change.

Regulatory Regimes

In the literature the role of industry in environmental protection and innovation is often discussed in terms of self-regulation versus government regulation. The notion of self-regulation is a contradiction in terms, as every regulation implies that certain standards have to be met by parts of an institutional system external to itself. Although regulation can take many forms—government edict, certification practices, market conditions, public response to information—it will always involve interaction among members of the system. The question is not whether voluntary programs create an alternative to traditional mandatory regulation based on standards, but how they support or link to it. A number of studies have concluded that single measures are overemphasized in theoretical discussions. In practice, the multi-threaded use of measures combining mandatory and voluntary elements, designed to meet common environmental policy goals, is more effective (Gouldson and Murphy 1998; Hemmelskamp, Rennings, and Leone 2000).

The effect of governmental actions on industrial innovation also spurs controversy in the economic literature (e.g., see Hemmelskamp, Rennings, and Leone 2000). In many cases strict regulation focusing on very specific (technical) solutions has produced excess costs (Kemp 1997). However, this is not an argument that economic support for innovation is unnecessary or counterproductive. In fact grants to stimulate specific types of innovation do raise the priority of environmental improvements in companies. This suggests that the details of implementation schemes for environmental regulation are important and a constructive dialogue between business and regulators is required.

Thus numerous perspectives reveal and emphasize the interdependency of actors, their knowledge, and their interactions in environmental regulation. Regulatory bodies and companies are the major actors, but others (e.g., customers, suppliers, consultants, communities of practice, and law enforcement personnel) can play important roles too. Regulatory regimes lie at the core of this framework; they are the integrated systems of social control that define the roles and orchestrate the discourse among all players (Jørgensen 1993).

The concept of regulatory regimes arises from neo-institutional theory (e.g., Powell and DiMaggio 1991) that explains how institutions develop stability based on routine practices and produce resistance to change. Another important inspiration comes from technology studies; Wiebe Bijker (1987, 1995) introduced the concept of "technological frames" to describe how a group organizes and develops specific knowledge and routines in relation to a technological (or environmental) artifact. In this discussion pollution and its consequent environmental problems constitute such an artefact. In the same strand of thought the concept of "technological regimes" introduced by Thomas Hughes (1987) informs the concept of "regulatory regimes" applied in this chapter.

This chapter uses "regulatory regime" rather than regulation as a core notion in order to capture the interdependent character of a broad set of factors within systems. Coherent interdependency among these factors will produce organizational stability and at the same time provide resistance against new ideas and practices. Effective selection of a regulatory method cannot be made on the basis of case-related efficiency, independent of the nature of the established institutions that regulate and will be regulated. This is especially true for industrial environmental innovation (Jørgensen 1993; Wallace 1995; Kemp 1997; Hemmelskamp, Rennings, and Leone 2000). Rather effective regulatory policy must be designed to work within the existing regulatory regime or imply changes in this regime.

Regulatory regimes create routines that may conflict with specific efforts in regulatory innovation. A regulatory regime includes a set of instruments to be used in enforcing the rules. These can vary from legal procedures for when and how the regulating body can dictate requirements to definitions of responsibility or negotiated agreements about the realization of certain action plans. But regimes also require specific professional competencies and skills distributed among the actors involved in the process. This creates an institutional scenario based not only on rules and routines but also on an entire world view supported by external professional networks and educational systems.

In the following sections the Danish cleaner technology programs are introduced and evaluated both in terms of the environmental performance of the technologies resulting from the innovation projects and in terms of how well the new technologies are diffused beyond the initial project partners. Subsequent analysis shows how efforts to stimulate cleaner technologies have interacted with the traditional command-and-control measures for environmental regulation. Based on this analysis I draw conclusions on the ability to promote cleaner technology within the current Danish environmental regulation, including the modification of this regulatory regime.

Danish Cleaner Technology Support Programs

The Danish government's Development Program for Cleaner Technology was designed to stimulate cleaner technology innovations and demonstrations; this initial program ran from 1987 through 1989. It was introduced and managed by the Danish Environmental Protection Agency (DEPA). The aim of this program was "to reduce the strain on the environment through preventive efforts with the aid of improved incentives to employ cleaner technology" (DEPA 1986).

This program was followed by a continued series of action plans, the second running from 1990 to 1992, the third from 1993 to 1997, and a clean technology program that concluded in the late 1990s. Since that time, government policy support has shifted from cleaner technologies and toward cleaner products.

In principle, these programs offered four different types of grants to applicants from industry, consultancy, and R&D institutions:

1. Develop innovative new technological solutions for industrial use.

2. Survey branches of industry for information on potential and currently deployed cleaner technologies.

3. Implement cleaner technologies in full-scale production and document results.

4. Establish facilities demonstrating the implementation of cleaner technologies.

The two first types of projects were allowed to receive almost full government funding. The latter two, however, would only receive grants for those parts of the project that were not standard production investments in the companies involved; only the innovative parts and the extra costs for documentation and information were fundable. The funding scheme was defined in this way to avoid any accusation of subsidizing what industry would have done anyway, while still making the program attractive to industry by reducing the risk of unsuccessful research and development.

This definition of what constitutes an innovation is important; it specifies the scope of the program. Unlike typical technological research programs, the criteria for novelty were relaxed. Innovation was not restricted to completely new technologies or even the initial implementation of new technologies but could include the implementation of a cleaner technology in an industrial sector or process where it had not been used before. In a conference in 1989 on cleaner technology, the Danish EPA explained the support scheme as follows:

> The goal is to promote new technologies or technologies that have not previously been employed in the branch concerned by providing economic and technical documentation that the technologies are realistic. The strategy is example-based, whereby demonstration plants or projects shall convince other enterprises of the value of cleaner technology solutions to a pollution problem. (MoE 1989)

Avoiding the common view that an industrial development is determined by radical "breakthrough innovations," studies of innovation have emphasized the role of continuous, incremental improvements in technology and organization. In some studies general patterns of innovation in which generic technologies influence a number of sectors and industries simultaneously have been termed "system" and "paradigm" shifts (Kemp 1997; Wolters et al. 1999). These distinctions are not used in this study, as there is no simple link between the classification of innovations and their environmental impacts. Both minor innovations and implementations in new settings, as well as radical innovations, can contribute to environmental improvements.

While the environmental potential of major innovations in new materials and processes should not be neglected for certain sectors, the focus of the Danish programs has not been limited to large-scale ideas. They have also supported a broad variety of innovations, including changes in and supplements to existing equipment, as well as the implementation of minor equipment and process changes when these changes can bring major environmental improvements.

A definition of cleaner technology was formulated at the very beginning of the programs, emphasizing prevention rather than end-of-pipe. This definition implies that a technology should be derived from a multimedia approach so that the transfer of emissions from, for instance, water to solid waste is not considered "cleaner." In addition the program steering committees have been composed to avoid the possibility that a project might simply transfer impacts away from the environment and onto the health and safety conditions of workers in the production and handling of materials.

During the first two programs (1987 to 1989 and 1990 to 1992), a total of 315 million Danish Kroner (approx. 45 US$mil) was granted to 365 different projects. Of this portfolio 169 were designed to produce cleaner technologies for testing and use in industry; these also received the majority of the available funds—193 million Danish Kroner (approx. 26 US$mil). Some grants went to projects surveying potentially valuable technologies, and others to reports and demonstrations of the results of the cleaner technology projects. In the third period (1993 to 1997), the government provided funding of approximately 75 million Danish Kroner per year (approx. 10 US$mil per year), making the total government grants for cleaner technology innovation and implementation around 95 million US$ in 11 years.

These programs have not been large compared to industry's investments in production facilities and innovation in the same period. In addition a number of environmentally enhancing technologies have been supported by other government programs dedicated to supporting research into renewable energy systems, waste handling, water treatment, and organic farming. The primary impact of the cleaner technology programs cannot be measured by their economic size only but should be viewed in terms of their effectiveness in directing industry's investments toward more environmentally sound manufacturing and products.

The cleaner technology programs were, in principle, open to all ideas and relevant applicants. However, to focus the attention of industry and to secure a measurable impact in particularly problematic pollution sources and industrial sectors, every program and action plan designated some high-priority areas. These priorities also reflected other strains of environmental policy in which well-defined pollution problems had not found satisfactory solutions through the use of traditional regulatory measures. In fact the cleaner technology program and the later action plans became important supports for the overall policy and operations of the Ministry of Environment and Energy. Projects were prioritized and coordinated with the goals and strategies of environmental policy; for example, projects involving agriculture or the major unsolved problems of the fish-processing industries were coordinated with the Action Plan for the Aquatic Environment. In addition the PVC action plan was supported by cleaner technology program projects for alternative materials and processes, and plans for handling waste from electronics production were supplemented by projects investigating ways to reduce waste and toxic releases from this industry.

In the first program, priority areas included the use and release of organic solvents (VOCs), the release of heavy metals, and the release of organic materials in wastewater (BODs) from the wood and furniture, metal plating, metal finishing, and meat and fish industries. In most cases these high-priority areas were concentrated in industries characterized by many small- and medium-sized companies using quite similar technologies and processes. Compared to larger and more specialized firms, these companies were less able to fund and prioritize innovative projects focused on reducing pollution. The fact that cleaner technologies might lead to reduced material costs was not enough for these industries to invest in such innovations without a push from the environmental authority and its financial support to carry part of the risk.

In later programs new areas were addressed, including the efficient use of natural fertilizers in agriculture, materials to substitute for soft PVC in products, reduction of VOC use in the printing industry, and coloring and processing chemicals in the textile industry. Policy initiatives requiring reductions in waste from different industries such as electronics and construction were coordinated with cleaner technology action plans to demonstrate the value of alternative processes and materials.

Three examples from the 169 innovation projects illustrate the type of cleaner technologies that have been supported:

1. A demonstration and implementation project showed the possibility of substituting water-based paint with high solids in place of VOC-based paint to protect large steel structures from corrosion. Such a substitution could result in a 75 percent reduction in VOC use and a faster flow of production. A grant of 1.3 million D.Kr. (160,000 US$) was made for this project. Widespread adoption of the new process depends on the enforcement of emissions reduction in environmental permits.

2. An innovation project (also involving a foreign supplier) invented a non-PVC plastic film to be used for blood bags and other medical materials. Adoption of these alternative materials could lead to a reduction in the use of PVC-based film of more than 500 tons. The project received a grant of 1.5 million D.Kr. (0.2 US$mil). As the companies receiving the grant are the only producers in Denmark, their competitors are foreign companies who are not forced to find substitutes for PVC. Competitive use of the more expensive new product depends on more stringent "green buyer" regulations.

3. Process modifications were developed for the fish-processing industry to reduce water consumption and BOD discharge in wastewater through dry collection of waste products. Machinery provided by foreign suppliers was modified and the new parts developed and sold to other companies by the participating engineering firm. Initially the machine suppliers were not interested in participating in the innovation project, as they viewed the demand as not typical of their world market. This situation subsequently changed. The impact has been a 60 percent reduction in water consumption and a 65 percent reduction in BOD loads in wastewater from fish processing. The grant was 2.1 million D.Kr. (300,000 US$); the payback time of the investment turned out to be less than one year.

Beginning in the early 1990s, and during the subsequent action plans, grants for cleaner product development and models for life cycle assessment[1] were introduced and received a growing part of the funding. Highest priority continued to be given to cleaner technologies that had the potential to be diffused among several companies.

In most of the cleaner technology projects the composition of the project team was taken quite seriously. A team had to include competencies in the development, implementation, and documentation of the technology and its performance. The potential for the diffusion of project results was taken care of by consultants or companies with an interest in marketing the innovations as products or services to other companies or with the capacity to bring suppliers of materials or machinery into the project (Andersen and Jørgensen 1997).

Innovation and Diffusion of Cleaner Technologies

Three evaluations of the cleaner technology program and action plans have been made by independent researchers. The second evaluation made in 1994 (Andersen and Jørgensen 1997) was the most complete and detailed. It included a study of the immediate outcome of the cleaner technology projects from 1987 to 1992, examined in detail the diffusion of the new technologies throughout Danish industry, and assessed their environmental impacts. A third evaluation (Remmen 2000; Remmen and Lassen 2000) focused only on parts of the 1993 to 1997 action plan covering the textile and electronics industries, transportation, and cleaner products. This evaluation included a study of the diffusion of the cleaner technologies, but did not address their environmental impacts. These evaluations and several studies of the implementation of cleaner technology options in environmental enforcement activities form the basis for the following analysis of the immediate results of the cleaner technology program and the identification of some limitations of the wider implementation strategy.

The results confirm that engaging industry and R&D institutions—on a voluntary basis—in developing cleaner technology solutions has been a very successful strategy. Out of 169 projects funded in the period from

1987 to 1992, 71 percent have led to useful technology options. Of these technologies, 50 percent were implemented successfully in full-scale production (Andersen and Jørgensen 1995).[2] Similar figures have been collected for the projects funded by the 1993 to 1997 action plans. The selective character of the evaluation for the later period precludes such a highly detailed overall picture (Remmen 2000; Remmen and Lassen 2000). Compared to the risks involved in making ideas for cleaner technologies both environmentally and economically feasible and the difficulties of overcoming real-life testing and scaling up of technologies, these results lead to the conclusion that these programs were successful in developing cleaner technology options.

In principle, the organization of support schemes for cleaner technology projects was independent of the local authorities' implementation of environmental permits. However, in practice, the regulatory status of companies applying for grants was sometimes one of the factors influencing companies to engage in a cleaner technology project. Slightly more than half of the projects were motivated by actual or potential environmental requirements (Andersen and Jørgensen 1995). These requirements, as defined in action plans, concerned regulations, for instance, for reductions in VOC discharges, the use of PVC, and discharges to the aquatic environment. A few cases addressed actual discharge criteria required for environmental permits or anticipated future requirements. The pricing of water consumption and wastewater discharges motivated companies, especially in the meat and fish industry, to develop cleaner technologies for their processes. However, in addition to satisfying regulatory requirements, the companies that sponsored successful new technologies also experienced increased sales, advertising advantages, lower costs, and in some cases even improved health and safety conditions for workers.

It is also evident that the economic motive alone—despite the cases showing lower costs and reductions in environmental taxes—was not sufficient to stimulate companies to undertake the costs and risk of developing cleaner technologies on their own. The support scheme enabled them to build development teams with suppliers and consulting engineers and to reduce economic risk. More important, it turned the attention of companies toward cleaner technology solutions.

The evaluations do not substantiate overall rational behavior by companies choosing among investment opportunities. In fact the risk aspects of most technology change make it quite difficult for companies to use simple economic measures to discriminate among technology development options. Factors influencing the selection process included the following:

- Environmental orientation of the company.
- Available knowledge about potential solutions.
- Pressures from environmental regulation, which were often rather weak but well defined.
- Access to risk-limiting funds and knowledge providers.

Together these elements influenced the willingness of a company to participate in cleaner technology programs (Andersen and Jørgensen 1995).

Arguments like "pollution prevention pays" have often been used to motivate companies. While relevant in some cases, for many this argument remains at the rhetorical level of management motivations. The economic benefits of cleaner technologies are in many cases not discretely measurable, as they are merged with other aspects of change. For example, production equipment is rather difficult to change; suppliers may not be motivated to provide cleaner technology solutions or even to accept changes without technical support. This was the case both in the fish processing and textile industries. In both instances the (mostly foreign) suppliers were not particularly interested in cleaner technology options developed in and for Denmark because they were oriented toward the global market. Many of these suppliers later copied some of the equipment improvements.

While the support schemes motivated industry to participate in the development of cleaner technologies, another voluntary approach introduced in the same period of time did not lead to the expected results. Environmentally beneficial improvements based on voluntary agreements between industry and government were less technologically radical than the results derived from the systematic introduction of cleaner technology solutions. In fact in most of the cases analyzed (e.g., the VOC and the PVC agreements) industry was well on its way to implementing the results when the final agreements were signed. The agreement negotiations may have influenced industry awareness, but the solutions obtained were rather "conservative" compared to other available cleaner technology options.

Substantial resources were used in the evaluations to document the diffusion of cleaner technology in industry. Diffusion is, of course, the result of a multitude of influences, only one of which being the cleaner technology programs themselves. But by tracing companies' knowledge about the program and their insights in specific cleaner technology options, a fairly accurate picture may be drawn. Environmental impact assessment of cleaner technology implementations, combining emission data with model calculations, can establish—still with large margins of uncertainty—the extent of their effects.

Although large differences exist between individual projects, the overall results point out that over 50 percent of all companies knew about the support schemes. There were, however, major differences as shown in table 7.1, in how much companies knew about the cleaner technologies developed through the programs (Andersen and Jørgensen 1995).

The differences in knowledge distribution and implementation may be confusing: fewer companies seem to know about the specific cleaner technology projects than the number that have introduced these technologies. The explanation is that knowledge about technologies is

Table 7.1
Proportion of companies aware of the cleaner technology options developed in the Danish programs, and proportion of companies that implemented cleaner technologies

	Percentage of companies knowing the specific CT options developed in the programs	Percentage of companies that implemented equivalent kinds of CT
Wood and furniture	25	27
Metal coating and finishing	37	73
Printing	14	60
Meat and fish processing	74	80

distributed in many different ways. Companies may have gained the information through consultants or through industrial organizations. Also brochures distributed by the DEPA have presented the potential technologies without identifying their sources.

The size of a company and its attitude toward environmental protection have a significant influence on the implementation of cleaner technology, which underscores the importance of knowledge about options and the interest in seeking solutions. Environmental permits can motivate the implementation of cleaner technology, but the study shows that their role is rather weak and more based on future expectations than actual requirements (Andersen and Jørgensen 1995).

Another important pattern may be found in analysing the strategic attempt to create development teams in each of the projects that would support further diffusion of the results after the project was finalized. When only one company was involved, and it had an economic interest in keeping the results to itself, no diffusion occurred. Technologies developed by project teams including consultants or suppliers with a commercial interest in marketing the results were diffused much more widely. When only laboratory testing of cleaner technology options was carried out, generally no diffusion was observed unless a follow-up project involving companies was initiated.

The major part of the cleaner technology projects, if realized, had the potential to reduce core pollutants in their specific sectors by 40 to 90 percent. The actual reductions achieved through industry implementation differ by sector and are much lower, due to both differences in the efficiency of certain implementations, and because very different numbers of companies in each sector have implemented the technologies.

In the wood and furniture industry emissions of VOC show an estimated reduction of 40 to 45 percent from 1988 to 1994. In the metal plating industry an account of actual reductions was not possible, although the potential for reduction was well documented to be about 40 percent in the use of acids, and about 80 percent for the disposal of hardened and heavy metals from the processes. Water-based paints now account for about 46 percent of total consumption in the metal finishing industry. The use of cleaning substances has been reduced remarkably, by more than 50 percent of VOC-based substances from 1987 to 1993.

In the printing industry VOC reductions have been approximately 40 percent with a potential of at least the same percentage through a current action plan. In the meat- and fish-processing industry the load of organic materials in the wastewater has been reduced by 25 percent, even though production has increased by about 40 percent. The use of fresh water has also diminished.

Overall, these figures reflect about 25 to 40 percent of the reductions that could have been achieved through the complete implementation of cleaner technologies by all companies. This limited impact is not caused by a failure of the cleaner technologies themselves but to their limited diffusion throughout industry. Although further improvements can be expected, interviews with companies show that without further external pressure or motivation, the noncompliant companies cannot be expected to implement cleaner technology solutions. We may conclude that as a voluntary approach and as an innovation program, the cleaner technology support measures have been quite successful, but as a diffusion program, the voluntary approach is limited unless regulatory measures for environmental improvements are in place.

Cleaner technology efforts have also had other impacts. In combination with the demands from environmental regulation and the introduction of environmental management schemes, cleaner technology efforts have spurred the creation of a new group of environmental professionals. Because knowledge and competence must be transferred among industry, consultancy and public authorities, this group of professionals has been crossing institutional boundaries in increasing numbers.

The Context of Danish Cleaner Technology Policy

The cleaner technology support programs—especially up to the mid-1990s—were viewed as a supplement to the traditional environmental protection measures that have been introduced since the establishment of the Danish Ministry of the Environment in 1973. The enforcement of the new laws and regulations for industry were left to local communities, which were delegated the responsibility for reducing pollution and improving the local and regional environment. They were given the authority to interpret the standards to fit the local environmental conditions (e.g., air or watershed quality and capacity), which gave them a great deal of discretion.

In the mid-1980s the serious ineffectiveness of local enforcement was recognized and a number of new policies were added. Certain enforcement activities were moved from the local communities to the regional authorities, which were believed to be better able to build the specific environmental competencies needed to deal with high risk industries. Regulation of smaller firms, especially those with well-understood production processes, was left to the local communities. The Danish system of environmental enforcement continued to be rather decentralized at the regional and local levels.

The baseline of all these activities is the command-and-control regime in environmental protection policy that sets standards for polluting discharges and the use of hazardous materials and gives companies environmental approvals, even though the construction of regulatory institutions and the division of labour between government and local authorities may reflect a uniquely Danish type of decentralization. The background for this regime is the science-based identification of strains on given components of the ecosystem. This fundamental perspective is often supplemented with attempts to define the ecological capacity of places and regions. This implies quite naturally a focus on local enforcement and local criteria.

Although rather successful in reducing emissions of well-known pollutants from specific industrial sources, command-and-control measures and their enforcement had also created some well-known regulatory disadvantages: a quite hostile relationship developed between companies and authorities, and a focus of both industry and regulators on end-of-pipe solutions.[3]

Less visible, but equally important, was the fact that the type of knowledge produced and used by regulatory authorities was different from the knowledge useful in industry to create less polluting technologies. The first type of knowledge focuses on the environmental strain on nature and the impact of certain emissions, while the latter focuses on the optimization of processes and changes that reduce or replace hazardous materials and auxiliary substances. This asymmetry in information also contributed to the production of suboptimal resource use.

Companies might develop their own understanding of environmental strains and potential damages, but the authorities were not able to enter into a constructive dialogue with industry about potential solutions.

To overcome the inadequacies of the existing system of regulation, an environmental reform was introduced during the late 1980s to create a new relationship between business and government. The idea was expressed in different discussion papers. The concluding report *Simple and Effective* (MoE 1988) promoted cleaner technology as a future means of environmental improvements:

> The challenge of the 1990s is the acknowledgement that environmental policy must and can be formulated so as to stimulate the development and use of new and cleaner technologies. The known technologies have taught us that economic effectiveness can very well go hand in hand with an active environmental policy. A flexible environmental policy must encourage industry to implement new production forms and to develop products that are more friendly to nature and the environment.

The first few years of experience with the cleaner technology programs helped pave the way for this environmental reform. The support schemes were constructed without any direct link to local enforcement activities in order to protect companies from any tightening of local standards that might follow innovations that promised increased environmental performance. Furthermore the cleaner technology program was made attractive by establishing it as a separate and voluntary arrangement. By building on a mutual interest in improving production processes, reducing costs of materials and waste, and developing greener products, the program was designed to create new forms of interaction and cooperation between industry and government.

The cleaner technology reform was initiated by new group of environmental professionals, including engineers and environmental planners, who had studied industrial processes and been active in the environmental movement. They were hired by the environmental agency to improve the quality of government policies. Fortunately these new policies were created during a period of increasing international support for cleaner technology, a climate hospitable to the idea of "modernizing" environmental protection.

In the longer term the upgrading of performance criteria facilitated by successful cleaner technology innovations was expected to contribute to the overall improvements in environmental performance and lead to tightened emission standards. Due to the success of the program, the environmental act was revised and the cleaner technology concept was included in the legislation. Although individual industries were not supposed to be forced to use any specific technology, the use of cleaner technology was made an independent objective of the new environmental act of 1992. Though it did not specify any measure to enforce the use of cleaner technologies, the act made the development of a preventive strategy based on the implementation of available cleaner technologies a central consideration both for industry and the regulatory authorities. The general duty of companies to reduce pollution was stated in the act as follows:

> In the design and operation of the plant, including choices of production processes, raw materials and auxiliary substances, measures shall be taken to minimize the use of resources, pollution and generation of waste. (MoE 1992)

In principle, the environmental act requires a minimization strategy based on the best available cleaner technology. The best-performance criteria are supposed to form the basis for the environmental permits given to companies. Even when a permit has already been granted, the development of an improved cleaner technology with lower emissions can be used to re-assess the permit and tighten the requirements for a specific company. The principle of using cleaner technology as the basis for setting emission standards should give each firm the freedom to choose how to reduce its own emissions. It was never understood to be an absolute requirement to use specific technologies but rather an injunction to follow the ALARA (As Low As Reasonably Achievable) standard concerning discharges from the production and disposal of products.

Regulatory Regimes at Play

Despite the success of the cleaner technology programs and the government's intention of making cleaner technology a core concept in Danish environmental policy, the two worlds of regulation—fixed emission standards and dynamic improvements based on cleaner technology—still exist in parallel as two separate systems. An integration of these systems to create a coherent set of regulatory practices has not emerged from the past 15 years of policy experience, despite a number of serious attempts.

The introduction of an environmental reform based on the cleaner technology approach in the early 1990s was expected to improve environmental regulation in three ways:

- Support innovations reducing the need for filtering of pollutants and disposal of hazardous waste.
- Support innovations improving industry's environmental performance beyond the criteria defined in emission standards and beyond the boundaries of the factory by reducing environmental loads from products during their complete life cycles.
- Create a cooperative relationship between industry and regulators enabling early responses to future environmental problems.

The overall vision was to improve the environmental performance of production and products beyond the limits of the emissions-based command-and-control regime.

What hindered the re-construction of the existing regulatory regime and the integration of the outcomes of the clean technology programs? The following analysis of the existing regulatory regime identifies these obstacles, and also point to the changes needed to secure the environmental improvements that are possible from cleaner technologies that have already been developed.

In traditional environmental regulation the objects of regulation are emissions of specific, measurable substances to certain well-defined ecosystem components. The standard conception is that pollution emanates from distinct sources and affects the ecological capacity of a bounded (local) set of surroundings. Consequently regulators need to have knowledge about the ecosystem and its capacity and to define emission limits and requirements for monitoring emissions, while they do not need to know about the specific technologies used in production. Rather, as regulation is aimed at continuous pollution control, regulators require information about pollutants to maintain regulation and control. A specific kind of professional is required to understand the environmental problems and to command the knowledge necessary to maintain the specific type of regulatory regime. For command-and-control regulation, the focus on ecological capacity makes environmental control the territory of chemists and biologists.

While the command-and-control regime is based on emission standards that are translated into local requirements that depend on the specific local and regional ecosystem conditions, performance standards based on the specific industrial processes and technologies form a different frame of reference. In technical terms this new frame may simply translate into more strict industrywide requirements. However, a transition to performance standards requires a change in the routines and competence of the regulator, who must be able to match the capabilities of implementing cleaner technologies in different branches of industry. It also demands improved negotiation skills to handle exceptional cases that require different technologies. In sum, the knowledge base of the regulator in the new regime will be different from the skills necessary in the command-and-control regime (Kroman, Binder, and Øhrgaard 1996; Andersen and Jørgensen 1997).

A different kind of professional is thus required. The object of regulation may even include the organization of production, the operators' skills and the companies' procedures for handling technologies. The focus on cleaner technologies moves production and environmental engineers into the core group of competent regulators. While the command-and-control regulator focuses on emission standards and ecosystem capacities, the cleaner technology regulator will have to focus on the best available and least polluting production solution.

Neither the Danish Environmental Protection Agency (DEPA) nor the local authorities have adequately pursued the process of knowledge creation and learning that is needed to sustain the achievements of cleaner technology innovation or to translate them into regulatory requirements. Even though all evaluations have emphasized the need to build and support a learning process in the regulating authorities at the local level, systematic training and competence building among local law enforcement officials has not taken place (Andersen and Jørgensen 1997; Remmen 2000).

Although the Industry Office of the DEPA tried to address this shortfall, its effort has not led to a change in the content of typical environmental approvals. The Industry Office's information strategy used the developed cleaner technologies as the basis for advice to local enforcement authorities on tightening the standards in environmental permits; cleaner technology options were included in guidelines for industry and local authorities of the different industry branches.[4] Some of these guidelines include tightened criteria for certain types of companies subject to mandatory environmental permits.

However, due to lack of resources the processing of the guidelines has been slow, and they were more or less obsolete when finally published. While industry organizations and bureaucracies were still debating the details of the guidelines, deadlines for environmental applications passed, forcing both companies and local enforcement authorities to use existing standards and experiences. Priorities also seem to have changed in the DEPA in recent years as no further guidelines are being produced. Thus local authorities will find it even more difficult to use the results of cleaner technology options as emission-setting criteria for granting future permits, as now each of them must collect and systematize knowledge from rather different and heterogeneous areas.

In addition to the efforts by the Industry Office, several studies and reports have been written (e.g., Kroman, Binder, and Øhrgaard 1996) to help local authorities implement and use cleaner technology options in environmental permits. Nonetheless, not much has happened, and almost all environmental permits still refer primarily to the general standards for emissions and discharges of pollutants. Very few permits for specific emission requirements in the 1990s are based on cleaner technology options, despite the possibilities as stated in the environmental act (Andersen and Jørgensen 1997: 175–76).

A survey produced in 1998 does show a widespread positive focus on cleaner technology, insofar as it has opened positive dialogues between authorities and companies and produced improvements for companies that have adopted these options (DEPA 1998). In this respect the voluntary aspect of the cleaner technology programs has had a positive impact, despite the fact that the enforcement of stricter standards based on translations of the cleaner technology options into specific standards is almost nonexistent. These positive attitudes, however, are not shared equally across firms. The results of a diffusion study showed that the noninnovative and environmentally more ignorant firms have not taken up the cleaner technology options at their own initiative. To harvest the benefits of environmental reform, enforcement must include the possibility of tightening emission standards for specific types of companies based on the availability and applicability of cleaner technology. Unless the performance levels reachable through cleaner technology can be translated into mandatory requirements, an even wider gulf will develop between an "environmental elite" of companies developing and implementing cleaner technologies and a large laggard group of slow adopters or even environmentally ignorant firms.

All in all, regulators have underestimated the work needed to introduce and sustain the new knowledge base for cleaner technology implementation, and more important, to overcome the established routines of regulators involved with the traditional command-and-control regime. Because of these difficulties, controversies and conflict arising from differing emission standards have given both industry and regulators an easy excuse to oppose the merger of the two regimes. Thus the point of minimizing environmental loads as stated in the Environmental Act is lost in the daily routines of formulating approvals and in legal procedures.

Resistance to a new way to set standards is rather passive and based on purportedly pragmatic bureaucratic arguments. However, this bureaucratic resistance may be based on a widespread scepticism among governmental environmental professionals who view themselves as the "protectors" of the environment based on scientific arguments. This view also reflects the perceived risk of too much negotiation and of the difficulty of assimilating a huge amount of knowledge about production processes available through cooperation with industry.

Touting of one method of environmental regulation as the "best available" is often based on an idealized view, in which the complexity of the regulatory regime has been underestimated. This was surely the case when the cleaner technology concept was introduced in Denmark in the late 1980s. The conflict between the command-and-control regime and the new cleaner technology regime installed through the environmental reform lies in the dual knowledge base that must be maintained by the authorities. It is very difficult to translate new cleaner technology solutions into specific emission criteria in the environmental permits an individual company must secure. Problems in the organization and distribution of knowledge between the DEPA and local authorities have not been resolved. The traditional command-and-control regime not only prevails but also seems even to resist adjustment to new standard-setting mechanisms based on cleaner technology and best available technology options.

In addition industry has contributed to delaying the use of best available technologies in regulation. It has been too easy to question the relevance of certain cleaner technology options in negotiations covering aggregate branches of industry; the idea of "shared responsibility" is undermined by traditional polarized negotiation positioning. Regardless of how understandable these negotiation strategies may be, they at the same time undermine productive interaction. This leaves industry and government with the traditional command-and-control regime that has so often been criticized.

Best Practice or Equal Rights to Pollute?

Even though environmental legislation and the formal goals of Danish environmental policy express support for a cleaner technology approach, legal institutions and routines have not been changed accordingly. Rather, they continue to define their own strict set of emission standards (Andersen and Jørgensen 1997). Consequently the need for extended knowledge and learning from experience discussed above has not been the only factor hindering the introduction of cleaner technology measures in environmental approvals. Several issues related to the legal system and legal rights have inhibited integrating cleaner technology approaches into the Danish environmental regime.

Legal procedures for appealing permit requirements still reference the stringent measures cited in the general emission standards. Thus approvals based on performance levels reached by other companies in employing cleaner technologies are easily attacked in the appeals court. Environmental activists have also challenged approvals giving companies wider latitude to achieve environmental goals based on rising level of trust between the authorities and the firm. For these reasons the Ministry has unofficially advised against the use of cleaner technology references and approvals based on future improvements promised and environmental plans defined by companies.

Even in the few cases in which local authorities have been experimenting with reference to best available technologies or requests for continued improvements during the period covered by the environmental permit, the DEPA has tried to avoid conflicts and advise against these experiments. The Agency has argued that the legal system does not support environmental permits that fail to refer to rather exact emission standards authorized by the ministry or that do not provide very specific arguments for further reductions.

Another core problem is the idea implicit in legal interpretations of compliance that all companies in all industries enjoy equal rights. While emission standards are supposed to reflect such equal rights, the level—equal or unequal—of emissions may reflect different technologies. Cleaner technologies are creating different capabilities for protecting the environment in different industries. The logic of enforcement based on the best technology available to each industry is based on the idea that environmental loads throughout all industrial sectors should be reduced as much as possible. This implies that the companies who are able to reduce emissions of a given pollutant at reasonable costs also have the greatest obligation to do so, even if companies in another industry, with a different set of processes, cannot reproduce these results.

The differentiation in emission standards resulting from cleaner technology options for different industry sectors provides an easy entrance for critique. It is very easy for companies to attack deliberate differentiation in enforcement during negotiations as well as in the courts, where the procedures and interpretations are dominated by nonenvironmental professionals. Although the Environmental Act states that priority should to be given to cleaner technology solutions, it does not define an appropriate new perspective on equality under the law enabling the courts to move beyond the command-and-control regime. The idea of best available technology as an informed basis for differentiated regulation does include a "cost-of-performance" measure to substitute for the equality in emissions, but it is difficult to defend in the legal system.

At present, the situation appears to be deadlocked. Reductions in support for developing cleaner technology and consolidating knowledge make it difficult to implement performance measures based on the best available technology. The infrequent use of performance measures in environmental approvals reduces the interest of regulators in maintaining a cleaner technology knowledge base.

An increasingly dominant focus on cleaner products in Danish environmental policy leaves even fewer instruments in the hands of the regulators. Products cannot be regulated by focusing on the company as the direct polluter that leaves the regulators without an object to regulate, except in a few cases such as hazardous materials for which the use can be restricted even in cases where pollution is indirect. Thus, promoting clean products is even more dependent on the diffusion of good practices across companies.

The shift in emphasis to clean products fuels the risk of leaving behind the cleaner technology approach as last year's fashion in environmental policy. In the world of politics it may be more expedient to introduce new concepts at high speed as symbolic manifestations of positive environmental intent. It does not, however, make the implementation of cleaner technology-based measures less relevant in environmental approvals. The technologies have proved themselves, but the hard work of integrating their implications into the regulatory framework remains to be done.

Notes

1. Life cycle assessment was seen as a methodology to support the design of products by providing data to minimise environmental loads and materials and energy used accounting for all phases of the products lifetime from the extraction of natural resources and production of supplies to the final disposal of the used product.

2. Eight percent of the analyzed projects were still at a so early stage of development that they were excluded from the judgment of the usefulness of the resulting cleaner technology.

3. The critique of the command-and-control regime is not based on the simplification that it is not working, or is counter to innovation, as it is often argued in the literature (e.g., Wallace 1995: 22). Traditional environmental regulation has produced significant results, but it is not optimal from the viewpoint of engaging

industry in further environmental improvements, and it has little impact beyond the boundaries of the factory.

4. So far a general overview of publications covering cleaner technology options has been published, together with a limited number of guidelines giving detailed advice to selected branches of industry. The guidelines cover car repair shops, car disposal, galvanizing and electroplating industry, asphalt industry, color and paint industry, and reuse of iron and metals.

References

Andersen, M. S., and U. Jørgensen. 1997. Evaluation of the Cleaner Technology Programme 1987–1992: Results, diffusion and future activities. *Environmental Review*, No. 14. Copenhagen: Danish Environmental Protection Agency. The Danish version was published in 1995.

Bijker, W. E. 1987. The social construction of Bakelite: Toward a theory of invention. In W. E. Bijker, T. P. Hughes, and T. Pinch, eds., *The Social Construction of Technological Systems*. Cambridge: MIT Press.

Bijker, W. E. 1995. *Of Bicycles, Bakelites, and Bulbs: Toward a Theory of Sociotechnical Change*. Cambridge: The MIT Press.

DEPA. 1986. *Udviklingsprogrammet for renere teknologi* (The Cleaner Technology Development Programme). Copenhagen: Danish Environmental Protection Agency.

DEPA. 1988. *Enkelt og effektivt* (Simple and Effective). Copenhagen: Danish Environmental Protection Agency.

DEPA. 1990. *Handlingsplanen for renere teknologi 1990–92* (The Action Plan for Cleaner Technology 1990–92). Copenhagen: Danish Environmental Protection Agency.

DEPA. 1998. *Inddragelse af renere teknologi i tilsyns- og godkendelsesarbejdet* (Incorporation of Cleaner Technology Options in Environmental Enforcement). Miljøprojekt 388. Copenhagen: Danish Environmental Protection Agency.

EU. 1999. *Directive on Integrated Polution Prevention and Control*. Brussels.

Georg, S., U. Jørgensen, and I. Røpke. 1993. Clean technology—Innovation and Environmental Regulation. In *Environmental and Resource Economics*, vol. 2. Dordrecht: Kluwer, pp. 533–50.

Gouldson, A., and J. Murphy. 1998. *Regulatory Realities: The Implementation and Impact of Industrial Environmental Regulation*. London: Earthscan.

Hemmelskamp, J., K. Rennings, and F. Leone, eds. 2000. *Innovation-Oriented Environmental Regulation: Theoretical Approaches and Empirical Analysis*. Heidelberg: Physica.

Hillary, R., and N. Thorsen. 1999. Regulatory and self-regulatory measures as routes to promote cleaner production. *Journal of Cleaner Production* 7: 1–11.

Hughes, T. P. 1987. The evolution of large technological systems. In W. E. Bijker, T. P. Hughes, and T. Pinch, eds., *The Social Construction of Technological Systems*. Cambridge: MIT Press.

Jørgensen, U. 1993. New regulatory regimes in environmental protection. Paper for the conference on Designing the Sustainable Enterprise, Boston. Lyngby: Unit of Technology Assessment, Technical University of Denmark.

Jørgensen, U. 1995. *Projektoversigt i tilknytning til evalueringen af støtteordningerne 1987–92* (Projectdirectory related to the Evaluation of the Support Programmes 1987–92). Arbejdsrapport 22. Copenhagen: Danish Environmental Protection Agency.

Kemp, R. 1997. *Environmental Policy and Technical Change: A Comparison of the Technological Impact of Policy Instruments*. Cheltenham: Elgar.

Kroman, L., J. C. Binder, and L. Øhrgaard. 1996. *Integration af renere teknologi i miljøsagsbehandlingen* (Integration of Cleaner Technology in Environmental Enforcement Activities). Miljøprojekt 331. Copenhagen: Danish Environmental Protection Agency.

MoE. 1988. *Enkelt og Effektivt—et Debatoplæg om Lovgivning og Administration* (Simple and effective—A discussion of legislation and administration). Copenhagen: Ministry of Environment and Energy, Denmark.

MoE. 1989. *Renere teknologi—konferencegrundlag* (Cleaner Technology—Conference Materials). Copenhagen: Ministry of Environment and Energy, Denmark.

MoE. 1992. *Miljøloven af 1.1.1992* (The Environment Act of 1.1.1992). Copenhagen: Ministry of Environment and Energy, Denmark.

Remmen, A. 2000. *Renere produkter—nye værktøjer, aktører og relationer* (Cleaner Products—New Methods, Actors, and Relations). Orientering 12. Copenhagen: Danish Environmental Protection Agency.

Remmen, A., and J. Lassen. 2000. *Renere teknologi via produkt- og kortlægningsprojekter* (Cleaner Technology highlighted through Product and Survey Projects). Miljøprojekt 522. Copenhagen: Danish Environmental Protection Agency.

Van Dijken, K. Y. Prince, T. J. Wolters, M. Frey, G. Mussati, P. Kalff, O. Hansen, S. Kerndrup, B. Søndergård, E. Lopes Rodrigues, and S. Meredith. 1999. *Adoption of Environmental Innovations: The Dynamics of Innovation as Interplay between Business Competence, Environmental Orientation and Network Involvement*. Dordrecht: Kluwer.

Wallace, D. 1995. *Environmental Policy and Industrial Innovation—Strategies in Europe, the US and Japan*. London: Earthscan.

II

Firm-Level Approaches

8

The Dutch Policy Program on Environmental Management: Policy Implementation in Networks

Theo de Bruijn and Kris Lulofs

Shared responsibility is a central pillar of Dutch environmental policy, which since the late 1980s has involved close cooperation among government, the business community, nongovernmental organizations (NGOs), and other actors (Bressers and Plettenburg 1996). The policy program on environmental management, which ran from 1989 to 1996, exemplifies this new approach. Environmental management has become a common practice in North America and Europe, where public and private organizations have made considerable progress in environmental management during the 1990s (De Bruijn, Groenewegen, and Grolin 1997). Companies now have many tools to choose from for managing their environmental impacts; nonetheless, implementation is still not easy. In many instances individual companies must undergo a drastic shift in vision, in addition to the organizational and managerial changes required (De Bruijn 2001).

While larger firms have been developing the capacity and capabilities needed over the last decades, SMEs (small- and medium-sized enterprises) often lack the skills, knowledge, and expertise as well as the finances and time to make the desired changes by themselves. Instead of working directly with SMEs, the Dutch government focused on facilitating and managing the formation of networks in which intermediary organizations acted as agents for change. The idea behind the approach was to generate mutual trust on which to base government-industry collaboration, to enhance capacity building within industry, to involve third parties, and to build networks in addition to aiming for direct environmental impacts. The program was conceived as a long-term strategy without precisely articulated goals in order to leave the process open to policy learning.

This chapter reviews the effectiveness of the Dutch policy approach. It is organized as follows: The first section presents the basic philosophy behind policy implementation in networks. The second section describes the central elements of the 1989 to 1996 Dutch policy program on environmental management and its effectiveness. The third section analyzes the effects of the program in terms of company attitudes and behavior. Section four explains why the program was relatively successful, focusing on the network approach and firm characteristics. The fifth section examines additional crucial factors, which include the wider policy approach and the fit with the general mediating policy style of the Netherlands, and the close involvement of industry in the program. The concluding section evaluates the potential effectiveness of the network approach for transforming SMEs.

Collaboration in Policy Networks

Relationships between state and society may be described in terms of hierarchies, markets, and networks (Van Heffen and Klok 2000). In network governance, policy choices emanate from highly organized social subsystems, such as the production system, rather than from a central state authority (Kohler-Koch 1998). Efficient and effective governance must recognize the specific rationalities of the subsystems. In order to mirror these rationalities and the creativeness of the subsystems, policy processes need to be open and decentralized. In network models the state plays a more limited role and concentrates on establishing supportive policy networks by bringing together the relevant actors in society (Kohler-Koch 1998; Young 2000).

Policy networks can facilitate a consultative style of government, reduce conflicts, de-politicize issues, and make policies predictable. Thus, actors in network governance have greater freedom to act on their own property although some actors may have specific authority to intervene (Van Heffen and Klok 2000).

During the last decade the network model has become quite popular, especially in environmental policy making (see O'Toole 1988; Bressers 1993) for two reasons. First, it has been very hard for governments using a hierarchical approach to develop meaningful environmental regulations for industry. Industry consists of many quite different sectors. It is virtually impossible for governments to have detailed knowledge of all sectors and the variations within each. Government must rely on industries to provide information on, for example, the amount of pollution they generate, abatement costs, and technological options. Lévêque (1996) has explored this shift from public regulation, in which public authorities set the environmental objectives and the measures to achieve them, to a system of co-regulation in which relationships between public authorities and firms are "pervasive and close." In his view, co-regulation is especially useful in the presence of uncertainty because it enables players to learn jointly how to achieve the objectives set by public authorities.

The second benefit of collaboration in policy networks lies in making expertise and support accessible to SMEs. Over the last decade many companies have shifted from simply complying with regulation to interpreting their environmental responsibilities more broadly (Young 2000). The corporate strategies of industry have come to include environmental issues, making environmental management a common, though not universal, practice in North America and Europe, at least in larger corporations with the resources to develop appropriate practices. For most SMEs, learning to become a more sustainable enterprise has been a more difficult struggle. Lacking external pressures and incentives as well as concrete support, SMEs will likely fall behind and not make the desired changes. Policy networks in which intermediary organizations play a central role may enable SMEs to access the information to develop their environmental capacities and capabilities.

In sum, network approaches harness the creativeness of industry to inform the policy-making processes. In this way solutions can be developed that fit the unique situations of different sectors of industry. At the same time network partners can reach out to SMEs and enable them to meet environmental policy goals.

Actors within policy networks have different roles to play in maximizing these benefits. The framework for analysis used in this chapter is based on the subjective rational actor model (Riker and Ordeshook 1973; Kiser and Ostrom 1982). This model explains the course of interaction processes in terms of the attitudes of the actors involved. An "attitude" is formed by the combination of the actor's motivation, resources, and position of power. The attitude therefore represents what an actor wants, his capacity to act accordingly, and his power to resist possibly conflicting preferences of other relevant actors.

The activities of intermediary organizations within a policy network affect all three dimensions of attitude formation through the performance of three functions: persuasion, support, and pressure. Network organizations perform a *persuasive function* in trying to influence the motives of the dominant coalition within companies. If firm-initiated environmental management is the goal, network organizations must convince this dominant coalition that it is valuable to improve their environmental performance through a management system. The *supporting function* of a network organization comes into play by increasing a company's ability to manage its environmental performance. This involves supplying model approaches, guidelines, and manuals and offering courses and training. The network organization also exercises a *pressuring function*, gently or vigorously pushing unwilling members of the target group in the desired direction. Effective pressure, of course, requires a position of power. The three functions are summarized in figure 8.1.

Under this model of network behavior, companies will show a positive attitude and carry out of the desired activities in direct proportion to the exercise of the three functions by the network. Government can play different roles in this context (Koppenjan et al. 1993; Mazmanian and Sabatier 1989). Acting as a "broker," government can promote integration by bringing together actors with converging goals and resources. It can act as a "mediator" when conflicts occur or as a "facilitator" in deploying incentives. By initiating actions, government can function as another "entrepreneur." In serious cases it can threaten or use its posi-

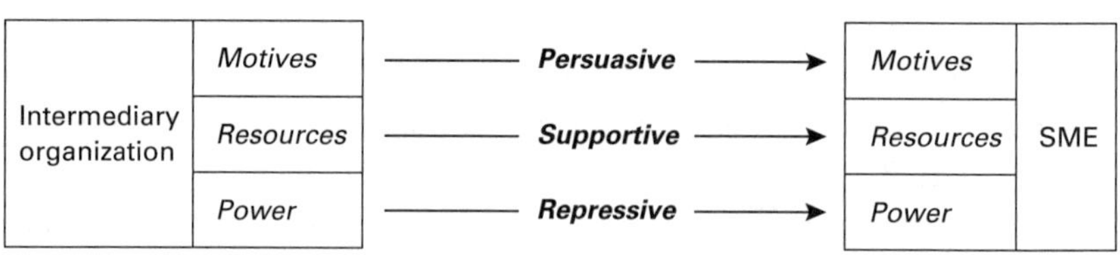

Figure 8.1
Three network functions between SMEs and intermediary organizations

tion in the hierarchy to pressure a corporation to change undesirable behavior.

None of these network actions, however, will produce the desired changes if the target group is self-referential and autopoietic (Katz and Kahn 1966). SMEs have proved to be a particularly difficult to reach target group, possibly because opinions within it are largely self-referential on environmental issues. The analytic objective of the empirical case analyzed below is to determine how well networks function for improving environmental management in SMEs.

The Dutch Policy Program on Environmental Management

In 1989 the Ministry of the Environment in the Netherlands (Dutch acronym: VROM) published the Memorandum on Environmental Management (TK 1988–1989, 20633, nr.3). The objective was to have companies in the Netherlands voluntarily introduce environmental management systems by 1995. No sanctions were set in the short run for companies that did not implement management systems, other than stating that they might be subject to more and severe enforcement.[1] The basic strategy of the program was to convince companies of the usefulness of environmental management by explaining its central concepts and then to offer concrete support during implementation.

In the memorandum, a distinction was made between the (few) larger companies and the main group of SMEs.[2] The memorandum included a set of activities based mainly on the distribution of specific knowledge through, for example, research and sample projects, for each industrial sector. This knowledge was to then be disseminated among companies through guidance and education. By supplying information, attempts were made to stimulate companies to actually introduce environmental management.

Supporters of the program believed that individual firms would implement appropriate measures if their costs and uncertainty were reduced by adequate support. In this cooperative process, the government would have the opportunity to develop regulatory strategies tailored to the realities of each type of industry. Intermediary organizations were asked to play the role of keeping in touch with the individual companies.

Policy implementation therefore took place in two distinct phases. In phase 1, VROM supported and developed intermediary organizations to translate the concept of environmental management into concrete actions for specific target groups. In phase 2, companies were expected to integrate these actions into their daily management practices (with the help of intermediary organizations).

Thus, instead of working directly with SMEs the (central) government tried to facilitate the formation of networks that could reach out to individual companies. Government personnel believed they would be more effective working in collaboration with the network organizations (e.g., trade associations) than acting alone (see O'Toole 1988). Policy designers depended on the network organizations' easy access to the target groups. To guide the implementation of this initiative, a special Program Office was established within VROM. It administered available funds (especially for phase one)[3] and was responsible for coordinating and evaluating all activities.[4]

The policy network consisted of trade associations, the industrial environmental agencies (Dutch acronym: BMDs; see below), municipalities, labor unions, consultancy agencies and the Ministry of the Environment.

The most important role of the trade associations was to offer their members (and preferably nonmembers as well) handbooks and courses, and to use their position of authority and power to force their members to manage their environmental impacts.

The 20 industrial environmental agencies were regional organizations established by Chambers of Commerce especially to introduce environmental management in SMEs. Their roles were mainly to motivate and support companies, and when complex problems arose to call in special help, for instance, from an innovation center or consulting firm. The BMDs were also expected to use the material developed by trade associations for their specific sectors.

Municipalities were expected to support the activities of the trade associations and the BMDs for the companies within their borders. They were also asked to fine-tune their regulatory strategies to encourage and support environmental management. In a note issued in 1995, VROM identified four sequential stages in the development of environmental management in a firm: (1) defensive, (2) reactive, (3) active, and

Table 8.1
Network organizations and their functions

Function actor	Persuasive	Supportive	Pressuring
Trade association	×	×	×
BMD	×	×	
Municipality	×	×	×
Labor unions	×		
Consultancy agencies		×	
VROM	×		

(4) pro-active. Local regulators were expected to adjust their approach towards firms accordingly, ranging from strict enforcement (phase 1), stimulation (phase 2), facilitation (phase 3) to trust (phase 4). Firms in phases 3 and 4 can also qualify for a framework permit (see also note 3).

Labor unions were expected to generate support on the shop floor, as workers play an important role in environmental management through the daily management of substance flows within the company. Labor unions were expected to raise the environmental awareness of these employees, who are one key to improving company performance.

Other players included consultancy agencies that had a role in assisting companies with the implementation process. Finally, the Ministry of the Environment placed itself in a motivating role and as co-financier of some of the other network organizations.

The functions expected to be performed by each actor in the Dutch policy program on EMS are summarized in table 8.1.

Implementation of the Policy Program and Its Effects

This section explores the two phases of implementing the program. During phase 1, a policy network was constructed. In phase 2, environmental management was integrated into company policy.

Phase 1: Construction of the Policy Network

Central to phase 1 was the construction of the policy network. Most of the funds available were devoted to sharing the costs of this process. The ministry could cover about half of the budget of a project, once a proposal submitted by an intermediary organization was approved. Between 1989 and 1996, 162 projects were implemented (VROM 1996). Fifty-five projects concerned initiatives by trade associations and some 20 BMDs were established.

Did these initiatives cover the three different functions that network organizations were intended to fulfill?[5] Of the 11 network functions anticipated (see table 8.1), only one failed to develop: the potential persuasive function of the labor unions. Based on these data, we conclude that a true network has been built over the years. To provide a fuller picture of network development in this program, the concrete activities of one trade association are discussed below.

The Royal Dutch Association for the Printing and Allied Industries (KVGO) represents the companies in the pre-press, printing, binding, and print finishing industries. KVGO acts an agent for its members and offers support on various issues. KVGO has a bureau including almost 80 people, some of them having environmental expertise. The printing industry is well organized: almost 90 percent of all companies are KVGO members.

The KVGO established a special foundation (Dutch acronym: SIMZ) to introduce environmental management. SIMZ developed several handbooks that could be of value to individual companies. The first was a handbook consisting of three sections on the organizational, legal, and technical aspects of environmental management with the greatest emphasis on organizational design and compliance (SIMZ 1993). The technical aspects were extensively covered in the next handbook, *Environmental Measures* (1993).

SIMZ also developed a model of environmental management including training and individual support that could be implemented by individual companies. Regional information meetings were held to attract the attention of companies. Special projects were organized with other partners, such as provinces and municipalities. SIMZ also carried out so-called pre-audits. This audit was voluntary for companies collaborating in the environmental management project. The procedures and contents resembled that of an official audit but were less costly.

Phase 2: Developments within Companies

Between 1990 and 1996, the authors closely followed the effects of these and other efforts by partners in the policy network. We analyzed and partly participated in environmental management projects initiated by network partners and by individual companies. We also maintained close contact with the Program Office of the Ministry. Four industrial sectors were investigated in depth: chemical industry, printing industry, synthetics processing industry, and concrete products industry.

Of 343 firms contacted, randomly selected from the four industrial sectors, 141 (41 percent) cooperated in this investigation. Our main research objective was to determine whether the new policy approach was successful. For this purpose we gathered data on 200 variables per company to measure how companies responded to incentives originating through the network and the internal barriers to change (see also De Bruijn and Lulofs 1996, 2000). Our analysis addressed the level of progress that each company had made and its attitude toward environmental management. We were especially interested in the effect of the network activities.

First, we assessed two aspects of the level of progress companies had made in implementing environmental management. The first of these was the extent to which companies had implemented environmental management systems. This is an indicator of progress at an organizational level, indicating a growing commitment and sense of responsibility. Because environmental management is more than just an organizational matter and the ultimate goal of the Policy Program on Environmental Management was direct environmental improvements, we also investigated the extent of technical measures a company had applied to environmentally relevant issues.

We combined the two aspects—systems development and technical measures—to indicate the level of progress a company had made in implementing environmental management.[6] The "inactive" companies have developed few or no elements of an environmental management system and have hardly implemented any concrete measures to minimize their environmental impacts. The "advanced" companies are the opposite. They have developed all or nearly all organizational elements of a management plan, and have also implemented a relatively large number

Table 8.2
Progress of environmental management in companies (in %)

Level of progress	Percentage of companies
Inactive	6
Orientating	39
Initiating	51
Advanced	4

Table 8.3
Progress of environmental management in companies (in %)

Attitude[a]	Percentage of companies
Negative	23
Indifferent	21
Positive	56

a. In our analysis we distinguished among five different attitudes. For this chapter we summarized these in three main categories.

of technical measures. Table 8.2 shows the distribution of the companies on this variable (progress). The data are reasonably comparable with those found by the official evaluation studies (e.g. Heida et al. 1996).

The implementation of environmental management requires changes within the organization, and thus in the patterns of behavior of its members. Moreover, the policy program also aimed at changing the attitude towards environmental management within a company. The dependent variable "attitude" toward changing rules of behavior can be placed on a continuum ranging from acceptance to rejection of change. As explained earlier in this chapter, in our model an attitude is formed by the actor's motivation, his resources, and his position of power. To measure the attitude of the companies, we used a list of 20 items, each representing a different dimension of an attitude.[7] Via the construction of scales for each dimension, we were able to determine the attitudes of the companies. Table 8.3 gives the overview.

Research expectations were formulated about the positive correlation between the attitude and the level of progress of companies. These expectations proved to be true (Kendall's tau-c = 0.42, T-value 5.84).

To illustrate the types of effects that this program had on companies, we highlight a printing company that we will call DRUK which is located in a small town in the Netherlands.[8] DRUK employs about 50 people, making it one of the larger printing companies in the sector. In 1993 DRUK received an invitation to a regional information meeting organized by KVGO. The environmental coordinator of this trade organization also called personally to ensure the participation of DRUK. Given its size, the company was seen as a prime candidate for joining the environmental management project. DRUK did decide to participate in the environmental management project of the KVGO and to implement environmental management, which it did by 1995 with the support of KVGO/SIMZ and a trainee. Two and a half years later DRUK became ISO 14001 certified.

On an organizational level the company has changed quite drastically. All working procedures have been screened and, if necessary, adjusted. All procedures have been outlined in a company-specific handbook on environmental management modeled after the handbook provided by SIMZ. Each year DRUK formulates an Environmental Action Program, based on a pollution prevention approach. In past years technical measures have been taken to limit the environmental effects of DRUK, particularly concerning emissions to water and air, and waste. Here are a few examples:

- The amount of waste has been reduced by around 15 percent, while another 5 percent is re-used outside DRUK. Remaining quantities are now properly disposed of.
- Energy efficiency has increased by 13 percent.
- Emissions of volatile organic solvents have been reduced by more than 20 percent.
- Emissions of photo chemicals to water have been reduced by more than 90 percent. Emissions of heavy metals to water have been reduced by almost 60 percent.
- A management system is now in place to control the most relevant areas of the production process.

Equally important as these concrete measures has been the raising of environmental consciousness throughout the company. Environmental concern is now fully incorporated into daily management. As DRUK embraced the principles and practice of environmental management, the mission of the company has become much broader. Next to profitability and quality, DRUK now recognizes that environmental concerns are legitimate and important goals, and has therefore incorporated environmental concerns into its decision-making processes.

All major decisions regarding future production processes include an environmental assessment. Examples include investment in a new printing press, the choice of ink, and the choice of cleaning products. The newly appointed environmental coordinator helps in gathering the necessary information, often from contacts at KVGO/SIMZ. Recognizing environmental consequences at an early stage of decision making decreases the risk of investments in equipment or material that do not meet current or anticipated environmental requirements. The amount of freedom to choose environmentally responsible alternatives is, however, often limited by the suppliers of equipment (see Le Blansch 1995).

Overall, the combination of growing external pressures, internal motivation by the management of DRUK, and support by the trade association enabled the company to become a leader in the sector within a few years. DRUK considers the environmental project as beneficial to its survival in the long run. It adds to its profile of being a frontrunner in the sector both in terms of quality and in environmental matters. This way the early adoption of EMS supported the niche position on the market that DRUK has created for itself over the years.

Explaining the Results

The example of DRUK and these results more broadly suggest that the stimulation policy has been quite successful. However, the main question is whether the policy program, implemented in a network configuration, is accountable for those outcomes. To this end we developed a theoretical model in which the interorganizational framework stood central. The basic assumption was that as network relations are exercised to a greater extent, companies will show a more positive attitude and consequently more environmental measures will be taken. In addition to network activities, this model assumes that company-specific circumstances could

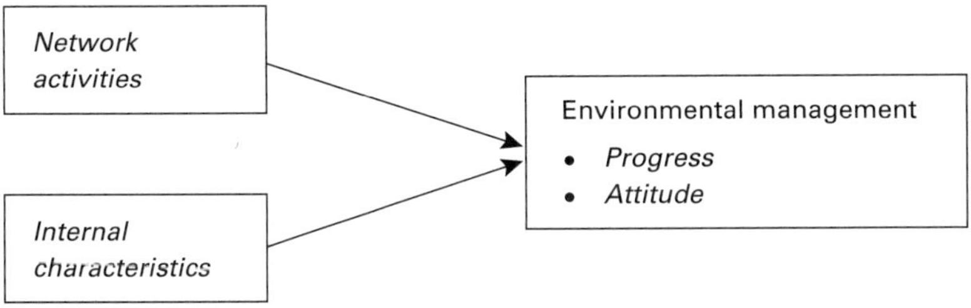

Figure 8.2
Research model

Table 8.4
Degree of penetration of the network activities (in %)

Network indicators	Percentage of companies
Familiar with terminology	92
In possession of the supporting material	62
Attended informative meetings	54
Implementation supported by the network	43

also explain why certain companies were further advanced and were much more amenable to environmental management than others. The research model is shown in figure 8.2.

To test this model, we first looked at the degree of penetration of the different activities carried out through the network. Network organizations (and especially trade associations) were supposed to develop model approaches, handbooks, courses, meetings, and so on. Table 8.4 summarizes the main findings on how companies perceived these activities. The percentages suggest that the activities in the network were certainly noticed by the companies.

Next we determined the frequency of contact between network organizations and the company. We then looked at the correlation between the different network functions and attitude. Each correlation proved to be positive (although not very strong) and significant (table 8.5). This means that as network organizations are more active, companies show a more positive attitude towards environmental management and a more advanced implementation of it.

Table 8.5
Correlation between network functions and company attitude

Network functions	Relation
Frequency of contacts	Kendall tau-c = 0.18, *T*-value 1.99
Persuasive function	Kendall tau-c = 0.27, *T*-value 3.08
Supportive function	Kendall tau-c = 0.24, *T*-value 2.29
Repressive function	Kendall tau-c = 0.16, *T*-value 1.71

These findings suggest that network activities seem to have influenced developments within companies. To get a better understanding of this relationship, we carried out several regression analyses.[9] We used indicators of network activities and internal characteristics as explanatory variables. Out of a multitude of variables, six proved to contribute significantly to the explanation. Among the network indicators, only those for the activities of trade associations and the Ministry of the Environment proved to contribute substantially to the explanation. The activities of other organizations in the network did not help explain developments within companies.

Turning to characteristics within companies, we found four significant variables. First, companies with a good internal communication structure achieved much better results. Such companies already had clear decision-making rules and regular job consultations. Second, companies that already had environmental expertise before the policy program was initiated proved to be much more advanced. Prior experience enabled them to respond to external pressures. A third explanatory variable was the level of profitability. Companies clearly need adequate resources to work on environmental management. The level of competition was our fourth explanatory variable. Companies facing very tough competition may have less time and resources to commit to environmental management.

Taken together, the network and internal variables can explain 58 percent of the variance in progress and attitude of the companies. Table 8.6 summarizes the variables that had significant explanatory power.

Trade associations proved to be by far the most influential actors in the network. Newly established organizations, for example, the BMDs, were not at all influential. The pro-active and influential role of trade

Table 8.6
Explaining the progress and attitude of companies

Explanatory variable	Explanatory power (%)
Network indicator	
Activities of Trade Association	14
Activities of Ministry of Environment	7
Internal characteristic	
Adequate communication structure	12
Environmental expertise	9
Profitability	10
Level of competition	6
Total	58

associations with regard to environmental affairs has also been observed in the United States (Nash 1999). Nonetheless, our research shows that other company specific factors can constrain developments. Stimulating companies through a network configuration is therefore not the ultimate solution. Companies with existing environmental expertise (e.g., the presence of an environmental coordinator) may be better equipped to recognize external pressures and better able to take advantage of the support the network offered. In sum, it is the interplay between external pressures and internal characteristics that enables companies to move forward.

The Program in Context: The Evolution of Dutch Environmental Policy

The policy program on environmental management and its implementation arrangement was a reasonable success. In this section we discuss how the context of this program contributed as much to its success as did the structure and contents of the program itself. The network plan of this program was not an isolated one, but highly integrated in a new approach to environmental policy that has developed in the Netherlands during the last decade. Other parts of this approach provided more mandatory imperatives for change that gave firms an incentive to improve their environmental management. Furthermore the program on EMS was a prime example of the new mediating policy style that has developed in Dutch environmental policies during the same period. This style fits well with the corporatist culture of the Netherlands in general. This "goodness of fit" also contributed to its success.

The evolution in Dutch environmental policy occurred because traditional policy approaches couldn't meet the environmental problems of the Netherlands. Governance in which voluntary programs go hand in hand with more coercive programs is much more appropriate and effective to this issue and this culture.

A New Approach to Environmental Policy Making

The Netherlands is one of the smallest European countries but has the highest population density.[10] Its economy includes a comparatively large amount of industry, intensive farming, and a fast-growing infrastructure. There is powerful competition for physical space among businesses, households, agriculture, traffic, and natural areas. As a result a system of physical planning was developed relatively early.

A large part of the Netherlands lies below sea level; several major European rivers (Rhine, Meuse) meet the sea in the Netherlands. From early in the nation's history, people were forced to collaborate, for instance, on Water Boards, to fight the danger of floods (see Raadschelders and Toonen 1993; Van Hall et al. 1999). Consensus-based community decision making underlies a planning tradition covering a wide range of social and economic aspects. A related Dutch characteristic is the long tradition of governmental consultation with various groups in society (VROM 1997). The general policy style of the Netherlands may therefore be described as corporatist, consensual, and pragmatic (see Van Waarden 1995).

In its initial stages the prevalent style in environmental policy making did not fit this corporatist culture. Legislation developed in the early 1970s revealed a distant, negative attitude toward industry and other target groups (Bressers and Plettenburg 1996). Media-specific environmental legislation was characterized by command-and-control mechanisms. Research indicated that this policy approach could not be properly implemented or enforced and was therefore substantially ineffective (see Schuddeboom 1994).

During the 1980s the decline of confidence in this approach coincided with a number of environmental accidents and crises. Numerous cases of land contamination were discovered, a very serious problem considering the scarcity of land and the hydrogeological characteristics of the Netherlands that make it easy for pollution to easily reach groundwater levels and threaten drinking water supplies. In order to remove or isolate contamination, a number of housing districts had to be torn down.

Another problem was the highly visible impact of acidification that, if not stopped, would cause the little forest that is left in the Netherlands to die and affect many historical monuments as well. Environmental worry was also excited by the discovery of dioxins in milk and in dairy products, which were traditionally seen as healthy, as a result of cows that had been grazing near waste incinerators.

These and other incidents led to a high level of public concern over environmental affairs in the late 1980s. They prepared the political stage for an integral study of the state of the environment in the Netherlands. The resulting report, called in Dutch *Zorgen voor morgen* (RIVM 1988),[11] in the context of high public concern became very influential. It showed that current policies were not only inadequate to the challenges of sustainability but also to some extent ineffective.

The publication of the report coincided with the growing lack of confidence mentioned already in the traditional policy approach with its emphasis on direct regulation (Bressers and Klok 1996). From this perspective it was inevitable that a complete rethink of the environmental policy strategy would take place (Bressers and Plettenburg 1996: 127). Rising standards for environmental quality and the lack of confidence in traditional approaches called for a strategy and style other than the hierarchical approach that accompanied the use of direct regulation. A new strategy was envisioned that would aim more specifically at eliciting private initiatives. This approach was to aim not only at achieving more broad based support for government policy, but also at recognizing that the know-how necessary to reduce environmental pollution is best known by the polluters themselves (see Lévêque 1996). In this perspective, industry is not just part of the problem but also part of the solution.

The first National Environmental Policy Plan (NEPP) published in 1989 was the policy response to the report. NEPP is the cornerstone of current Dutch environmental policy.[12] It calls for radical changes that would make environmental problems manageable within the next 25 years. For example, it would reduce emissions of the most heavily polluting substances by 80 to 90 percent. A fundamental principle underlying NEPP is that the responsibility for reaching such ambitious environmental targets lies primarily with the target group itself (Suurland 1994).

This new strategy therefore leans heavily on integrating more collaborative approaches into the rest of the policy system, thus bringing environmental policy making more into line with the basically mediating national style (Liefferink 1997: 224). Over the years the hierarchical stance with its distant, negative attitude toward target groups has evolved into a new approach designed to encourage collaboration, voluntary action, and self-regulation (Bressers and Plettenburg 1996: 116). The policy program on environmental management was a prime example of the new approach and marked a major shift in the philosophy of environmental governance and regulation.

Voluntary Action, Public Regulation, and Co-regulation

The policy program on environmental management was voluntary, but certainly not without engagement. It was deeply embedded in the total system of environmental policy in the Netherlands. In this system public regulation, co-regulation, and voluntary action are all-important, mutually sustaining, ingredients (see Lévêque 1996). The government decides on overall targets (as laid down in the NEPP); industry gets a say in what measures can best achieve the targets. The gains for industry are increased flexibility and efficiency. In the Dutch system of co-regulation, free riders will be forced in the end by the regulators to meet the same standards as participating companies. From the government perspective, this approach allows regulators to concentrate on a much smaller number of laggards. Furthermore the speed of change is no longer determined by the laggards, as co-regulation makes it possible for the public sector to absorb the knowledge of the private sector.[13]

The policy program on Environmental Management was a first step in a careful strategy aiming at capacity building within industry. At the same time more mandatory programs provided the imperative for change. In the mid-1990s with the program on environmental management still running, regulators introduced the next step—the integration of environmental friendly product-design and life-cycle approaches into environmental management systems.

Industry Involvement in the Policy Program on Environmental Management

The Policy Program on Environmental Management was developed in close collaboration with industry. In fact the debate on environmental management was initiated by industry itself. In the mid-1980s the employers' association VNO/NCW published their first brochure on the desirability and necessity of environmental management, an idea they had picked up from developments in the United States (VNO/NCW 1986). Their initial goal was to strive for deregulation in exchange for the implementation of environmental management by industry. The basic argument was that industry was more capable of setting meaningful standards and requirements than regulators. One of the prime means for achieving this was the implementation of environmental management systems. VROM responded favorably to the idea in general, although the opportunity to further engage industry, instead of deregulation, was their prime motive. The idea of environmental management systems fitted their wish for greater collaboration and their vision of shared responsibility. A joint committee was established to supervise the implementation of several experiments to clarify the concept of environmental management, at that time still a rather vague and abstract idea. In 1988 this committee reported its findings (Committee Environmental Management Systems 1988).

Most of the ideas in this report made their way into the memorandum that VROM issued in 1989. As described above, one of the key ideas of the memorandum was to actively involve industry and its representatives in the implementation of the policy program. Industry was heavily involved during the whole policy-making process from the initial development of the idea all the way through implementation and was thus able to determine the agenda to a large extent. For this reason the policy program was supported by industry, with trade associations as credible messengers to SMEs.

In conclusion, one can observe a strong trend toward integration and shared responsibility in the Dutch environmental policy system beginning in the late 1980s (OECD 1995: 32; Liefferink 1997: 218). Today Dutch environmental policy is being created through a process of close cooperation between government, the business community, nongovernmental organizations (NGOs), and other actors (Bressers and Plettenburg 1996). The introduction of the strong neo-corporatist traits of Dutch society into the environmental field introduced a multi-level approach in which top-down and bottom-up interaction becomes interconnected. Beside the macro-level (national government) and the micro-level (SMEs), the meso-level (intermediary organizations) is involved, facilitating access to actors at the micro-level.

The multi-level nature of the system is not particularly new. What is innovative is how the levels communicate, adjust to each other, and produce and implement agreements. Meso-level actors (e.g., trade associations) bring their influence, resources, and power on actors at micro-level (individual firms). Various programs contribute to the new strategy. The policy program on environmental management and the target-group policy (see chapter 2 in this volume) are two important elements. As the target-group policy specifies what needs to be done and provides the imperative for change, companies are expected to learn to implement these requirements via environmental management. Implementing environmental management systems thus increases the capacity for change. Moreover, given their close connection with the target-group policy, trade association projects often included both organizational and managerial perspectives as well as concrete measures for improving environmental performance. Typical workbooks developed by trade associations thus basically can be seen as a tool on how to *systematically* meet the highly ambitious standards established by the government.

Conclusions

Our empirical case shows that the new policy approach based on shared responsibility, with the policy program on environmental management as one of the prime elements, has had positive results. Specific knowledge useful to companies has been developed as a result of the activities of the policy networks. It has not only increased the capacity for change within industry but also contributed to improved performance thanks to the close link with the target-group policy.[14]

Will this policy approach make *the* difference in reaching for sustainable production in SMEs? Will it lead to radical changes within the industrial production system? Programs such as the one we have evaluated might enable companies to move in that direction, especially in cases where a program is supported by more mandatory elements, as our program was. Such external drivers are needed to mobilize increased capacity.

We must acknowledge, however, that until now, improvements in Dutch companies' environmental performance have been satisfactory but not transformational. The greatest benefit of the Dutch policy approach lies in providing flexibility. The current policy mix gives proactive companies room to innovate and supports them actively. At the same time the permit system can tackle laggards and secure developments in other companies. Therefore we feel that the network approach is a promising new supplement to reach out to SMEs in a way that truly is, in the slogan the ministry used to promote its environmental management policy, "voluntary but not without obligations."

The study identifies several factors that contributed to its success, and should be taken into account in future use:

- *Context of neo-corporatism.* Network approaches might be particularly successful in countries with a neo-corporatist culture. It is, to some extent, normal practice for trade associations in the Netherlands to collaborate with governmental agencies. They had already performed intermediary functions between industry and governments. The formation of the policy networks was therefore fairly easy, and of course facilitated by cost-sharing (50 percent co-financing).
- *Careful network design.* Organizations with a strong potential to influence SMEs should be selected for integration into the network. Organizations should be selected on the basis of their ability to perform three functions: a persuasive function, a supporting function, and/or a pressure function. A balanced mix is essential. Access to the target group is key to the success of intermediary organizations in networks. Important determinants of accessibility include homogeneity of the target group, the

number of companies involved, and the existence of a strong representative body that can act and negotiate on its behalf of the target group. In the case analyzed here, trade associations proved to be the most valuable intermediary actors in the policy network. They were good partners for governments and were trusted by the companies involved. Our case indicates therefore that it might be better to use strong, existing networks and partners than to create new intermediary organizations.

• *Substantial management of the network.* Networks are not an "easy choice"; they do not operate by themselves. Their value is found in bringing together actors in the subsystem to work toward practical solutions to problems that can only be addressed in general, nonoperational terms. This requires monitoring of the network design and the activities of the network. Government agencies, in our case the special Program Office within VROM, need to manage the networks.

• *Building sufficient pressure.* For voluntary approaches it is crucial that the target group perceives that changes of behavior in the long run will be unavoidable. A program must be embedded in a long-term strategy for change. The strategy should include fallback options in case the voluntary approach does not work. The level of public awareness also is a key factor, but not one that is easily manipulated. In an atmosphere of consensus that change is necessary, companies will move more easily in the desired direction.

• *Step-by-step approaches within a long-term strategy.* Environmental management as promoted by the Dutch policy program is a clear example of a step-by-step approach within a comprehensive, long-term agenda based on an integral analysis of the state of the environment. The analysis and agenda are renewed every few years, but goals extend decades into the future. This makes change less dependent on the dynamics of short-term politics. For industry, this stability and predictability of policies may well be the biggest benefit.

These recommendations underline our conclusions that the Dutch policy program on Environmental Management was well thought out in terms of implementation structure, instrumentation, and content. However, it did not achieve success in isolation. We feel that the context in which the program was implemented made companies more susceptible to accepting the challenges in the voluntary program, and thus contributed significantly to its success.

Notes

1. In 1995 the ministry published a note in which the importance of environmental management for the relationship between regulators and companies was stressed. Companies who had implemented a certified management system could qualify for a so-called framework permit (VROM 1995). In a framework permit regulators only set the general standards to be reached while the company provides the details regarding how it will reach those standards.

2. The memorandum mentioned 10,000 to 12,000 larger companies, and 250,000 SMEs.

3. Only some 22.5 million euros were available. By its size the policy program was only modest. A third part of the funds went to trade associations to support their branch projects, another third to the establishment of BMDs (see below). The remaining funds went to individual cases and to research and communication programs.

4. The archive established by this office provided much of the data for the research reported in this chapter.

5. Respondents evaluating the organizations were asked to determine the existence of a specific function on a ten-point scale. The average score was used to test the establishment of the various functions.

6. VROM had defined an EMS as consisting of eight elements that had to be implemented in a coherent way fitting the context of a specific firm. For the organizational dimension we measured how many elements had been implemented appropriately. We also asked for an overview of technical measures. These overviews were encoded in order to distinguish between firms that had taken relatively many measures and the ones with relatively few measures.

7. To give an example for our case: via the list of items we measured whether a company was in favor or against implementing environmental management, its expertise on it (sufficient or insufficient) and whether it could resist pressures by the trade association and other network partners to implement EMS.

8. The company has been made anonymous. The description of DRUK is based on actual (and updated) data of one specific company that we investigated, with some complementary data from other, comparable companies. This way DRUK can be seen as an example of a typical Dutch SME trying to include more environmental matters into its management and core processes. See also De Bruijn (2001).

9. Through regression analyses one can determine the relation between a dependent variable and several independent variables ($Y = a + \beta_1 X_1 + \beta_2 X_2 + \cdots + \beta_n X_n + e$). This gives an estimate of the explanatory power of independent

variables. We excluded variables from the regression equation that explained less than 1 percent of the variance of the dependent variables. The level of significance of β was 0.10.

10. The total area of the Netherlands is only some 41,000 km^2 (EU-15: 3,193,000 km^2). The population density in the Netherlands is about 380 citizens per square kilometer; the EU average is 117 (Eurostat).

11. The Dutch word *zorgen* has a double meaning. The title of the report indicates "Concerns about tomorrow" as well as "Taking care of tomorrow."

12. See chapter 2 by Hofman and Schrama in this volume.

13. We are not proclaiming that the Dutch system, in which the target-group policy stands central, is an overall success, although certain successes cannot be denied. Again, also see chapter 2 by Hofman and Schrama in this volume.

14. One of the official evaluation studies commissioned by VROM also confirmed the positive correlation between having an environmental management system and actual performance (Heida et al. 1996). This study found that in general, companies first began implementing some environmental measures and later proceeded to build management systems.

References

Bressers, J. Th. A. 1993. Beleidsnetwerken en instrumentenkeuze. (Policy networks and the choice of instruments). *Beleidswetenschap* (Journal of Policy Science) (4): 309–30.

Bressers, J. T. A., and P. J. Klok. 1996. Ontwikkelingen in het Nederlandse milieubeleid: Doelrationaliteit of cultuurverschuiving. (Developments in the Dutch environmental policy). *Bestuurswetenschap* (Journal of Government) 10(5): 445–60.

Bressers, J. Th. A., and L. A. Plettenburg. 1996. The Netherlands. In M. Jänicke and H. Weidner, eds., *National Environmental Policies. A Comparative Study of Capacity-Building*. Berlin: Springer.

Bruijn, T. J. N. M. de. 2001. Environmental management in a printing company: Transformation patterns in the Dutch industry. In K. Green, P. Groenewegen, and P. Hofman, eds., *Ahead of the Curve: Cases of Innovation in Environmental Management*. Dordrecht: Kluwer Academic Publishers, pp. 41–62.

Bruijn, T. J. N. M. de, P. Groenewegen, and J. Grolin. 1997. Global restructuring: A place for ecology. *Business Strategy and the Environment* 6: 173–84.

Bruijn, T. J. N. M. de, and K. R. D. Lulofs. 1996. *Bevordering van milieumanagement in organisaties* (Stimulating environmental management in organizations). Enschede: Twente University Press.

Bruijn, T. J. N. M. de, and K. R. D. Lulofs. 2000. Driving SMEs towards environmental management: Policy implementation in networks. In R. Hillary, ed.,

Small and Medium-sized Enterprises and the Environment: Business Imperatives. Sheffield: Greenleaf, pp. 263–74.

CEC. 1993. *Towards Sustainability: A European Community Programme of Policy and Action in Relation to the Environment and Sustainable Development.* Luxembourg.

Committee Environmental Management Systems. 1988. *Milieuzorg in samenspel (Environmental management in joint action).* The Hague: Ministry of Environment.

Hall, A. van, Th. G. Drupsteen, and H. J. M. Havekes. 1999. *De staat van water (The state of water).* Lelystad: Koninklijke Vermande.

Handboek milieumaatregelen grafische industrie (Handbook of environmental measures for the printing industry). 1993–1996. The Hague: Fo-Industie.

Hartman, C. L., P. S. Hofman, and E. R. Stafford. 1999. Partnerships: A Path to Sustainability. *Business Strategy and the Environment* 8(5): 255–66.

Heffen, O. van, and P.-J. Klok. 2000. Institutionalism: State models and policy processes. In O. van Heffen, W. J. M. Kickert, and J. J. A. Thomassen, eds., *Governance in Modern Society: Effects, Change and Formation of Government Institutions.* Dordrecht: Kluwer, pp. 153–77.

Heida, J. F., A. H. Hupkes, J. Iansen-Rogers, S. Karreman, J. van der Kolk, H. Senders, P. Stoppelenburg, and A. Vloet. 1996. *Evaluatie bedrijfsmilieuzorgsystemen 1996* (Evaluation of environmental management systems of 1996). The Hague/Tilburg: KPMG/IVA.

Katz, D., and R. L. Kahn. 1966. *The Social Psychology of Organizations.* New York: Wiley.

Kiser, L. L., and E. Ostrom. 1982. The Three Worlds of Action: A metatheoretical synthesis of institutional approaches. In E. Ostrom, ed., *Strategies of Political Inquiry.* Beverly Hills: Sage, pp. 179–222.

Kohler-Koch, B. 1998. *The evolution and transformation of European governance.* Political Science Series 58. Vienna: Institute for Advanced Studies.

Koppenjan, J. F. M., J. A. de Bruijn, and W. J. M. Kickert. 1993. *Netwerkmanagement in het openbaar bestuur: Over de mogelijkheden van overheidssturing in beleidsnetwerken.* (Network management in public policy: About possibilities for governance in policy networks). Den Haag: Vuga.

Le Blansch, K. 1995. Offset printing companies and the environment. *Business Strategy and the Environment* 4(4): 220–28.

Lévêque, F., ed. 1996. *Environmental Policy in Europe: Industry, Competition and the Policy Process.* Cheltenham/Brookfield: Edward Elgar.

Liefferink, D. 1997. The Netherlands: A net exporter of environmental policy concepts. In M. S. Andersen and D. Liefferink, eds., *European Environmental Policy: The Pioneers.* Manchester: Manchester University Press, pp. 210–50.

Mazmanian, D. A., and P. A. Sabatier. 1989. *Implementation and Public Policy.* Glencoe, IL: Scott, Foresman.

Nash, J. 1999. The emergence of trade associations as agents of environmental performance improvement. 1999 Greening of Industry Network Conference. Best Papers Proceedings. Chapel Hill: University of North-Carolina.

NEPP (National Environmental Policy Plan). Tweede Kamer (Second Chamber) 1988–1989, 21137 1–2.

OECD. 1995. *Environmental Performance Review Netherlands*. Paris: OECD.

O'Toole, L. J. 1988. Strategies for intergovernmental management: Implementing programs in inter-organizational networks. *International Journal of Public Adminsitration* (4): 417–41.

Raadschelders, J. C. N., and Th. A. J. Toonen. 1993. *Waterschappen in Nederland* (Water Boards in the Netherlands). Hilversum: Verloren.

Riker, W. H., and P. C. Ordeshook. 1973. *An Introduction to Positive Political Theory*. Englewood Cliffs, NJ: Prentice-Hall.

RIVM. 1988. *Zorgen voor morgen. Nationale milieuverkenning 1985–2010* (National Environmental Outlook 1985–2010). Alphen aan den Rijn: Samsom H.D. Tjeenk Willink.

Schuddeboom, J. 1994. *Milieubeleid in de praktijk* (Environmental policy in practice). Alphen aan den Rijn: Samsom H.D. Tjeenk Willink.

SIMZ. 1993. *Handboek Interne Milieuzorg* (Handbook for environmental management). Amstelveen: SIMZ.

Suurland, J. 1994. Voluntary agreements with industry: The case of Dutch covenants. *European Environment* 4(4): 3–7.

Tweede Kamer der Staten-Generaal. *Bedrijfsinterne Milieuzorg* (Second chamber, Memorandum on Environmental management) TK, 1988–1989, 20633, nr.3.

VNO/NCW. 1986. *Milieuzorg in bedrijven* (Environmental management in companies). The Hague: VNO/NCW.

VROM (Ministry of Environment). 1995. *Bedrijfsinterne milieuzorg als basis voor een andere relatie tussen bedrijven en overheden* (Environmental management as a base for changing relationships between industry and governments). The Hague: VROM.

VROM (Ministry of Environment). 1996. *Informatiebulletin Bedrijfsinterne Milieuzorg* (Information bulletin on environmental management). Special edition. The Hague: VROM.

VROM (Ministry of Environment). 1997. *Environmental Management: A General View*. The Hague: VROM.

Waarden, F. van. 1995. Persistence of national policy styles: A study of their institutional foundations. In B. Unger and F. van Waarden, eds., *Convergence or Diversity? Internationalization and Economic Policy Response*. Gateshead: Athenaeum Press.

Young, S. C., ed. 2000. *The Emergence of Ecological Modernisation: Integrating the Environment and the Economy?* London: Routledge.

9

Voluntary Regulation and Industrial Capacities for Environmental Improvement: The Case of the EU Eco-Audit Regulation (EMAS) in the United Kingdom

Andrew Gouldson

Considerable debate has arisen within discussions of regulatory reform over the relationship between mandatory, voluntary, and economic or market-based forms of regulation. This chapter examines the influence that one framework for voluntary environmental regulation, the EU's Eco-Management and Audit Scheme (EMAS), has had on the capacities for environmental improvement in the United Kingdom.

Adopted in 1993, the Scheme now operates in each of the fifteen member states of the European Union. It encourages companies to go beyond minimum legal compliance and to continually improve their environmental performance. Organizations that choose to participate in the scheme must implement an environmental management system (EMS) that meets the requirements of the International Standard ISO 14001 and produce an independently verified public report on their environmental performance. EMAS provides an opportunity for organizations that meet a range of conditions to gain a "seal of approval" that a firm can use both internally as it validates the integrity of management controls and improvement programs and externally as it enables organizations to provide verified information on their environmental performance to regulators and to their wider range of stakeholders. By the end of 2001, over 3,700 sites within the European Union were registered as EMAS organizations.

Although the specific impacts of voluntary regulations such as EMAS are difficult to isolate, the empirical analysis presented in this chapter suggests that EMAS has helped companies establish and develop their capacities for environmental improvement. However, it also suggests that the extent to which these companies develop and draw upon these capacities depends on the wider range of incentives and imperatives that they encounter. Regulations and economic incentives therefore drive the uptake and shape the influence of voluntary initiatives such as EMAS (see also Millar 1994; Hillary 1995; Bayliss et al. 1998). By helping to establish the capacities for change, voluntary regulations such as EMAS complement other command-and-control regulations and market-based incentives. Furthermore, by helping establish the capacity for change, voluntary regulations such as EMAS may enable governments to impose stricter standards, to establish more challenging incentives and disincentives, and to adopt more "responsive" approaches to implementation (see Ayres and Braithwaite 1992; Gunningham and Grabosky 1998).

These factors are significant for two main reasons. First, they suggest that voluntary regulations that build the capacity for change can only have a positive impact as part of a wider policy mix that also establishes imperatives and incentives. Second, it suggests that positive feedback can exist in the inter-relations between the imperatives, incentives, and capacities for change that can create an evolutionary "regulatory space" (Hancher and Moran 1989) within which change is particularly achievable. For these reasons the inter-relations between voluntary, mandatory, and economic forms of regulation are of central importance.

This chapter begins by examining the origins of voluntary EMS-based standards in the United Kingdom. It then moves on to consider the extent to which voluntary management-systems-based standards such as EMAS have been accepted by industry, regulatory agencies, and other stakeholders in the United Kingdom. Next it introduces a study that seeks to examine the influence of EMAS on the forms and levels of innovation and technological change in participating firms. While recognizing that it is difficult to isolate the influence of one instrument within a complex policy mix, it considers the motives for the voluntary adoption of EMSs and the influence of EMAS on both the levels and the form of environment-related innovations developed and/or adopted by participating companies.

Generally, the study shows that the motives for participation in the scheme were perceived to be limited and that, although the development of any EMS could be considered significant, the influence of EMAS over and above the influence of any preexisting EMSs such as ISO 14001 has thus far also proved to be limited. However, it is noted that the EMAS Regulation has been revised since the time of the study and that some regulators are beginning to investigate the potential for "regulatory relief" to be given to EMAS-registered sites. As a result it is concluded that the uptake and influence of EMAS might be more significant in the future.

EMAS Regulation in the United Kingdom

EMAS was adopted as an EU Regulation in 1993. Unlike EU Directives, which establish general principles or objectives but allow member states some flexibility in the ways in which they operationalize or achieve them, EU Regulations specify particular measures that all member states must take and/or standards that they must comply with. Consequently national governments have had to comply with the terms of the Regulation without transposing it into national or subnational legal systems. Thus, formally at least, the scope for the Regulation to be tailored to fit into the diverse national legal contexts has been very limited.

This approach has ensured the establishment of similar structures to administer the EMAS in all EU member states. Each is required to appoint a Competent Body to promote, develop, and administer EMAS and to establish a structure for accrediting those environmental verifiers (mainly consultancies) that visit participating sites to ascertain whether or not they comply with the various requirements of the scheme. However, the EU does not, and perhaps cannot, ensure that the actors nominated to operate within these structures will interpret, apply, or respond to the requirements of the Regulation in the same way. As a result significant differences have emerged among the member states (see Watzold 2001). The UK experiences cannot therefore be taken to be wholly representative of the experiences that other member states have had with EMAS.

Within the United Kingdom, the government initially, in 1993, appointed a central government ministry, then called the Department of the Environment, as the Competent Body for EMAS. However, in 1998 this function was transferred to a private sector organization, the Institute for Environmental Management and Assessment (IEMA). Administratively, the IEMA must ensure that any sites that apply to be registered in EMAS have been adjudged to comply with the requirements of the scheme by an Accredited Environmental Verifier (AEV). As the scheme demands legislative compliance as a minimum, the IEMA also liases with environmental regulators to ensure that registered sites comply with all applicable environmental regulations. Responsibility for the accreditation of environmental verifiers under EMAS was allocated by the Department of Trade and Industry to the United Kingdom Accreditation Service (UKAS). An individual or organization desiring AEV status applies to UKAS, which then considers qualifications, competence, and experience of the applicant. UKAS also witnesses and audits the verification work conducted by AEVs to ensure that they interpret the EMAS according to the details of *The Accreditation of Environmental Verifiers for EMAS* (UKAS 1995). These structures and the role of the various organizations are summarized in table 9.1.

Legislative and Institutional Context for the Implementation of EMAS in the United Kingdom

Prior to the adoption of EMAS, industry in the United Kingdom had been familiar with voluntary management-systems-based standards for many years. Driven particularly by concerns about quality management, government departments, local authorities, and larger companies began to demand that their suppliers adopt systems in the ISO 9000 series; these diffused rapidly during the 1970s and 1980s. Thus ISO 9000 "trickled down" from larger organizations through the supply chain to their smaller suppliers. By the beginning of the 1990s, compliance with standards in the ISO 9000 series had become virtually a prerequisite for doing business in many sectors.

Following the upsurge of environmental concern and the introduction of various new environmental regulations in the late 1980s and the early 1990s, industry in the United Kingdom began to show interest in developing an environmental version of the already familiar quality management systems. In fact it was industry that called for, and that led, the development of the British Standard on Environmental Management Systems (BS7750). This standard, developed by the British Standards Institution in collaboration with various UK firms, was piloted in various

Table 9.1
EMAS structures and responsibilities in the United Kingdom

Member state	Responsible for appointing the Competent Body and the Accreditation Body; also for promoting the scheme and for ensuring implementation. The UK Department of the Environment, Food and Rural Affairs (DEFRA) fulfills this function in consultation with representatives from Northern Ireland, Scotland, and Wales. DEFRA is a government department created in 2001—prior to this period responsibility for EMAS rested with the Department of Environment (from inception to 1997) and the Department of Environment, Transport and the Regions (DETR) (from 1997 to 2001).
Competent body	Appointed by DEFRA. Responsible for registering, suspending or deleting sites. Contributes to promotion of the scheme by providing advice to DEFRA and participating in promotional events. The Competent Body is the Institute of Environmental Management and Assessment.
Accreditation body	Appointed by DETR. Responsible for ensuring the competence of environmental verifiers, through witnessed assessments and ongoing supervision of their activities. This function is performed by the UK Accreditation Service (UKAS).
Competent enforcement authority	Responsible for informing the Competent Body if a site is in breach of legal requirements—effectively has a veto on all applications. Depending on site location, the regulators are the Environment Agency, Scottish Environment Protection Agency, the Northern Ireland Environment and Heritage Service, the Department of Trade and Industry and the local authority(s).
Environmental verifier	Responsible for validating that the site's policy and management system comply with the requirements of the regulation and that the information in the environmental statement is accurate and reliable.

industrial sectors in the early 1990s. Launched in 1994, it was in many ways the pilot scheme for the subsequent development of an international standard that closely approximated it. As a result BS7750 was withdrawn in 1997 and replaced by the International Standard on Environmental Management Systems (ISO 14001).

Given the familiarity of UK industry with voluntary management-systems-based standards, and particularly with quality and environmental management systems in the form of ISO standards, it is perhaps not surprising that the development and application of EMSs in the United Kingdom has been relatively rapid. By the end of 2001, over 2,000 organizations in the United Kingdom had ISO 14001 certification, while approximately 120 sites had EMAS registration (ISO World 2001). Watzold et al. (2001) suggest that among manufacturing organizations with more than 20 employees approximately 30,000 firms are potential participants in these schemes within the United Kingdom. Within this category, participation rates are approximately 6.7 percent for ISO 14001 and 0.4 percent for EMAS. In the United Kingdom at least, low participation rates in EMAS appear to have stemmed from a reluctance to seek external validation for existing EMSs and from the perception that EMAS has thus far offered only limited "value added" over ISO 14001 (see below).

These experiences seem to be more widely reflected; preference for ISO 14001 over EMAS is evident in most of the other EU member states, except Germany where EMAS participation is much higher. Over 2,600 German sites are registered for EMAS, representing 68 percent of all EMAS-registered sites for the European Union and a participation rate of nearly 7 percent among manufacturing organizations with more than twenty employees. This high participation rate is seen by some to be largely the result of the "regulatory relief" given to EMAS-registered sites in Germany (Watzold et al. 2001). Such "responsive," "smart," or "tiered" approaches to regulation that regulate those sites with a demonstrated capacity to manage environmental performance with a lighter touch than those without (see Ayres and Braithwaite 1992; Gunningham and Grabowsky 1998: or Nash, chapter 10 in this volume, respectively) have yet to be widely adopted throughout the European Union. However, various national regulators have begun to consider the pros and cons of such approaches. A decision that regulatory relief will be given to EMAS sites, but not those with a nonvalidated EMS or with ISO 14001, will give companies a much greater incentive to participate in the scheme.

Participation in and Acceptance of EMAS in the United Kingdom

Voluntary approaches to environmental management and improvement such as EMAS have been broadly supported by industry in the United Kingdom for some years. For example, the Confederation of British Industry (CBI) has attempted to "... promote voluntary efforts by business to enhance environmental performance and to ensure that the policy and regulatory framework within which business operates is consistent with the need to gain competitive advantage" (Cridland 1994: 234). The CBI has long argued that the government should define broad environmental goals and priorities and set minimum standards but, if possible, let industry achieve these objectives in the most efficient way (e.g., see CBI 1994a). According to the CBI (1994b: 30), voluntary action "... should always be the first recourse of government when seeking environmental improvement" because in this way the likelihood of securing competitive advantage from environmental action is maximized. On this basis, voluntary schemes such as ISO 14001 or EMAS have received broad support from industry.

However, in the early stages of EMAS industry representatives were concerned about the bureaucratic nature of such schemes and about how well they actually deliver environmental improvement without overly restricting flexibility and undermining competitiveness (Webber 1994). While these concerns may have diminished over time, many companies remain sceptical about the value of external validation as they perceive the demands of schemes such as ISO 14001 and EMAS to be overly bureaucratic with claims that these approaches merely impose a "paper chase" within firms being common.

Because of its desire to reduce the level of regulation and to minimize spending, the UK government has also accepted and supported the development of voluntary schemes such as ISO 14001 and EMAS. However, government has also been slow to develop formal links between these schemes and the framework of mandatory regulation. This reflects a reluctance on behalf of government and the regulatory agencies to hand over responsibility for significant areas of public policy to the private sector; both government and the public are concerned about the credibility of voluntary regulations and the accountability of voluntary regulators (Jenkins 1995). It also reflects practical difficulties, relating for example to the ability of regulators to recognize and respond to changes in conditions in regulated sites and a scepticism about the inherent value of EMSs (see also Nash, chapter 10 in this volume).

Indeed, regulators in the UK have suggested that the mere presence of an EMS does not guarantee any particular level of performance or even the capacity to manage performance. They have suggested that they should respond not simply to the presence of ISO 14001 or EMAS but rather to documented improvements in environmental performance that these systems might allow over time. Regulatory agencies in the United Kingdom have therefore been reluctant to offer the "regulatory relief" so closely associated with the wider uptake of EMAS in countries such as Germany (Watzold 2001).

Voluntary measures to protect the environment have generally been welcomed by the public and by pressure groups. However, there is a suspicion that the verification structures that are developed to assure the quality and demonstrate the integrity of voluntary regulations such as EMAS may be more susceptible to "regulatory capture" than those that seek to ensure compliance with mandatory regulation (Jenkins 1995; Gouldson and Murphy 1998) in that they may further the interests of the regulators and the regulated industries rather than those of the broader public or of the environment. There has also been only limited public interest in the environmental reports published by EMAS-registered companies. Hillary (1998) notes that a significant majority of requests for EMAS reports has come from researchers and students and very few requests from the general public or other stakeholders. In part this has been because EMAS has required rather restrictive reporting practices; EMAS sites have limited ability to target different stakeholders with different forms of communication (see below).

To summarize: Support for schemes such as EMAS within government has been broad but largely rhetorical. In the regulatory arena, support has stopped short of formal recognition in the form of the responsive or tiered approaches to implementation that might regulate EMAS-registered sites more lightly. Despite broad industrial support many companies are concerned about the bureaucratic nature of management systems and sceptical about the benefits of external validation. While the public and pressure groups tend to support the adoption of environmental management initiatives in industry, they are cautious about the integrity of schemes such as ISO 14001 and EMAS and the associated implementation and enforcement mechanisms.

Methodology

Against this background the European Commission, through the Institute for Prospective Technological Studies, commissioned a project to examine the impact of EMAS on innovation in different member states (see acknowledgment below). Although EMAS does not explicitly set out to stimulate innovation, the study was motivated by the perception that EMAS might help build capacity for technological change in industry. If so, it could function as part of a broader mix of policy instruments introduced to force, encourage, or enable environmentally desirable forms of technological innovation.

For the purposes of this project, within the United Kingdom 19 of the 72 sites then registered for EMAS were interviewed (see table 9.2) The companies chosen for interview were selected from various different industrial sectors and included four small- and medium-sized enterprises (SMEs) with less than 250 employees worldwide as well as a range of larger companies. All interviews were conducted in April and May of 1999 with the person responsible for environmental management on the registered site. In the case of the larger companies this person tended to be the environmental manager, although in some instances this function was combined with health and safety management. In the case of the smaller companies, the interviewee was usually the managing director or the site or process manager. Interviews were also conducted with representatives of the competent body for EMAS and of the accreditation body responsible for awarding Accredited Environmental Verifier status.

Table 9.2
EMAS registered sites interviewed in the United Kingdom

Company	Group	Employees	Turnover	Sector	NACE code
National Power—Drax	International	Site c.600	£1.2 billion	Power generation	40.1
Curtis Fine Papers	National	Site c.500	—	Paper making	21.1
Xerox UK	International	Site c.2000	£250 million	Office equipment manufacture	30.0
National Power—Eggborough	International	Site c.250	£200 million	Power generation	40.1
Auto Smart	No	Site c.80	£10 million	Chemical manufacture	24.5
Nuclear Electric	International	Site c.500	£200 million	Power generation	40.1
Intercolor Ltd.	International	Site c.25	—	Printing ink manufacture	24.3
The Intex Group (UK) Ltd.	No	Site c.100	£3.5 million	IT equipment recycling	37.2
INA Bearing Co. Ltd.	International	Site c.600	—	Bearing manufacture	29.1
Tioxide Europe Ltd.	International	Site c.400	—	Titanim dioxide manufacture	24.1
Biffa Waste Services Ltd.	International	UK c.2000	£250 million	Waste management services	90.0
Volvo Bus and Truck Assembly	International	Site c.450	£80–90 million	Bus and truck assembly	34.1
Philips Components	International	Site c.350	£30 million	TV screen manufacture	26.1
Soapworks Ltd.	International	Site c.100	—	Soap manufacture	24.5
Exhall Plating Ltd.	No	Site c.60	£2 million	Metal finishing	28.5
Alcan Smelting & Power UK Ltd.	International	Site c.500	—	Primary aluminium production	27.4 & 40.1
Kautex Textron (UK) Ltd.	International	Site c.170	—	Plastic blow moulding	25.2
Huntsman Polyurethane	International	Site c.100	£50 million	Polyurethane manufacture	24.6
The Beacon Press Ltd.	No	Site c.150	£6 million	Printing	22.2

All of the interviewees found it difficult to separate the influence of the EMS, of ISO 14001 certification, and of EMAS registration. For this reason the discussion below considers the combined impact of each on the levels and the forms of innovation in EMAS-registered companies. Therefore EMAS may be only one of a number of factors leading to change. This is a critical point; many of the people interviewed suggested that the beneficial impacts identified would have been realized in the absence of the EMAS registration.

Other important issues concern causality. Schemes such as EMAS may stimulate change by creating conditions that allow firms to respond to other factors that induce change. For example, some studies have shown that while the presence of an EMS establishes the capacity for change, the use of that capacity varies according to the economic incentives and regulatory imperatives facing companies (see Gouldson and Murphy 1998). As this study focuses on EMAS particularly, it is possible that its influence is overstated where innovations that had been stimulated by a wider range of interacting factors were attributed by respondents to EMAS alone. Consequently this analysis presents changes that have been associated with, though not necessarily caused solely by EMAS.

While this may appear to suggest that EMAS has had only a limited direct impact, it may be that EMAS plays a role as one of a broader range of factors on both the supply side and the demand side that stimulate and enable innovation is compatible with the views of Rothwell (1992), Soete and Arundel (1995), and Kemp et al. (1998). These authors underscore the value of a systems-based approach to innovation that relies on supply-side conditions to push and on demand-side conditions to pull innovation and of interaction between the various actors and stages of the innovation process.

Findings

This section covers different approaches to EMAS registration, motives for registration, the influence of EMAS on levels and forms of innovation, and its impacts on economic and environmental performance and on external communications.

Approaches to EMAS Registration

Although some of the companies interviewed had sought EMAS registration from the outset, the majority had developed an EMS for internal reasons before deciding to pursue ISO 14001 certification and then ultimately EMAS registration. All but one of the eighteen companies selected for interview from the list of EMAS-registered sites also had ISO 14001 certification. Although some companies had pursued ISO 14001 certification and EMAS registration simultaneously, in only a small minority of instances had EMAS registration been sought before or in the absence of ISO 14001 certification.

Many interviewees said that their firms decided first to develop an effective EMS and to gauge its influence on economic and environmental performance before considering participation in schemes such as ISO 14001 or EMAS. For most of the companies the merits of external validation were therefore considered only after an EMS had been installed. Although in some instances the desire to have the EMS certified led firms to pursue EMAS registration, most sought EMAS registration only after ISO 14001 certification. Importantly, there was not a general perception that EMAS demanded the application of a more rigorous EMS than ISO 14001. After developing the EMS and seeking ISO 14001 certification, the final stage for most companies was to prepare an environmental report and seek EMAS registration.

Motives for Registration

The first stage for many companies, the development of an EMS, was motivated by both internal and external factors. Internally, EMS development reflected a desire to improve risk management, process efficiency and control, employee awareness, and monitoring and performance measurement. Externally, EMS development was stimulated by the need to build competitive advantage by responding to market and stakeholder pressures and by the desire to present a positive environmental image, particularly by building trust and gaining the confidence of customers, regulators, and other stakeholders. Although most of the interviewees suggested that there were good reasons for developing an EMS, some observed that the time and cost implications of EMS development could be particularly significant if it appeared that an effective EMS would demand a change in the culture of the organization. Nonetheless, company managers commonly believed that the development and application of an effective EMS made good business sense. Consequently its development was supported initially both by the directors and senior managers of firms and by the plant managers and the environment, health, and safety managers of specific sites.

Managers seeking external validation of a firm's EMS hoped that the process of certification would hone the performance of the EMS, leading to both improved environmental management and to efficiency gains, and that certification would confirm the integrity of the EMS as an effective environmental management tool. Thus certification was motivated by the desire to check the integrity of the EMS for internal reasons and to enhance the credibility of the EMS for external reasons. While most of the companies interviewed suggested that there were no particular reasons for not certifying the EMS, all were aware of the cost and time requirements associated with the certification process.

Perceptions of the impacts and of the costs and benefits of EMAS registration therefore were mostly based on how much value EMAS added over that realised through the presence of an uncertified EMS or through ISO 14001 certification. In this respect many of the interviewees from EMAS-registered companies thought their firms were "leaders" or "front-runners." By adopting ISO 14001 at an early stage they had already done more than the vast majority of companies in the United Kingdom. In securing EMAS registration, they had done even more than the majority of companies with ISO 14001 registration. However, except in companies where "environmental excellence" was an integral facet of business strategy, managers were sceptical about the added value of EMAS. In fact a significant number claimed that EMAS added few benefits beyond those attained through the EMS and ISO 14001 certification.

The internal benefits that were collectively realized by the EMS, ISO 14001, and EMAS included better monitoring and measurement and improved information flows, higher levels of awareness among managers and employees, more effective process control, more effective cost management, reduced exposure to risk, and improved business planning. Together, these upgrades improved environmental and economic performance. Externally, certification and registration were seen to have enhanced the image of companies, particularly with suppliers, customers, regulators, and stakeholders. In some instances, certification or registration allowed companies to forge links with new customers and to be short-listed in competitive tendering processes; in others, certification or registration helped secure new contracts.

Many of the managers interviewed pointed out that as EMAS did not demand any significant changes to the EMS needed to gain ISO 14001 certification; the only added value of EMAS when viewed in isolation was that it enabled a verified statement to be used as a marketing aid. Even in this respect some interviewees complained that the format for the statement and the data to be presented within it was too rigid and hampered effective communication between the firm and stakeholders. They thought that the value EMAS could add was diminished because the format of the environmental statement was created by those who knew little about the demands of particular stakeholders or how to communicate with them. Thus there was a common perception that the statement was of only limited value.

There was some awareness of the costs of EMAS registration, notably in relation to the time, effort and expenses of setting up and maintaining the system, monitoring its performance and training staff, and with the accreditation, printing, and distribution of the EMAS statement. While most of the companies interviewed thought that these costs were not excessive, interviewees also said that the costs tended to be "front loaded" and that once registration had been achieved the costs would fall and the benefits would increase. This suggests that despite the prospect of benefits in the medium term, a short-term perspective may deter registration in some instances.

Influence of EMAS on Levels of Innovation

Innovation was promoted by a range of different actors and for a range of different reasons within the companies interviewed. Interviewees generally claimed that innovation was promoted by everyone from the board and the managing director through the marketing, the production and health, safety, and environment departments, to individual employees. Many of the large companies studied had their own in-house technical support services and research and development departments.

In the small- and medium-sized companies, innovation was seen as an integral and informal part of day-to-day life as companies adapted to evolving business conditions.

In general, innovation was stimulated by a range of different factors. Firms try to maintain or enhance competitive advantage by responding to the demands of existing customers, by identifying new market opportunities, and by exploring new approaches to production. These new approaches can improve efficiency and profitability while reducing environmental impacts and exposure to the risks of non-compliance. While few firms experienced serious barriers to innovation, there was recognition that it required time and resources and could be inhibited by instability, uncertainty, and risk aversity. Several companies also acknowledged that the scarcity of investment funds favored projects with demonstrable benefits and short rates of return. Intangible or uncertain projects or those with rates of return in excess of two to three years were commonly not supported.

There was a general belief that EMSs had raised awareness of the need to innovate, certainly within environment-related functions but also in the company more generally. Awareness had been raised partly by the level of corporate commitment needed to develop and maintain environmental policies, programs, and management systems and partly by the impact on managers of the information collected through reviews, monitoring programs, and audits. In many cases this information challenged preconceptions and created awareness of areas in which companies were exposed to risks and in which there was potential for economic or environmental improvement. Some companies also believed that EMSs had raised awareness of the need for innovation in their trading partners. For example, some of the larger companies claimed that they had begun to provide information and training and to build partnerships throughout the supply chain, increasing their suppliers' and customers' awareness of the need for innovation.

The capacities of these companies to innovate were also raised by the levels of experience and expertise that had accumulated through the process of EMS development. The growth of tacit understanding both of production processes and of improvement options appeared to be particularly important, as did the recognition that there were many opportunities for improvement throughout the company. The introduction of a company-wide EMS also served to break down the functional boundaries within the firm and to encourage systems change. Again, some of the larger companies thought that EMAS-prompted improvements in their capacity to innovate had stimulated new interactions with trading partners. For example, the new systems facilitated monitoring of the quality of material inputs and enhanced communication throughout the supply chain. However, EMAS-stimulated improvements were not seen to be significant compared to the effects of EMSs within registered companies.

Interviewees indicated that employees within the environment-related functions of their companies generally thought that developing an EMS and participating in EMAS had reduced the costs and the risks of innovation. However, this perception did not extend to other parts of the company. The managers of other company functions did not see any relation between environmental innovation in their own firm and in their trading partners. In the environment-related functions of registered companies, greater access to information, expertise, and understanding enabled companies to explore the potential and monitor the performance of new technologies with greater certainty. Furthermore, the information and insight gained through the EMS typically highlighted a range of low-tech and organizational opportunities for environmental improvement that required little, if any, financial investment and generated rapid and sometimes significant economic returns. This encouraged emphasis on the potential for organizational innovations of an incremental nature. However, it was often clear that opportunities for incremental environmental improvement, and the economic returns associated with them, tended to diminish over time. As further improvement opportunities became more scarce, companies expected that they would have to explore the potential of more radical change and that such changes would be associated both with higher costs and greater risk and uncertainty.

Although it appears that the presence of an EMS and/or participation in EMAS has stimulated incremental innovations that had led to improvements in process management and efficiency, in general, it was suggested that these initiatives had not changed the speed of innovation. While this may contradict the claim that initiatives such as EMAS had increased the level of innovation to some degree, many companies claimed that many of these changes were happening anyway but that the EMS had made the change process easier. This supports the view that EMS-based standards such as EMAS can enable innovation by allowing companies to respond more easily to regulatory demands or market opportunities.

Influence of EMAS on Forms of Innovation

Interviewees observed that EMAS had not typically been associated with the invention of new technologies. Instead, with the possible exception of investments in monitoring technologies, it had stimulated the adoption of existing technologies and techniques within registered companies. This is true both in the development of new and in the market for existing environmental technologies.

The majority of the innovations affiliated with an EMS or EMAS registration had been developed internally within the registered companies. Most of these were organizational or "low-tech" changes, for example, incremental changes to the calibration of process technology. Clearly, the wider adoption of EMSs may not generate the "demand-pull" that would stimulate the manufacturers of environmental technologies to develop new products.

Some differences emerged for other aspects of environmental management. It was generally accepted that EMAS had been associated with the development and wider adoption of new technologies and techniques for the use of energy and materials, though here most of the managers felt that the improvements that had been secured since EMAS registration would have been realized anyway, that further advances in energy saving were feasible and that incremental change would continue into the future. However, many managers also acknowledged that many of the easy options had already been exploited and that new initiatives would probably encounter diminishing marginal returns. By increasing the capacity to innovate, the presence of an EMS and/or participation in EMAS appeared to have extended the range of technically and economically feasible options, thereby delaying the point at which initiatives that relied on incremental change alone would encounter diminishing marginal returns.

Some interviewees believed that EMAS had been associated with the development and wider adoption of new technologies and techniques in waste management. Here again they felt that EMAS-associated changes were more organizational than technological. It is interesting to note that some interviewees thought EMAS had been associated with a shift to more radical forms of change in the search for improved approaches to waste management. While in many cases waste management initiatives were underway already, a significant number of the people interviewed felt that EMAS had been associated with new waste management initiatives, commonly because it forced firms to look at the sources of waste and the methods of waste management in a different way and to pursue continuous improvement.

Some companies accepted that EMAS had been associated with the development and wider adoption of new technologies and techniques for pollution prevention and control. It was not generally felt that EMAS had been associated with a shift from end-of-pipe to integrated clean technologies for pollution prevention and control. However, it was felt that EMAS had been associated with a change in the emphasis of pollution prevention and control so that the potential of organizational change was explored as well as that of technological change. In this respect some of the initiatives stimulated by the presence of an EMS and/or participation in EMAS were not already underway. Once more, however, there was not a general feeling that any shift from incremental to radical change had occurred.

Impacts on Economic and Environmental Performance

It was generally accepted that the presence of an EMS and/or participation in EMAS had led to improved monitoring, enhanced awareness of and commitment to environmental concerns, reduced energy and materials consumption, more effective pollution prevention and control, and reduced exposure to risk. In many of these instances there was a general perception that the environmental benefits associated with EMAS were not realized in the short term as much as in the medium to long term. In other areas, for example, the environmental aspects of purchasing policies, product design, and distribution activities, there was a general perception that EMAS had not been associated with environmental improvement at all.

Not all of the managers interviewed felt that the influence of EMSs and EMAS had generated economic benefits through improved decision making, reduced exposure to risk, cost reduction, or improved access to new markets. There was some consensus that the presence of an EMS and/or participation in EMAS had not led to economic benefits through improvements in product quality. As these issues were some of the main motivations for participating in the EMAS scheme, the economic aspirations that many of the companies had for the EMAS scheme in advance of their registration were not perceived to have been fully realized.

Impacts on External Communications

While interviewees rated the impacts of EMAS to be low in most areas, there was one important exception: It was widely believed that EMAS had improved the public image of registered companies and their relations with stakeholders. While most interviewees felt that their firms had already established effective and open communication channels with their stakeholders, as a result of their participation in EMAS communications were improved not only with suppliers, clients, customers, shareholders, investors, and insurers and but also with the public authorities, regulators, and the local community. Although it was generally felt that the EMAS statement was an effective communication tool, there were some common concerns that the format required for the EMAS statement actually inhibited effective communication.

Conclusions

Analysis of these research results suggests some conclusions about the impacts of EMAS, and about how revisions to the EMAS Regulation should enhance the rate of adoption within EU countries.

Impacts of EMAS

In general, companies in the United Kingdom have sought to improve their environmental performance by adopting and developing effective EMSs and, in some instances, by subsequently seeking ISO 14001 certification for them. Only a relatively small number of companies have then proceeded to seek EMAS registration by preparing and publishing a verified environmental report. Consequently many of the managers interviewed for this study suggested that many of the impacts associated with having an EMS would have been realized even if they had not sought EMAS registration. Thus, when the EMAS scheme is viewed in isolation and its effects separated from those associated with the presence of an EMS and ISO 14001 certification, many companies perceived that EMAS had generated little added value. A major exception in this respect is the benefit realized from publishing a verified environmental report. Although most of the managers interviewed suggested that the format required by EMAS for the environmental report was somewhat restrictive, they also acknowledged that it had been associated with improvements in their public image and in their relationships with various stakeholders.

Interviewees from most companies found it difficult to separate the influence of EMAS on innovation and on the economic and environmental performance from the influence of their EMS and ISO 14001 certification. Nonetheless, the combined impacts of the EMS, of ISO 14001 certification, and of EMAS registration together were significant in some respects. It was generally accepted that EMAS had been associated with increases in the level of awareness of the need to innovate and in the capacity of companies to do so. They also observed that EMAS played a role in reducing the costs and the risks associated with innovation. Consequently EMAS helped establish the conditions enabling registered companies to recognize and respond innovatively to various market incentives, stakeholder pressures, and regulatory imperatives. In this sense EMAS was generally associated with an increase in the level of innovation because it increased the capacity for change.

Although EMAS had not generally been associated with the invention of new technologies and techniques, it did go hand in hand with the development and wider adoption of existing technologies and techniques. In most instances existing technologies and techniques were developed in-house and were of a low-tech or an organizational nature. Most of the innovations associated with EMAS tended to be incremental rather than radical. It was commonly acknowledged that as the opportunities for further incremental change become scarce, the environmental and economic returns associated with the presence of an EMS and/or participation in EMAS can be expected to diminish. Although to date the innovations associated with EMAS had commonly led to improvements in both environmental and economic performance, it is not clear that opportunities for further improvement will continue to be available into the future.

In conclusion, it appears that even the small minority of EMAS-registered companies in the United Kingdom have expressed doubts about the effectiveness of EMAS in stimulating innovations that improve environmental or economic performance. While it is reasonably clear that the presence of an effective EMS raises the capacity of companies to innovate, most of the managers interviewed were less convinced about the benefits resulting from EMAS registration, particularly as their firms were already registered to ISO 14001. It is probably fair to assume that the vast majority of companies that are not EMAS registered and who were not interviewed as part of this project are either less aware or are less convinced about the potential benefits of EMAS.

Policy Learning? Revisions to the EMAS Regulation

The basis for many of these criticisms has recently been removed or reduced through recent changes to the EMAS Regulation. Within the original Regulation provision was made for a review of the progress of the scheme five years after its adoption. The review suggested that EMAS has had a useful role to play in raising the capacity for environmental improvement in participating organizations and in helping them provide credible, validated information to their various stakeholders. However, reflecting the findings outlined above, the review also recognized that to date the incentives for participation have not been high enough to encourage widespread participation. Based on the review, the revised EMAS Regulation, which was adopted in early 2001, amends the original regulation to encourage wider adoption of the scheme. Its proponents hope that the introduction of some new measures will encourage organizations with ISO 14001 to seek EMAS registration in order to gain greater credibility and recognition for their achievements (Baxter 2001).

With this in mind, the revised Regulation recognizes the value of public reporting and the need to provide credible information that is externally validated. Many people think this is especially significant given the proliferation of new environmental policy instruments, particularly those economic instruments and voluntary and negotiated agreements that depend on the provision of reliable information. The revised Regulation also makes it easier for participating companies to achieve external recognition by extracting data from a wider body of information that is EMAS-validated and using it within their communications and marketing strategies. It allows participating sites to adopt a new logo that can be used to illustrate that the organization is registered and that any information on environmental performance that is being provided in various forms of communication has been externally validated. This logo cannot be used directly on products however because of fears that it may clash with another form of voluntary regulation, namely the EU eco-labeling scheme (Baxter 2001). Finally, in addition to the discussion of the nature of the Regulation itself, there is ongoing debate whether the provision of regulatory relief for EMAS-registered sites would encourage companies to participate in the scheme in the European Union and its different member states (Watzold 2001).

In conclusion then, it appears that if EMAS could be more effectively promoted and more widely adopted, for example, by emphasizing the extra integrity and the added value of EMAS over ISO 14001 and by more formally linking EMAS with the broader framework of regulation, EMAS might become more widely adopted in UK industry. If this were to happen, the capacity of companies in the United Kingdom to innovate in response to the various market incentives, stakeholder pressures, and regulatory imperatives would be increased as would the extent to which existing technologies and techniques are developed and adopted within registered companies.

Acknowledgments

The empirical sections of this chapter are based on the results of the UK case study undertaken by the author as part of a broader project on the Impact of EMAS on Innovation in Europe. This study was undertaken for the European Commission Joint Research Centre and the Institute for Prospective Technological Studies in association with Diana Bradford at the Centre for Exploitation of Science and Technology, London, Andrea Marsanich at the Fondazione Eni Enrico Mattei (FEEM), Milan, and Heinz Kottmann and the Instituts für ökologische Wirtschaftsforschung (IOW), Berlin.

The constructive comments made by the editors of the volume and by Martin Baxter of the Institute for Environmental Management and Assessment in the United Kingdom were also very helpful and much appreciated.

References

Ayres, R., and I. Braithwaite. 1992. *Responsive Regulation: Transcending the Dergulation Debate*. Oxford: Oxford University Press.

Baxter, M. 2001. New EMAS brings greater flexibility. *The Environmentalist—Magazine of the IEMA*, Issue no. 3 (January).

Bayliss, R., L. Connell, and A. Flynn. 1998. Company size, environmental regulation and ecological modernization: Further analysis at the level of the firm. *Business Strategy and the Environment* 7: 285–96.

Confederation of British Industry. 1994a. *Agenda for Voluntary Action*. London: CBI.

Confederation of British Industry. 1994b. *Environment Costs: The Effects on Competitiveness of the Environment, Health and Safety*. London: CBI.

Cridland, J. 1994. Meeting the cost of environmental regulation. In A. Boyle, ed., *Environmental Regulation and Economic Growth*. Oxford: Clarendon Press.

ENDS. 1996a. DoE widens small firm grants for EMAS registration. *The ENDS Report* 260, p. 33.

ENDS. 1996b. HMIP urged to find incentives for firms to improve. *The ENDS Report* 252, p. 34.

Gouldson, A., and J. Murphy. 1998. *Regulatory Realities: The Implementation and Impact of Industrial Environmental Regulation*. London: Earthscan.

Gunningham, N., and P. Grabosky. 1998. *Smart Regulation: Designing Environmental Policy*. Oxford: Clarendon Press.

Hancher, L., and M. Moran. 1989. *Capitalism, Culture and Economic Regulation*. Oxford: Clarendon Press.

Hillary, R. 1995. *Small Firms and the Environment: A Groundwork Status Report*. Birmingham: Groundwork.

ISO World. 2001. The number of ISO 14001/EMAS registration of the world. www.ecology.or.jp/isoworld—last accessed 2/7/2001.

Jenkins, T. 1995. *A Superficial Attraction: The Voluntary Approach and Sustainable Development*. London: Friends of the Earth.

Millar, C. 1994. Motivation to Participate in the Eco-Management and Audit Scheme. *Eco-Management and Auditing* 1(pt. 4): 9–12.

Rothwell, R. 1992. Developments towards the fifth generation of innovation. *Technology Analysis and Strategic Management* 1(4): 73–75.

Soete, L., and A. Arundel. 1995. European innovation policy for environmentally sustainable development: Application of a systems model of technical change. *Journal of Public Policy* 2(2): 285–385.

UKAS. 1995. *The Accreditation of Environmental Verifiers for EMAS*. London: United Kingdom Accreditation Service.

Watzold, F., A. Bultmann, M. Eames, and S. Schucht. 2001. EMAS and regulatory relief in Europe: Lessons from national experience. *European Environment* 11(1): 37–48.

Webber, G. 1994. Environmental management systems: Benefit or burden? *Eco-Management and Auditing* 1(pt. 3): 2–6.

10

Tiered Environmental Regulation: Lessons from the StarTrack Program

Jennifer Nash

Many observers of environmental policy have concluded that the existing regulatory system does little to encourage firms to achieve performance that goes beyond compliance with environmental regulations. It punishes the bad but fails to motivate the good. This shortcoming is particularly significant given that many environmental problems lie beyond the scope of existing environmental regulations. Many firms are capable of doing more than merely complying with environmental laws, which are written for the general case. Environmental policy should find ways to make superior performance attractive to those businesses that are capable of achieving it.

Tiered systems of environmental regulation are a new approach agencies have developed to encourage companies to strive toward higher levels of environmental protection. Under a tiered approach, regulators invite facilities to institute programs that go beyond regulatory requirements in return for a range of benefits. To date, tiered systems have been adopted in twelve states and by the US Environmental Protection Agency (EPA).

The question addressed in this chapter is whether tiered regulatory systems are likely to lead managers to achieve higher levels of environmental protection than they would otherwise choose. To answer this question, I draw on the experience of the StarTrack program, a pilot program that created a special regulatory track for facilities with strong environmental performance. The program was run by EPA's Region I (New England) office during the late 1990s.

I conclude that StarTrack did little to motivate environmental protection in participating firms. In the view of the private sector managers who took part, the program was mainly a paperwork exercise they undertook to garner EPA recognition of established environmental management practices. StarTrack facilities' environmental performance did improve during their participation in the program, but not as a result of their participation. Facilities that met program entry criteria were managed by people who had already invested in environmental performance improvement, and were committed to continuing to do so. The benefits EPA provided program participants were meager, and less than the agency had promised. As a result StarTrack attracted only a handful of participants.

StarTrack officially closed its doors in June 2000 when EPA launched its National Environmental Performance Track. EPA considers the Performance Track the "culmination" of a number of reinvention efforts including the Common Sense Initiative, Project XL, the national Environmental Leadership Program, and StarTrack (EPA 2000). The National Environmental Performance Track is based on StarTrack, and shares many of its features. Participation rates are higher, but implementation has raised a complex set of issues. Defining what constitutes strong environmental performance, and how such performance should be rewarded, have proven to be daunting tasks for agencies.

The Rationale for Tiered Systems

Complying with the law is easier for some firms than it is for others. Similarly, going beyond what the law requires poses lower costs, and greater benefits, for some firms than for others. These simple observations form the basis for tiered regulatory systems. Tiered systems have been used in many areas of public policy. Regulators have distinguished between small and large firms, setting more lenient standards for small organizations for which the costs of compliance are often higher. Small hydroelectric facilities have been exempted from licensing requirements, for example, and small airlines have been subject to fewer reporting requirements. Agencies have also drawn distinctions on the basis of facility age. Old plants are often required to meet less stringent standards than new facilities. Tiered systems reduce the costs of achieving a regulatory goal by focusing more stringent regulation on those facilities for which the costs of compliance are relatively low (Sullivan 1982).

While tiered regulation is not new, the approach EPA and states have proposed as part of efforts to "reinvent" environmental regulation is novel. Under the new idea of tiered regulation, firms for which the costs of environmental protection are relatively low, or for which the benefits of strong environmental performance are relatively high, are granted leniency in some procedural aspects of compliance[1] in return for achieving higher levels of environmental performance.

What constitutes a higher level of environmental performance varies in different jurisdictions. Often to be admitted into a tiered system firms must have operated without violating environmental regulations for several years and have taken steps to reduce unregulated environmental impacts. One requirement common to nearly all agency tiered programs is that facilities must have adopted an environmental management system or EMS. An EMS is a set of rules and resources that managers adopt in order to meet corporate environmental goals (Coglianese and Nash 2001). Agencies have made EMSs a criterion for deciding which facilities receive special treatment. Agencies have argued that when facility managers take responsibility for identifying and reducing their environmental impacts with an EMS, the strategies they adopt are likely to be less costly and more effective than they would be under standards imposed by government.

Tiered environmental regulation is intended to create a "parallel system for environmental protection" (Aspen Institute 1996: 9). Major policy groups such as the National Academy of Public Administration (1996) and the Aspen Institute (1996) have endorsed the idea of imposing lower regulatory burdens upon firms willing and able to attain higher levels of environmental protection. The gains offered by tiered environmental regulation are purportedly many. The Aspen Institute lists ten categories of benefits, including "increased environmental gains," "increased innovation," "integration of pollution prevention and continuous improvement into core decisions," and "increased technological advances" (1996: 9).

Three assumptions form the basis of tiered regulatory approaches. The first has already been mentioned. Environmental protection poses different costs and benefits for firms. A second assumption is that agencies can distinguish top performers from lagging or below-average firms. A third—and critical—assumption is that recognizing and rewarding top performers will motivate new behavior on the part of firms and agencies. Each of these assumptions is discussed briefly below.

Firms Respond Differently to Environmental Demands

According to EPA, the variance in firms' responses can be depicted as a bell-shaped curve (EPA 1999). At one tail of the curve are "laggards," managed by people who view environmental protection as a cost to be externalized. Without close agency oversight, these managers will intentionally or negligently violate environmental laws (Spence 2001). At the opposite tail are what EPA calls "top performers." These firms are both motivated and able to implement environmental programs beyond what is required under the law. They are run by managers who view environmental protection as a business need and have linked environmental performance with their overall business strategy (Hoffman 2000). Under a tiered system, agencies regulate laggards through a command and control approach while top performers are provided incentives for achieving superior environmental performance.

Agencies Will Know Strong Environmental Performance When They See It

A second assumption is that agencies can distinguish top performers from laggards. This second assumption poses problems for regulators. Historically tiered systems have been successful when they have discriminated on the basis of obvious differences—between large and small facilities, or old and new (Sullivan 1982). But environmental performance is difficult to measure. A firm may be a leader when it comes to recycling but a laggard in its air toxic control efforts. Furthermore a firm's past performance does not guarantee that it will continue to be a top performer in the future. Strong environmental programs may be curtailed during economic downturns, and a vigilant and well-intentioned manager may be replaced by someone unfamiliar with regulatory requirements. These scenarios present substantial risk to agency officials. A well-publicized incident at a facility deemed by EPA to be a top performer could jeopardize the careers of program managers and embolden the agency's political opponents (Wilson 1980).

In an effort to avoid such risk, agencies have established detailed criteria for entry into performance tier programs. Legal compliance has been the baseline for entry. To be admitted to StarTrack, a facility was required to show that it had not been the target of any agency enforcement action within the past three years and had in place an auditing program to identify and correct violations of regulations should they occur. It had to demonstrate a history of cooperating with environmental regulatory agencies and of preventing pollution.

In addition StarTrack companies were required to commit themselves to implementing an EMS (EPA 1998). The EMS was intended as a way to "lock in" environmental performance commitments and prevent backsliding (Coglianese 2001). Third-party verification of compliance and EMS audits was also required. Finally, facilities had to publish an annual environmental performance report that included results from compliance and EMS audits as well as information about the facility's environmental impacts (EPA 1998).

Tiered Systems Will Improve Environmental Protection

A final assumption underlying tiered regulatory systems is that these approaches will lead to higher levels of environmental protection. These higher levels are to be achieved in three ways. First, tiered systems are supposed to help agencies allocate their inspection resources more efficiently. Agencies do not have sufficient resources to inspect all facilities on a regular basis. Estimates are that over a two-year period, less than 1 percent of facilities with federal permits were inspected to ensure compliance with air, water, and waste management regulations (Hale 1998). StarTrack was originally designed to test the effectiveness of independent third parties as substitutes for direct agency inspections. StarTrack's original name was the "Third Party Certification Program." It was intended to privatize some of the oversight normally assumed by agencies, thereby allowing government to target its scarce resources to poor performers.

Second, agencies maintain that the performance of participating facilities should improve as they implement program requirements. The components of StarTrack—auditing, environmental management systems, external verification, and public performance reporting—create feedback loops whereby managers must stop, examine their environmental performance, share information with external constituencies, and fine-tune their performance. The combination of tools utilized in StarTrack aims to create within facilities "the conditions for self-critical reflection about behavior and how to improve it" (Fiorino 1999: 449). The process by which government encourages self-critical reflection is sometimes called "reflexive law" (Fiorino 1999; Orts 1995).[2] Reflexive law does not dictate how firms are to achieve environmental goals, but establishes the conditions under which managers will seek to work toward these goals in their own way. While programs like StarTrack focus on management practices such as auditing and reporting, they are intended to help businesses come up with new practices, including technological innovation (Orts 1995).

Finally, tiered regulatory systems are intended to improve the performance of facilities that do not qualify now but aspire to qualify in the future in order to enjoy program benefits. EPA and state officials speak of shifting the entire bell-shaped performance curve in the direction of environmental protection. "The ultimate goal," according to a review of the StarTrack program undertaken by a President's Council for Sustainable Development Task Force, is "to encourage a broad range of companies to adopt comprehensive EMSs and auditing schedules that will ensure ... continuous improvements in environmental performance" (PCSD, undated document: 2).

Agencies recognize that in order to motivate new behavior they must offer incentives. Agencies have offered a variety of incentives in tiered programs: regulatory flexibility, recognition, and relationship changes. The idea that regulatory flexibility and agency recognition will motivate superior environmental performance grows primarily out of a literature authored by business leaders (Smart 1992; Bowers and Mills 1996; Schmidheiny 1992). These accounts are mainly anecdotal and informed by roundtable discussions and expert surveys (Davies and Mazurek 1996).

Existing studies fail to explain the role of regulatory flexibility in environmental protection. Are agencies offering flexibility because they believe that the inflexibility of command and control regulation is a significant obstacle to firms reducing their environmental impacts? Or do agencies offer flexibility because it is their chief negotiating asset, one that they will reluctantly provide in order to reap the presumably greater performance benefits achieved through tiered regulation? Existing research, furthermore, tells us little about the *level* of incentive necessary to motivate change (Davies and Mazurek 1996). Flexibility when it comes to record keeping may be a significant incentive if it means never having to file another air quality report, but insignificant if it only reduces the frequency of reports from quarterly to semi-annually. Recognition may be a powerful motivator if it comes in the form of a press conference with the President; it may mean little if it is only a letter of thanks from the EPA administrator.

To StarTrack facilities, EPA *promised* penalty amnesty, modified inspection priority, and faster than normal processing of permit applications. The only incentive EPA was able consistently to deliver, beyond what it offered to any facility whether or not it participated in StarTrack, was recognition. EPA listed the names of StarTrack firms on its Web site, and recognized their achievements at annual meetings.

Evidence of Change in StarTrack Facilities

In 1999, I was one of four researchers that undertook an investigation of the StarTrack program.[3] Our goal was to test the most critical assumption that forms the basis of tiered regulatory systems: Did this program motivate change? Did StarTrack improve the efficiency of government enforcement programs, and did it motivate improvements in firms' environmental performance beyond what is required by regulation?

By 1999, fifteen facilities participated in StarTrack. They represented a wide variety of industries including building materials, bulk gas supply, defense, electronics, paper, and plastic manufacturers. A facility of the US Postal Service and of the US Coast Guard also participated. Six commercial facilities participated in the program throughout its entire tenure, from 1996 until 2000. To understand the impact of StarTrack over time, we contacted these six facilities and asked to meet with managers to discuss their impressions of how the program was working. Two facilities declined to participate. We conducted interviews at the remaining four sites.

We conducted in-depth face-to-face meetings with four facility managers at each site. Each interview lasted approximately two hours. In addition we asked each person we interviewed to complete a written questionnaire comparing environmental management practices in their facility before and during StarTrack. We toured each plant to see pollution prevention projects first hand. We also examined StarTrack application materials for each facility, performance reports, compliance and EMS audits, and third party verification reports. We spoke at length with EPA Region I staff members responsible for StarTrack implementation, state agency environmental staff members who had participated in StarTrack audits, and EPA headquarters personnel. We interviewed environmental advocacy groups that had participated in the program as well as local officials. In addition, we conducted interviews at two facilities that we thought would meet the StarTrack eligibility requirements but whose managers had chosen not to participate in the program. Research took place during the spring, summer, and fall of 1999.

In the discussion below I consider how each of the major programmatic components of StarTrack contributed to change in government and private firms.

Environmental Management Systems

EPA required that StarTrack facilities commit to adopting EMS modeled on ISO 14001, the international environmental management standard.[4] EPA also required facilities to incorporate pollution prevention and continuous improvement in environmental performance as EMS goals.

Both facilities B and C were under corporate directives to adopt ISO 14001. Facility managers believed that participating in StarTrack would demonstrate to corporate officers that they were taking this corporate directive seriously. Managers at a third facility, facility D, decided to become certified to ISO 14001 in part to distinguish themselves with corporate management. These managers had already become certified to the Irish environmental management standard. They wanted to be the first facility in the company to step forward to adopt the international standard.

Managers at these facilities sought the advice of regulators and third-party consultants in their certification process. Receiving such help was part of their motivation for joining StarTrack.

Adoption of ISO 14001 helped to strengthen pollution prevention programs at two of the three facilities we studied, according to facility managers. For example, each of the four managers we interviewed at facility D described certification to ISO 14001 as the most significant event that had shaped environmental management in their plant over the past three years. These managers said that ISO 14001 was the tool they used to identify environmental aspects, set environmental performance goals, and design products with fewer environmental impacts. They said they used their ISO 14001 certification to differentiate their products from those of competitors, telling customers, "Our product was made with zero compliance issues."

According to those managers, implementation of ISO 14001 had led workers to seek out environmental opportunities. Managers mentioned several pollution prevention projects they had recently implemented, all of which had yielded direct benefits that exceeded their costs. The first project was a chemical lab consolidation that helped to lower pollution control costs and make environmental management practices more uniform. The second project involved changing the containers used for dispensing chemicals from beakers to squeeze bottles. Squeeze bottles reduced emissions of volatile organic compounds as well as the volume of chemicals procured for production. The third project reduced energy demand through changing lighting fixtures and adding sensors.

Managers at facility B held quite a different view of the value of the international environmental management standard. For them, ISO 14001 was a tedious administrative exercise. Initially managers at this plant included environmental objectives in their EMS that were outside the scope of regulatory requirements. One such objective was to reduce emissions from the facility's boilers. However, managers soon discovered that continuous reductions in boiler emissions would be costly and could potentially place them at a competitive disadvantage. Managers decided to change the EMS objectives to focus more narrowly on compliance. In addition the paperwork associated with documenting workers' responsibilities—which the facility environmental manager called "administrivia"—was overwhelming.

Managers at this facility were skeptical about their potential to prevent pollution. While in past years they had successfully reduced emissions and energy consumption, recently the facility's pollution-prevention performance had plateaued. Because of the high cost of many of the products the facility produced (e.g., $100,000 for a single circuit board), managers were conservative about undertaking changes in their production process that might reduce environmental impacts. They considered current levels of chemical use and waste generation extremely low. They told us they would rather waste small quantities of chemicals than experiment with new processes that might yield lower quality.

Environmental management systems are intended to "lock in" (Coglianese 2001) managers' environmental commitments, even as personnel, product lines, and leadership within a company change. Two of the four facilities we studied had undergone important management changes over the course of their StarTrack participation. Both had been purchased by larger companies. At facility C the environmental policy, which emphasized pollution prevention, and managers' commitment to EMS implementation, remained constant despite the ownership change. The new owner was a European-based company with a reputation for strong environmental performance. Managers we spoke with explained that the highly competitive nature of their facility's industry required them to focus on environmental performance. Product differentiation based on environmental attributes was potentially a crucial aspect of business success. In addition, minimizing production costs was an important determinant of profit margins. Managers knew that reducing their wastewater discharges would help to keep down the amount of waste product needing to be processed, and overall operating costs.

At the facility B, however, management changes had led to a de-emphasis of beyond-compliance initiatives. The new facility environmental manager we interviewed told us that the ambitious environmental objectives contained in the EMS were the decision of the previous environmental manager, who aggressively sought out opportunities to establish the facility as an environmental leader. "It's time to get back to basics," he explained, meaning a compliance-focused environmental management program.

Managers at facility A had not implemented a facilitywide EMS. They had hired a consultant to determine the effort required to bring the current EMS, which only addressed a portion of the plant's operations, up to the level of ISO 14001. The consultant found that the equivalent of one person working full time for a year would be required. Managers took some but not all of the steps necessary to close this gap. While regulatory compliance and pollution prevention were management priorities, implementation of a facilitywide EMS was not. The facility environmental manager explained that every worker was expected to comply with environmental regulations and to strive for the lowest possible releases. "The people on the manufacturing floor know if we have a spill what to do ... but they do not know we have an EMS," he reported. We were also told that this facility had the lowest discharge rate of any in the company and was also among the most profitable.

Auditing

Auditing was a major emphasis of StarTrack. StarTrack required that facilities conduct compliance and EMS audits, and have audit results verified by an EPA-certified third party. Compliance auditing was already established in these facilities prior to participating in StarTrack. (An established compliance-auditing program was a requirement for entry.) StarTrack opened up the auditing process, involving agencies and other external organizations in conducting the audits and reviewing the results.

Prior to StarTrack, corporate environmental managers were generally responsible for facility compliance audits. Corporate managers inspected these facilities, usually on an annual basis, to determine their compliance with environmental regulations as well as corporate policies. Under StarTrack, EPA and state environmental agencies participated in these audits as observers as well. Often between five to ten agency personnel would take part. What had been a corporate undertaking became a joint corporate-agency activity, with facility managers joining in. In addition representatives of environmental advocacy groups were invited to participate in the audits.[5] According to EPA, such observation was necessary to check the reliability of the auditors and verifiers.

In one case agency personnel *took the place* of corporate auditors. The environmental manager at facility A requested that internal compliance audits during its second year of StarTrack participation be conducted by the state environmental agency. This manager preferred the state agency audit over corporate review, noting that corporate auditors were very difficult to please and looked at many aspects of performance, not just compliance and EMS programs. Managers at facility B echoed this view. They told us that corporate auditors had graded their environmental management as a "C" while EPA hailed the facility as an environmental management leader. From EPA's perspective this facility's environmental management practices were superior to those of comparable firms and conformed to StarTrack requirements. Corporate officers, however, were dissatisfied with facility managers' progress toward meeting corporate safety and energy conservation goals.

Third-party EMS audit reports provide some evidence to back up facility managers' claims about the rigorousness of corporate audits. An initial third-party audit at facility A, for example, conducted before corporate managers decided not to audit the facility themselves, found "no nonconformances" between the facility's stated policy with respect to employee training and observed training practices. The third party notes, however, that corporate auditors "conducted a more detailed review of the training program," identifying a number of "gaps." Corporate auditors determined that certain departments were exempting themselves from environmental training and directed facility managers to take steps to correct the problem.

Prior to StarTrack, audit results were internal documents, shared among facility and corporate managers. EPA required audit results to be included in facility performance reports, which were public documents. Anyone who asked could receive a copy. EPA's intention in requiring facilities to disclose audit results was to provide the public an opportunity to see that participating businesses had in fact adopted strong environmental programs. Should a participating facility slacken in its commitments, disclosure might generate adverse publicity that could pressure managers to do better. Facility managers we talked with considered public disclosure of audit results to be one of the most challenging aspects of the program. One manager claimed that it required a "transformation" of their relationship with the agency from "suspicion" to "absolute trust."

While StarTrack changed the composition of the audit teams to include outsiders from government and made audit results public information, participants' compliance performance changed little over the course of their involvement in the program. All facilities reported violations of environmental regulations in both their 1997 and 1999 performance reports. None of these violations were serious or egregious. Both before and during the program, compliance at these facilities was strong, although violations were consistently found. While StarTrack did not appear to improve compliance performance, it also produced no slackening in managers' attention to compliance. At facility A, for example, managers repeatedly mentioned compliance improvement as a top goal. Managers at this plant consistently ran their operations at 10 to 20 percent of permitted levels, but still looked for ways to reduce emissions further. Similarly managers at facility B spoke of compliance as "job #1" of the environmental management department.

Third-Party Review

Third-party review was required during the first year of a facility's participation in the program, and every three years thereafter. Third-party review was intended to ensure that compliance and EMS audits were conducted in accordance with StarTrack protocols and that findings were accurate. Third-party reports, on file in EPA offices, show that reviewers identified numerous deficiencies in facilities' EMSs. For example, the third-party review of facility C's EMS in 1997, before ISO 14001 certification, found that the environmental management program was largely informal and undocumented. Environmental programs did not grow out of a systematic evaluation of the facility's environmental impacts. A third-party review in 1999 found that those issues had been corrected. At that time, however, the reviewer found that temporary workers were not adequately informed about the organization's environmental policy and objectives. Third-party reviews brought to managers' attention issues that might otherwise have been overlooked.

Environmental Performance Reports

Two of the four facilities we studied had no formal mechanism for reporting their environmental performance to the public prior to their participation in StarTrack. The two others published performance reports, but not with the level of detail required under the program. According to the facility managers we spoke with, performance reporting contributed little to their understanding of the environmental impacts of their facilities. These impacts were already brought to managers' attention during compliance and EMS audits, in which agencies and others participated. Preparing a facility performance report required the time of the plant StarTrack representative. It was perceived as a cost of participation.

In theory, public reporting of a facility's environmental performance should spark performance improvements through political checks on private behavior. Knowing that the results of a facility's compliance audits will become public information, managers should be motivated to eliminate all potential violations (Sunstein 1999). In other words, performance reporting should improve behavior by disclosing information that could potentially embarrass managers. However, according to EPA, no StarTrack facility report was ever requested by a member of the public. The reason, perhaps, was that StarTrack facilities were perceived by the public to be strong environmental performers. Their performance was not a substantial concern. Environmental advocacy groups took little notice of a program dedicated to improving the performance of facilities whose track records were already good.

Relationship Changes

Changes in facility-EPA relationships have been noted as a positive feature of many reinvention initiatives (Beardsley 1996). StarTrack also appeared to improve facility-EPA interaction. Before participating in StarTrack, agency and private facility managers generally did not communicate directly except when completing administrative tasks, such as filing and approving permits, or in adversarial interaction, as when an agency inspected and fined a facility. StarTrack opened up new opportunities for interaction between EPA and private sector managers.

During the course of the interviews we were able to witness such interaction firsthand. Facility A's managers had instituted an open door policy with state environmental officials, and a representative of the state agency sat in on our interviews at the site. At one point during a discussion about pollution prevention projects the facility environmental manager announced what he considered to be a promising new approach to managing waste. By combining several waste streams, he hoped to create an inert material that could be used on site for landscaping, potentially saving millions of dollars in disposal costs. The state agency representative immediately responded by raising a host of regulatory concerns. Without a program like StarTrack, the environmental manager might have been reluctant to share his plans at such a preliminary stage. At best, the dialogue would probably have occurred in a more formal manner, consuming more agency and facility resources.

The auditing process established under StarTrack also helped to build positive agency-facility interaction. Facility managers told us that when agencies observed compliance and EMS audits, they developed a deeper appreciation of facility environmental programs. They saw how hard managers were working to meet regulatory requirements. They told us that agencies went away from the audits with a new respect for the way private sector managers were addressing their compliance responsibilities.

Those we talked with from EPA also emphasized aspects of the program that undermined positive relationships. EPA designed the program to motivate "beyond-compliance performance," however it did not define what it meant by that term. EPA officials we interviewed maintained that StarTrack facilities were not necessarily superlative performers. These officials mentioned environmental programs that were not in place in these facilities, such as design for environment programs and life-cycle assessment programs. EPA's problems delivering promised benefits, discussed below, further strained relationships with participating facilities.

Incentives

According to EPA representatives, the major weakness of StarTrack was its inability to deliver promised benefits to participating facility managers. The only benefit consistently delivered was recognition. The EPA

Regional Office presented each facility with a plaque and listed its name on the EPA Web site. StarTrack facilities were invited to an annual conference in which their achievements were lauded by EPA officials. While welcome, private sector managers described this benefit as "weak." The environmental manager of facility A described the form he would have liked recognition to take. "I'd like to be able to put an EPA seal of approval [on the side of the containers coming out of my plant]," he explained. "Customers might notice that."

Other promised benefits did not materialize. For example, the EPA Region I Web site listed "express-lane permitting" as a benefit of StarTrack. The facility managers we spoke with would have valued this benefit, had EPA and state environmental agencies delivered it. But none of the managers felt that their permits were processed any faster than before joining the program. State environmental agencies are responsible for reviewing facility permit applications. While these agencies were parties in StarTrack agreements, they saw StarTrack as an EPA program in which they were not fully invested. Permit applications went through regular state administrative channels, resulting in usual processing times.[6]

EPA also promised what it called "modified inspection priority" as a benefit of StarTrack. Facility managers interpreted that to mean inspection relief. The agreements signed by EPA, the state agency, and facility managers stipulated that agencies "will not, during the [StarTrack project], conduct any routine regulatory inspections" (EPA 1996). However, EPA staff members told us that agencies could inspect StarTrack facilities if they chose to. Region I had recommended "spot checks of participants' performance and adherence to program requirements" (EPA 1998). Two of the facilities we studied reported that they had not been inspected since joining the program, while the other two had been subject to inspections.

StarTrack agreements signed by facility managers and federal and state agencies stipulated that EPA and states would not initiate any enforcement action for violations discovered at StarTrack facilities if they were corrected within 60 days.[7] EPA said this benefit should be valued by facility managers since it guaranteed that operations would not be disrupted by agency actions. Region I's policy with respect to StarTrack facilities was consistent with EPA's Audit Policy, however.[8] It provided no greater level of enforcement forbearance to StarTrack facilities.

Because EPA was not able to develop meaningful incentives, participation in the program fell below agency expectations. Only 15 facilities, two of them agencies of the federal government, chose to join. In addition EPA found it had little ability to insist that participating facilities fully implement the program. StarTrack companies did not always file performance reports on time. The agency was not always satisfied with the consultants facilities picked to serve as third-party verifiers. Yet EPA staff members felt they were in no position to complain about those issues since they were unable to uphold their end of the agreements.

Lessons from the StarTrack Experience

EPA launched the StarTrack program to test the role of tiered environmental regulation in motivating "top" environmental performance on the part of facilities and agencies. What lessons can be learned from this experience? There is no evidence that performance declined at StarTrack facilities or that managers attempted to exploit their closer relationships with EPA to subvert environmental protection goals. StarTrack entry criteria were effective in selecting facilities run by people committed to environmental protection.

Environmental performance improved in StarTrack facilities—but not as a result of the program. Corporate policies drove facility environmental management practices. Corporate auditing programs were perceived by facility managers as rigorous and thorough; agency inspections were if anything less comprehensive. Corporate policies dictated EMS adoption, and when this was not the case, agency EMS requirements were mostly ignored.

Corporate requirements for environmental management systems, auditing, and external verification appear to have fostered self-monitoring and self-correction in StarTrack facilities. These requirements encouraged managers to stop, observe, and reflect on their actions, and undertake corrections to bring action into line with policy objectives. Of course, self-monitoring and self-correction did not always lead managers to strive for higher levels of environmental protection. In one facility the implementation of an ISO 14001 EMS bogged managers down in exhausting "administrivia." Even after careful deliberation, not all StarTrack facilities believed that further efforts to prevent pollution were in their financial interest. Furthermore the pollution prevention projects managers described to us were more in the nature of housekeeping improvements than major innovations.

These observations clarify the role of EMSs in the environmental performance of the StarTrack facilities we studied. These systems served primarily to reinforce corporate environmental objectives. EMSs were not in themselves a necessary or sufficient condition for environmental performance improvement. Facility A made substantial progress without implementing a facilitywide EMS; Facility B's environmental performance stalled after implementing a formal system. The finding that EMS serve mainly as reinforcing mechanisms has been confirmed in other recent studies (King and Lenox 2000; Matthews 2001; SPRU 2001). These studies have failed to establish a correlation between EMS adoption and strong environmental performance.

StarTrack appeared to increase, rather than reduce, agency oversight of participating facilities, which were considered by regulators to be strong environmental performers. This is a negative result for a program dedicated to improving the efficiency of agency compliance oversight programs. In StarTrack, agency personnel participated as observers of compliance and EMS audits, a level of oversight greater than before the program. According to one state environmental agency representative, preparing for the StarTrack audits, participating in them, and assessing their results required one full-time employee. Four StarTrack facilities were located in this state.

While EPA initiated StarTrack to test the potential of tiered environmental regulation, the test was not really a fair one. EPA's inability to define and deliver meaningful incentives undermined the program's capacity to motivate change on the part of private sector managers. Participating facilities did not slacken in their existing commitments, but they did not set higher standards of performance as a result of their StarTrack participation. Several facilities had not fully implemented StarTrack requirements, but EPA was in a weak position to demand more since it had not delivered the benefits it had promised. There was no waiting list of facilities striving to qualify for membership in the program. The managers we spoke with from nonparticipating facilities expressed little interest in joining a program that offered only intangible benefits.

Perhaps these negative conclusions are premature. The inability to deliver promised benefits could have been due to StarTrack's being a regional, rather than national, program, without the full support and authority of EPA headquarters. In June 2000 EPA announced a national program that built on the StarTrack experience, the National Environmental Performance Track. The Performance Track offers a further opportunity to explore the potential of tiered regulatory systems.

Will the National Environmental Performance Track Yield Greater Benefits?

The National Environmental Performance Track is a program of EPA's Office of Policy, Economics, and Innovation. Like StarTrack, the National Environmental Performance Track is intended to "recognize and encourage top environmental performers ... [by] captur[ing] opportunities for reducing costs and spurring technological innovation" (EPA 2000). The Performance Track builds on and consolidates previous voluntary initiatives such as the Common Sense Initiative, Project XL, the national Environmental Leadership Program, and StarTrack.

Program Requirements

Programmatic requirements are more specific than StarTrack's. Such specificity should provide helpful guidance to facility managers seeking to meet EPA expectations. To be admitted, a facility must have implemented an EMS, not just have made a commitment to doing so. EPA does not require facilities to have modeled their EMS on ISO 14001 as it did under StarTrack. Instead, it specifies what it considers essential EMS features.[9]

While StarTrack required that facilities "continuously improve" their environmental performance, program documents did not explain what was meant by that term. Performance Track provides more information to facility managers about its expectations for performance improvement.

Under Performance Track, facilities must commit to making improvements in at least two of the following categories: energy use, water use, materials use, air emissions, waste generation, water discharges, accidental releases, preservation and restoration, and product performance. Within each of those categories, EPA has identified aspects of performance facilities must target; in the "waste generation" category, for example, facility managers must pledge to reduce solid waste, hazardous waste, or toxic releases to land. A minimum of four aspects must be addressed. Progress toward meeting these targets must be reported to EPA and to the public.

Other program requirements are less stringent than StarTrack's. To gain entry into the Performance Track, a facility must have an established program for compliance and EMS auditing. Managers are not expected to invite EPA and state environmental officials to observe these audits, however. Facility managers must include in their annual reports to EPA a summary of audit findings, but not the audit reports themselves. Third-party review of compliance and EMS audits is not required.

How Performance Track Could Work

Under the most optimistic scenario, how might the program work? Conceivably, as facility managers became aware of Performance Track, they could use the EMS, performance reporting, and other requirements of Performance Track as a benchmark for their own management practices. While some managers might see Performance Track as primarily a paperwork exercise to formalize existing commitments, others might use the application process as an opportunity to go beyond previous plans. They might pledge to make new improvements in the areas specified by the program, reducing emissions to air or water, wastes, or their use of natural resources. Progress toward meeting these commitments could be regularly checked through the EMS process, and problems along the way could be identified and corrected. Managers could systematically disclose their progress in annual environmental performance reports, which could serve as the basis for a community dialogue on facilities' environmental impacts.

Ultimately participation in Performance Track could result in discernible improvements in the environmental performance of participating facilities, particularly in areas that lie beyond the scope of regulation. As managers gained more knowledge of the program, and observed the benefits enjoyed by program participants, interest in the program would grow. More and more facilities would apply.

Conceivably Performance Track could also lead to changes in the behavior of agencies. For the facilities participating in the program, EPA would no longer have to spend time monitoring their environmental performance. These facilities would be essentially self-regulating, and EPA could rely on facility performance reports to ensure compliance. Knowing that an EMS was in place in these plants that addressed compliance as well as other critical performance considerations, EPA could direct its regulatory resources elsewhere. Relationships between EPA and facility managers would become more open and cooperative.

If Performance Track became a normal part of corporate management, EPA and states could move toward greater reliance on performance-based regulation. The existing regulatory system could move away from technology-based standards toward a system built around the performance commitments articulated through Performance Track. These commitments would represent private sector managers' assessment of how best to reduce their most significant environmental impacts. These commitments would become the relevant benchmarks for agencies as they regulated these facilities. Managers would be allowed to meet these commitments in whatever ways they considered most effective from both a performance and cost perspective.

Potential Pitfalls

As of August 2003 more than 300 facilities had been accepted into the EPA Performance Track. EPA program managers have worked hard to develop a package of benefits that will motivate facilities to take part. They have promised participants recognition, specifically, use of a Performance Track logo,[10] and visibility in EPA publications, Web sites, and events. They have also promised access to "state-of-the-art" information such as a best practices database.[11]

A new regulation, to be issued in final form in late 2003, will provide Performance Track members with modest regulatory benefits. The regulation will allow Performance Track members that generate large quantities of hazardous waste to store their wastes on site for 180 days without a federal Resource Conservation and Recovery Act (RCRA) permit. It will also reduce the frequency of reports required under the Maximum Achievable Control Technology provisions of the Clean Air Act, and allow Publicly Owned Treatment Works that participate in the program to report using the Internet rather than local newspapers. These charges, while welcomed by Performance Track members, may marginally reduce the costs of regulatory compliance for member facilities.

Subtle changes in facility-agency relationships may be the chief benefit for program participants. Recently, for example, facilities that have joined Performance Track formed a "Participants Association" to help guide EPA program implementation. By participating in Performance Track, managers have established a new channel for communication that may result in their playing a role in the development of more meaningful benefits for program participants in the future.

Because program benefits are modest, Performance Track is likely to appeal to managers of facilities that already meet program requirements. Information available about Performance Track participants indicates that most have well-established environmental management programs. Nearly two-thirds had achieved certification to ISO 14001 prior to submitting an application. Nearly 80 percent have in place externally verified environmental management systems, a further indication that EMSs have been strongly integrated into organizational routines. The program does not appear to be attracting facilities that would need to improve their environmental performance in order to meet entry requirements.

Without meaningful incentives, the commitments managers adopt as part of their Performance Track applications are likely to represent only small improvements over what they would have done in the absence of the program. Facility environmental management systems are unlikely on their own to motivate changes in environmental performance. Facility managers will follow directives from corporate environmental officers, and their EMSs will serve to bring environmental practices into line with established corporate objectives. These objectives may or may not encourage managers to strive toward excellence.

Conclusions

While the idea of tiered environmental regulation is relatively simple, in practice implementation has raised a complex set of issues. A facility's environmental performance has many dimensions and past behavior does not necessarily predict future conduct. As a result agencies have established detailed criteria for entry into the tier of "top performers," requiring firms to have strong records of compliance and pollution prevention and to have implemented formal environmental management systems that disclose environmental performance to the public.

These requirements do not ensure, however, that facilities will exhibit the behavior agencies are seeking. The adoption of an EMS is a weak proxy for top environmental performance. A facility with a formal EMS may fail to strive for performance beyond what is required under the law, and a facility without a formal system may be an exemplary environmental performer.

The primary benefits agencies have offered facilities that participate in tiered programs are recognition and relationship changes. The package of regulatory changes EPA is working to implement for Performance Track facilities is modest. Furthermore EPA has not articulated the rationale for these regulatory benefits. Do the demands of compliance upon private sector managers stand in the way of beyond-compliance performance, thereby justifying agency flexibility? On the contrary, the StarTrack experience suggests that stringent regulatory requirements may motivate firms to push beyond compliance. For StarTrack facility managers, compliance was the foundation of a strong environmental management program.

While EPA, states, and numerous policy groups are advocating tiered environmental regulatory systems, the role of these systems in environmental protection is not yet clear. Environmental management systems, a central component of these programs, may build capacity for environmental improvement but do not ensure strong environmental performance on their own. Agency reluctance to provide meaningful benefits has made it difficult to test the potential of tiered approaches. The assumption that tiered regulatory systems will change the performance of facilities and agencies has not yet been established.

Notes

1. Tiered systems generally involve no relaxation of substantive standards of performance.

2. A primary example that shares many elements with StarTrack is the European Eco-Management and Auditing Scheme (EMAS).

3. The other researchers were John Ehrenfeld, Jeffrey MacDonagh-Dumler, and Pascal Thorens from the Technology, Business, and Environment Program at the Massachusetts Institute of Technology.

4. EPA did not require StarTrack facilities to adopt ISO 14001 but rather that auditors compare their EMSs to this standard and identify gaps. Managers were expected to make consistent progress toward closing those gaps. Many EPA managers believe the ISO standard does not require firms to go far enough in terms of compliance, pollution prevention, or information disclosure. While the agency has endorsed the use of environmental management systems by firms, it maintains that the ISO standard is one of many possible guides.

5. During the study period, one StarTrack facility (not one of the four we studied) had its compliance and environmental management systems audited by a group that included a representative of an environmental advocacy organization. This representative commended facility managers for the openness and thoroughness of the audit process.

6. For example, when facility B's environmental manager learned he would need two separate permits for manufacturing operations in separate buildings (historically regulated jointly under a single permit) he expected that his permit application would go to the "top of the pile" at his state environmental agency. Instead, he received the same treatment as before joining StarTrack.

7. Enforcement action would still be taken in the case of criminal violations, violations that resulted in serious environmental harm, or violations that resulted in significant economic benefit to the facility.

8. EPA audit policy states that "where violations are found through voluntary environmental audits or efforts that reflect a regulated entity's due diligence, and are promptly disclosed and expeditiously corrected, EPA will not seek gravity-based (i.e., noneconomic benefit) penalties and will generally not recommend criminal prosecution against the regulated entity" (EPA 1995).

9. EPA requires that the facility have established an environmental policy that includes commitments to compliance, pollution prevention, continuous improvement in environmental performance, including areas not subject to regulations, and sharing information with the community. Facility managers must identify their significant environmental aspects and set measurable objectives and targets for reducing them. They must assign responsibility for meeting these targets and train all employees about their environmental responsibilities.

10. The logo may be used on facility-specific brochures and other printed materials. It is not to be used on products.

11. For a list of proposed benefits, see *www.epa.gov/performancetrack/benefits/admin.htm.*

References

Aspen Institute. 1996. *The Alternative Path: A Cleaner, Cheaper Way to Protect and Enhance the Environment.* Washington, DC: The Aspen Institute.

Bowers, D., and K. Mills. 1996. How the Aspen process shapes environmental strategy. *Corporate Environmental Strategy* 3(4): 5–11.

Beardsley, D. P. 1996. *Incentives for Environmental Improvement: An Assessment of Selected Innovative Programs in the United States and Europe.* Washington, DC: Global Environmental Management Initiative.

Coglianese, C. 2001. Policies to promote systematic environmental management. In C. Coglianese and J. Nash, eds., *Regulating from the Inside: Can Environmental Management Systems Achieve Policy Goals?* Washington DC: Resources for the Future.

Coglianese, C., and J. Nash. 2001. Toward a management-based environmental policy? In C. Coglianese and J. Nash eds., *Regulating from the Inside: Can Environmental Management Systems Achieve Policy Goals?* Washington DC: Resources for the Future.

Davies, T., and J. Mazurek. 1996. *Industry Incentives for Environmental Improvement: Evaluation of U.S. Federal Initiatives.* Washinton DC: Global Environmental Management Initiative.

Environmental Protection Agency (EPA), Office of Enforcement and Compliance Assurance. 1995. *Audit Policy: Incentives for Self-Policing.* Washington, DC: December 22, 1995.

Environmental Protection Agency (EPA), Region I. 1996. "Third Party Certification Project Leadership Agreement between International Paper Company—Jay, Maine Facility, Maine Department of Environmental Protection, Town of Jay, Maine, and United States Environmental Protection Agency—New England." Boston, June 26, 1996.

Environmental Protection Agency (EPA), Region I. 1998. "Guidance for Environmental Management Systems Audit, Environmental Management Systems Audit Report and EMS Implementation Plan." Boston, March 9, 1998.

Environmental Protection Agency (EPA). 1999. *Aiming for Excellence: Actions to Encourage Stewardship and Accelerate Environmental Progress.* Report of the Environmental Innovations Task Force. Washington, DC.

Environmental Protection Agency (EPA). 2000. *A Summary of EPA's Performance Track Proposal.* Washington, DC: EPA.

Fiorino, D. J. 1999. Rethinking environmental regulation: Perspectives on law and governance. *Harvard Environmental Law Review* 23(2): 441–69.

Hale, R. 1998. *The National Expansion of StarTrack: Issues Regarding the Delegation of Environmental Compliance Oversight*. Boston, MA: US Environmental Protection Agency, Region I.

Hoffman, A. J. 2000. *Competitive Environmental Strategy*. Washington, DC: Island Press.

King, A. A., and M. Lenox. 2000. Industry self-regulation without sanctions: The chemical industry's responsible care program. *Academy of Management Journal* 43(4): 698–716.

Matthews, D. H. 2001. Assessment and design of industrial environmental management systems. PhD thesis. Department of Civil and Environmental Engineering, Carnegie Mellon University, Pittsburgh.

Nash, J., J. Ehrenfeld, J. MacDonagh-Dumler, and P. Thorens. 2000. *ISO 14001 and StarTrack: Assessing Their Potential as Tools in Environmental Protection*. Washington DC: National Academy of Public Administration.

National Academy of Public Administration. 1996. *Setting Priorities, Getting Results: A New Direction for EPA*. Washington DC: National Academy of Public Administration.

Orts, E. W. 1995. Reflexive Environmental Law. *Northwestern University Law Review* 89(4): 1227–1340.

President's Council for Sustainable Development. No date. PCSD Environmental Task Force Review of the StarTrack Program, U.S. Environmental Prtection Agency, New England—Region I. On file with the author.

Schmidheiny, S. 1992. *Changing Course: A Global Business Perspective on Development and the Environment*. Cambridge: MIT Press.

Smart, B. 1992. *Beyond Compliance: A New Industry View of the Environment*. Washington, DC: World Resources Institute.

Spence, D. B. 2001. Can the second generation learn from the first? Understanding the Politics of Regulatory Reform. *Capital University Law Review* 29: 205–22.

SPRU. 2001. *Measuring the Environmental Performance of Industry (MEPI)*. EC Environment and Climate Research Programme, Research Theme 4, Human Dimensions of Environmental Change, Contract ENV4-CT97-0655.

Sullivan, T. J. 1982. Tailoring government response to diversity. In E. Bardach and R. A. Kagan, eds., *Social Regluation: Strategies for Reform*. San Francisco: Institute for Contemporary Studies.

Sunstein, C. R. 1999. Information regulation and informational standing: Akins and beyond. *University of Pennsylvania Law Review* 147(3): 613–76.

Wilson, J. Q. 1980. The politics of regulation. In James Q. Wilson, ed., *The Politics of Regulation*. New York: Basic Books.

11

Cooperative Environmental Regulation: Examining Project XL

Alfred A. Marcus, Donald A. Geffen, and Ken Sexton

Governments, businesses, public interest groups, affected communities, and academics are searching for new and innovative ways to achieve common environmental goals. They are striving to find approaches and techniques that provide "cleaner, cheaper, and smarter" solutions to environmental problems. The Environmental Protection Agency's (EPA's) Project XL (eXcellence and Leadership) has been one of the most ambitious and potentially consequential of these experiments (Hirsch 2001; Marcus, Geffen, and Sexton 2001). Project XL was conceived to be a centerpiece of the Clinton administration's effort "to develop innovative alternatives to the current regulatory system" (Clinton and Gore 1995). It was a primary vehicle by which EPA would "enter into partnerships with businesses, environmentalists, states, and communities to test alternative strategies for single facilities, industrial sectors, or geographic areas" (Clinton and Gore 1995). The end result was to be "a coordinated series of demonstration projects designed to provide the opportunity to implement alternative management strategies" (Clinton and Gore 1995). According to the Clinton administration, "The knowledge gained from such bold experimentation" would "lay the groundwork for developing a new environmental management system for the 21st century" (Clinton and Gore 1995).

The EPA selected 3M's world-class tape manufacturing facility in Hutchinson, Minnesota, as one of the first pilot projects to be carried out under the auspices of Project XL.[1] The purpose of the 3M pilot was to obtain a permit that incorporated many of the features of a new approach to environmental management, including superior environmental performance; flexible, facilitywide performance-based environmental goals; pollution prevention; an innovative environmental management system; reduction in compliance and enforcement costs, and explicit measures to ensure accountability. The agreement was to have been forged with community stakeholder and interested party participation, and the overarching aim was to achieve better environmental results than could have been achieved under existing requirements. Caring and committed people with different beliefs, backgrounds, and opinions worked collaboratively and in good faith for 18 months to forge a cooperative, mutually agreeable pilot that was both environmentally responsible and economically sound. Their goal was to achieve a classic "win–win" outcome in which both the environment and the company would be better off. Despite these good intentions, rather than getting a pilot project up and running, the negotiations among EPA, 3M, and the Minnesota Pollution Control Agency (MPCA) faltered and broke down.

This chapter compares the substance and process of the proposed 3M project with three XL pilots that were approved—Intel, Weyerhaeuser, and Merck. This discussion focuses on the gap between the rhetoric of reinventing environmental regulation and the reality of implementing changes to the existing system. Substantive differences explain why issues were contentious but not why some pilots were approved and others were not. The key substantive issues are related to disagreements about the definition of superior environmental performance (SEP). The main process difference was the direct involvement of the state government in the 3M pilot, which had no parallel in the other cases.

Ultimately the story of Project XL, as reflected in these four cases, reveals the challenges and complexities inherent in attempts to move beyond the familiar paradigm that has shaped US environmental programs and policies for more than two decades. The conventional paradigm has promoted adversarial relationships among stakeholders, and involved command-and-control strategies, end-of-pipe approaches, narrow media-based statutes, overly proscriptive laws and regulations, means-based rather than outcome-based standards and rules, and limited use of economic incentives. The new paradigm, in contrast, embraces cooperative approaches, and involves collaborative business-government partnerships, pollution prevention, holistic multimedia approaches, place-based decision making, flexible, easy-to-adjust rules, results-based standards, increased use of market mechanisms, and meaningful stakeholder involvement (Sexton, Murdock, and Marcus 2002).

The setup of this chapter is as follows. First we describe the design of Project XL in general. Then we move on to looking at four early pilots more in detail by comparing substance and process differences between them. After examining some changes in XL policy we draw our conclusions on the value of the program.

Project XL

On March 16, 1995, the Clinton administration announced a series of regulatory reinvention initiatives calling for wide-scale experimentation with new methods of environmental management. Though Project XL was only one among a number of programs that the White House had proposed to test innovative ideas, the White House referred to it as the "crown jewel." According to administration materials (Clinton and Gore 1995), the expected benefits of Project XL were as follows:

1. Increased flexibility to adopt innovative solutions to environmental problems.
2. Increased and more cost-effective environmental protection.
3. Improved compliance and increased use of innovative technologies.
4. Expanded use of waste minimization and pollution prevention strategies.
5. A more cooperative relationship between regulators, the facility, and the community.

The government would establish a high bar of environmental performance for excellent companies and they would be given the flexibility to decide how they were going to jump over it. For companies in this category, the government was willing to "throw out the rulebook" (Steinzor 1996). The Clinton administration envisioned a quid pro quo, in which regulated businesses would be granted greater regulatory flexibility in return for achieving superior environmental performance. If businesses applying for an XL permit achieved results better "than expected under existing law" (Clinton and Gore 1995), EPA would then be in a position to waive existing requirements.

As EPA originally conceived Project XL, individual companies that wanted to participate would design and submit proposals to the agency. If a proposal was accepted, the company would draft a permit and Final Project Agreement (FPA) working in collaboration with the state pollution control agency, the EPA regional office, local stakeholders, and EPA national offices. The goals of the EPA (Federal Register Notice 1995) in asking for proposals under Project XL were as follows:

- Environmental performance superior to that which could be achieved under current and reasonably anticipated future regulation.
- Cost savings and paperwork reductions for regulators and the affected firm.
- Stakeholder support.
- Innovative environmental management strategies.
- Multimedia pollution prevention.
- Transferability, feasibility, and clearly defined objectives, measures of success, and time frames.
- Easily understandable information, including performance data, made available by participating firms to stakeholders.
- Consideration of risk to worker health and safety.
- Cooperative relations between regulators, the facility, and the community.

Concerns focused on the definition of SEP. EPA summarized these concerns as a desire to have the agency "push harder on the environmental benefits" and be "an advocate for environmental excellence." EPA's XL managers wanted to make clear that superior implied more than just better than the regulatory minimum, since it was clear that many companies already perform better than required by regulation.

EPA was aware of the ambiguity in the White House's statements. It was certain that some people might interpret SEP as just doing better than the regulatory minimum. The burden on the applicant was to convince EPA that a proposed pilot could achieve SEP. EPA staff admitted that it had difficulty communicating its concept of SEP (Interviews 1997–98). Other parties read the same words differently. They each took parts of what the President said and formulated their own definition of what he meant.

It was difficult for EPA to get across a simple, clear, consistent, and uniform message that was not easily misinterpreted. People who attended the President's announcement of XL in the Rose Garden walked away with diverging impressions of what SEP meant and of the essential elements of XL. Individual groups returned to their own particular settings and circumstances and further developed their idiosyncratic interpretations in isolation from each other. EPA could not enforce a uniform understanding on diverse and scattered participants, and according to EPA staff, the gulf between what EPA meant by SEP and what other parties interpreted it to be widened over time and distance (Interviews 1997–98).

Early XL Pilots

By the end of 1996 EPA had received 45 XL proposals (EPA 1996) and two signed agreements had been reached. Forty-three projects had been deferred or rejected, were in review, or were in the process of negotiation.[2] The proposed 3M-Hutchinson pilot was one of XL's first (PPC 1996a), and represented a serious effort to forge an XL agreement by a company with an excellent environmental record. Nevertheless, the 3M-Hutchinson negotiations broke down in December of 1996 and 3M withdrew its proposal.

Of the first seven projects that EPA approved, the ones most similar to the 3M proposal were Intel, Weyerhaeuser, and Merck. Intel signed an FPA for its chip fabrication site in Chandler, Arizona, in November of 1996. Weyerhaeuser signed an FPA for its pulp manufacturing facility in Oglethorpe, Georgia, in January of 1997. In December 1997 Merck signed an agreement for its pharmaceutical manufacturing plant near Elkton, Virginia (the Stonewall Plant). One area where they were especially similar was the relatively high transaction costs of the negotiation. Blackman and Mazurek (1999) estimate that the average cost of forging these kinds of XL agreements was nearly $700,000 each. The companies bore the bulk of these costs—about $500,000 apiece—while EPA's expenses for each agreement averaged about $200,000.

Intel, Weyerhaeuser, and Merck were similar to 3M in other ways. All four of these companies were large publicly traded corporations known for innovation. They innovated both in the way they conducted their business and in the way they conducted their environmental affairs. Intel and Merck, in particular, are similar to 3M in their need to rapidly introduce new products. Getting to the market quickly is a critical part of their business models (Kessler and Chakrabarti 1996). These companies are under intense pressure to stay ahead of the competition. Thus they have a strong incentive to forge cooperative environmental agreements quickly. It is reasonable to expect that negotiations with these types of companies should have proceeded expeditiously.

At the outset EPA managers had hoped that projects would move from initial selection to implementation in six months or less. However, the time Intel, Weyerhaeuser, and Merck took to forge agreements (see table 11.1) ranged from 16 to 26 months, according to estimates (Blackman and Mazurek 1999). 3M gave up trying after approximately 18 months.

The 3M, Intel, Weyerhaeuser, and Merck Cases

Some background on these cases (see table 11.1) is necessary to set the stage for a comparison. The 3M pilot involved the company's premier tape (video and sticky) manufacturing facility located in Hutchinson, Minnesota, an industrial town about 60 miles west of the Twin Cities. The Intel pilot involved a facility that was supposed to become the company's newest Pentium microprocessor fabrication facility. It would be located in Chandler, Arizona, 14 miles from downtown Phoenix. The Weyerhaeuser pilot dealt with a pulp manufacturing facility in Oglethorpe, Georgia, which made millions of tons of absorbent fluff used in plastic diapers (enough pulp to make 30 million diapers daily). The Merck pilot concerned a pharmaceutical production site in Elkton, Virginia, adjacent to the Shenandoah National Forest. Its employees were engaged in a range of activities such as fermentation, solvent extraction, organic chemical synthesis, and finishing operations.

At Oglethorpe, GA, between 1995 and 1997, a total of about 4,500 tons of particulate matter, sulfur dioxide, carbon monoxide, volatile organic compounds, and nitrogen oxides were emitted into the air annually. At Hutchinson, MN, a total of about 3,000 tons of these pollutants

Table 11.1
Overview of four XL pilot projects

Type of facility	Type of pollution addressed by XL pilot	Location	Site characteristics	Time needed for negotiating an agreement
Weyerhaeuser Pulp Mill	Serious water pollution emissions; about 4,500 tons of criteria air pollutants emitted annually; and solid waste problems	Oglethorpe, Georgia	Recreational activity in Lake Brashear and Flint River water basin	16 months: agreement signed in January 1997
3M Tape Manufacturing Facility	About 3,000 tons of criteria air pollutants emitted annually	Hutchinson, Minnesota	Industrial town about 60 miles west of the Twin Cities	No agreement: after 18 months 3M withdraws from Project XL in December 1996
Merck Pharmaceutical Production Site	About 1,500 tons of criteria air pollutants emitted annually	Elkton, Virginia	Adjacent to Shenandoah National Forest and subject to special air pollution regulations as a Class I region	26 months: agreement signed in December 1997
Intel Pentium Microprocessor Fabrication Facility	A limit of 150 tons of criteria air pollutants proposed for the site	Chandler, Arizona	In the Phoenix metro area there are smog problems and water shortages	17 months: agreement signed in November 1996

were emitted into the air annually between 1995 and 1996, and at Elkton, VA, a total of about 1,500 tons of these pollutants were emitted into the air annually between 1995 and 1997. Intel's proposal was to limit total emissions of these pollutants to less than 150 tons into the air annually at Chandler, AR.

The Weyerhaeuser site at Oglethorpe and the 3M site at Hutchinson were in towns relatively removed from metro areas (Oglethorpe is in Macon County, about 100 miles southwest of Atlanta), and both were in attainment areas for national ambient air quality standards (NAAQS). They were therefore subject to less stringent regulation than the Merck plant in Elkton, which was subject to special restrictions under the Clean Air Act because it was adjacent to a National Park. The Intel site at Chandler is located in the Phoenix metro area, which experiences relatively high levels of automobile pollution and has chronic water shortages. Total air emissions from the Intel plant were relatively small and qualified as a minor source under national air pollution control laws.

The Intel pilot was relatively straightforward, and did not involve waiving any national environmental rules or regulations. Moreover it did not require a site-specific agreement, which was the legal arrangement used in the Weyerhaeuser and Merck cases and contemplated in the 3M case. This was not a case of a company seeking breakthrough environmental benefits in exchange for the possibility that it might have to violate federal laws. Still, the current regulatory system generally is more stringent on new pollution than old. Thus the Intel request, while apparently minor, was nevertheless a proposal to add new pollution in a metro area where pollution levels were already worrisome.

Although all four cases are comparable in certain respects, the most direct comparisons are between 3M and Weyerhaeuser, on the one hand, and Intel and Merck, on the other. 3M and Weyerhaeuser were examples of proposals to emit large amounts of pollution into less restricted areas, while Intel and Merck were examples of proposals to emit small amounts of pollution into more restricted areas. In terms of the magnitude and scope of the pilot proposal as well as its political and regulatory sensitivity, the Weyerhaeuser proposal should have presented the most problems. However, Weyerhaeuser obtained a signed agreement more quickly than either Intel or Merck (and 3M never got that far).

Substantive Issues

As already discussed, XL got off on the wrong foot when the White House and EPA defined SEP differently. The White House seemed to offer greater flexibility and other regulatory benefits to companies in exchange for environmental performance that was superior to what would have otherwise occurred under "existing" regulation (Clinton and Gore 1995). EPA, on the other hand, said it would entertain proposals from companies for regulatory changes only if the changes produced environmental results that were superior to what the companies would have achieved under "existing and reasonably anticipated future" regulation (Federal Register Notice 1995).

Ten substantive issues played important roles in defining SEP in these four cases (See table 11.2). The first issue has to do with differences between actual emissions (from current operations) and allowable emissions (allowed by existing laws). According to our calculations, 3M-Hutchinson had the largest difference in this parameter.

The second issue has to do with the difference between the emission caps in the XL permits and the limits in current regulations. By this criterion, 3M-Hutchinson would be subject to more restrictive legal requirements under its XL permit than the other pilots.

The third issue has to do with differences between the caps in the XL permit and current emissions. Both 3M's and Weyerhaeuser's air pollution potentially could increase under the XL agreement, but the gap between caps and current air emissions was greater in the 3M case. 3M also did have immediate plans for expansion at its Hutchinson plant and used emission-per-unit-of-production prominently in its proposal (MPCA 1996c, d, e).

The fourth issue was the difference between current-pollution-per-unit-of-production and pollution-per-unit-of-production under the XL permit. 3M, which anticipated higher production at Hutchinson, promised a 37 percent reduction in tons per unit of production over a ten-year time span.

The fifth issue has to do with the time period used to establish a baseline for superior environmental performance (i.e., SEP compared to what). EPA insisted that 3M use the most recent year prior to the XL permit being in place, which would have been 1995. In contrast, EPA

Table 11.2
Comparison of superior environmental performance for four XL pilot projects

	3M (proposed XL agreement)	Weyerhaeuser (signed XL agreement)	Merck (signed XL agreement)	Intel (signed XL agreement)
1. Difference between actual and allowable air emissions	91.8% lower	71.0% lower	44.3% lower	A start-up, thus no current actual emissions[a]
2. Difference between the cap in the XL permit and currently allowable air emissions	82.0% lower[b]	60.0% lower	55.5% lower[c]	51% lower
3. Difference between the cap in the XL permit and current air emissions	54.9% higher	26.4% higher	20.0% lower[d]	A start-up, thus no current actual emissions[e]
4. Difference between tons of pollution per unit of production in the XL agreement and current tons of pollution per unit of production	37% lower	Water usage: 2.9% higher Water pollution: 7.5% lower	Not calculated	Not calculated
5. Baseline year (or years) used in making comparisons	1995	1993–95	1992–93	A start-up—no baseline available, therefore current federal Requirements for minor sources used
6. 10 year XL permit allowed	No	Yes	Yes	Yes

7. Promise of technological breakthrough in 10 years	No	Yes	No	No
8. Economic factors considered	No	Yes	No	No
9. Comprehensiveness of XL agreement	Limited to air pollution	Air, water, solid waste, forest practices	Limited to air pollution	Plan for reduced water usage
10. Some credit for previous voluntary controls	No (but moving in this direction)	Yes	No	No

a. The 1994 permit for the facility was 75 percent lower than the federal requirements.

b. For volatile organic compounds, approximately 65.45 percent lower if the allowed emissions after maximum achievable technology (MACT) is determined and from 15.1 percent lower to 36 percent higher after ten years with no grandfathered coaters and MACT.

c. If 650 tons per year margin for growth in VOCs is taken into account, then the difference is 31.4 percent lower.

d. If 650 tons per year margin for growth in VOCs is taken into account, then the difference is 23.2 percent higher.

e. The Project XL permit was 4.4 percent lower than the 1994 permit.

allowed Weyerhaeuser to use averages from 1993 to 1995 and Merck to use averages from 1992 to 1993.

The sixth issue was whether a ten-year time period for determining SEP would be allowed. In the Weyerhaeuser, Merck, and Intel cases, EPA agreed that such a permit was acceptable. However, in 3M's case, EPA maintained that under the provisions of the Clean Air Act, it could not sign a ten-year agreement.

The seventh issue was whether a promise of a future technological breakthrough could be used as evidence of SEP. In Weyerhaeuser's case, EPA agreed to such a promise but in no other case was it even brought up. EPA's intent was to use the hoped-for technical advance to coax, and ultimately compel, other pulp manufacturers to do what Weyerhaeuser would demonstrate was possible.

The eighth issue was whether economic considerations mattered in evaluating and granting the XL permit. In the Weyerhaeuser case, EPA relied on economic arguments (the mill employed 500 people and made a large contribution to the Georgia economy) to justify its support for the agreement. In the 3M case, EPA explicitly refused to consider economic impact (PPC 1996b).

The ninth issue was whether a more comprehensive (e.g., multimedia, broad-based, holistic) agreement constituted greater evidence of or potential for SEP. The Weyerhaeuser agreement was the most comprehensive and complicated of the four cases.

The final issue involved determining whether any credit in the XL agreement would be given for previous voluntary controls. In most cases EPA rejected this option, but it did give Weyerhaeuser some credit for earlier reductions and it was moving in the direction of giving 3M some credit before negotiations broke off.

Process Issues

When XL started, EPA was uncertain and unclear about what standards would be used for determining SEP. The Agency attempted to negotiate solutions on a case-by-case basis, but it was also uncertain how these negotiations would be conducted, who was to be involved, and how an agreement was to be reached.

In drafting XL agreements, the 3M case was the only one of the four that included direct early involvement of the state government. The Minnesota Pollution Control Agency (MPCA) was the only state agency formally selected by EPA to be an XL participant (EPA 1996). Thus the MPCA considered itself to be a full partner with 3M in drafting the Hutchinson agreement. 3M had a good working relationship with the MPCA, going back several years to the time when MPCA had agreed to apply an innovative "pollution bubble" approach to controlling emissions at 3M's St. Paul plant (another tape-making facility). Thus MPCA dived into XL with 3M without first developing an understanding (e.g., they did not execute a memoradum of understanding) with EPA Region V that specified roles and responsibilities. In the other three cases relevant EPA regional offices were directly involved from the start. As Dawson (1998: 41) writes about the forging of the Weyerhaeuser agreement, "The EPA (i.e., its regional office in Atlanta) devoted a great deal of resources to the program ... The Georgia Environmental Protection Division did not devote too much time or resources...."

3M worked directly with the MPCA, while Weyerhaeuser, Merck, and Intel formed broad-based stakeholder coalitions in which the EPA region and other interested parties participated, (Sexton, Murdock, and Marcus 2002). For example, EPA's regional office had representatives on the Merck stakeholder group, as did many other organizations. The company's stakeholder group consisted of EPA, the Virginia Department of Environmental Quality (VADEQ), the US Department of Interior (DOI) National Park Service (NPS), and community representatives, all of whom had a part to play in drafting the agreement (Federal Register 1997).

Merck may have intended to have its stakeholder group play a strictly advisory role, but they brought effective pressure to bear and pushed the negotiation to the point where they were nearly full partners in crafting the agreement (Sexton, Murdock, and Marcus 2002).

Intel followed a similar process. Tim Mohan (1997: 27), who managed Intel's application for an XL permit, writes: "Intel conducted perhaps the most inclusive stakeholder process (involving all relevant federal, state, and local agencies, environmental organizations, tribal government, community representatives, and Intel employees) of any industrial permitting project in history." The group met for eleven months and had over one hundred meetings, including eight meetings through which the public was fully informed in advance and asked to participate. The Intel agreement was developed jointly by representatives from Intel, ten different agencies, and five Chandler residents.

The 3M case followed a different paradigm in coalition formation than the others (Odell 1999). Weyerhaeuser, Merck, and Intel used a process that centered on "influentials" (e.g., the relevant EPA region, national environmental groups) and tried to include a broad array of stakeholders. 3M, on the other had, employed a "friends first" process (i.e., MPCA, a hand-picked advisory group) and initially excluded EPA and national environmental organizations.[3] Consequently neither EPA nor national environmental groups were supportive of the 3M proposal.

Though opposition to the proposed XL agreements existed in all cases (primarily from national environmental groups or some local stakeholders), it did not prevent agreement from being reached in the Weyerhaeuser, Merck, and Intel cases. A critical difference between these three cases and 3M was the relationship with EPA. In the last weeks of May 1996, at literally the same time that 3M and the MPCA were first bringing their proposal to the attention of EPA Region V and asking that it be put on public notice (PPC 1996b), Weyerhaeuser, Merck, and Intel had a different kind of meeting with national environmental organizations, arranged by EPA national headquarters.

The first critical exchange with EPA about the contents of the 3M proposal occurred at a meeting at regional headquarters in Chicago (PPC 1996b) and did not go well. EPA's spokesperson (PPC 1996b) argued that before Region V could approve the XL permit, MPCA and 3M had to show environmental benefits, to ensure compliance, and to justify granting operating flexibility to the Hutchinson facility. EPA appreciated 3M's willingness to step forward and be innovative, but wondered to what extent the environment would be better off under the proposed permit. EPA officials expressed their view that EPA's obligation to Congress and the public was to guarantee environmental protection, not economic performance. According to EPA officials, the determining factor in whether EPA granted the XL permit would be MPCA's and 3M's ability to demonstrate, without qualification, a superior level of environmental improvement.

The meetings in Washington, DC, that EPA organized for Weyerhaeuser (EPA 1996a), Intel, and Merck were more positive in spirit and tone. These meetings were an outgrowth of contact between EPA's highest ranking policy and planning official, David Gardiner, and key players for national environmental organizations. Gardiner had met with them as part of an ongoing process to inform the national NGO community about the specifics of XL pilots as they moved toward approval. An important auxiliary objective was to improve communication about XL among EPA, XL proponents, and national environmental groups.

The EPA-arranged meetings involved an introduction of the participants, comments by Gardiner, summary presentations of the proposed projects by the staff of the companies, and a facilitated question and answer session. The facilitator was from the Keystone Center, which had helped other parties reach environmental agreements. The meetings provided a relatively stress-free environment to air opinions and exchange views. There were two separate meetings, one on May 15 at which Weyerhaeuser and Intel presented their proposals and one on May 23 at which Merck presented its proposal. Similar meetings with outside facilitators and joint company/EPA presentations did not take place in the 3M case. Instead, 3M, the MPCA, and the XL advisory group had a much more contentious meeting with EPA's Region V office.

The EPA-arranged meetings did not quiet all criticism from the environmental community. The National Resources Defense Council (NRDC), for instance, opposed not only the 3M agreement but also wrote detailed critiques of the Intel and Merck agreements. The most crucial difference between 3M and the other three cases was that the EPA joined in the chorus of criticism aimed at the 3M-Hutchinson proposal while EPA (1996d, e) was generally more supportive of the Weyerhaeuser, Merck, and Intel agreements, going so far as to defend them to critics. This is not to say that EPA did not harbor some reservations about the other three proposals, particularly the Weyerhaeuser pilot (Dawson 1998).

Summary of Substance and Process Differences

In sum, there were differences in the substance of these four cases with respect to how SEP was defined and in the processes used to negotiate an agreement. The facilities were in different locations, emitted different amounts and types of pollution, were subject to different legal and regulatory requirements, and the actors took different actions as part of the negotiating process.

3M was the only company that began by working exclusively with a state agency, the MPCA, which meant that unlike the other cases, the EPA Region did not play an active role in the negotiations. Unlike the others, 3M did not have an EPA-arranged meeting with national environmental organizations and other interested parties prior to requesting that its proposed agreement be put on Public Notice. Ultimately 3M did not gain EPA headquarter support and could not overcome the continued opposition from national environmental organizations like the NRDC. Both substance and process problems combined to thwart the best efforts of 3M and MPCA to craft an XL proposal that EPA would approve.

Subsequent Changes in XL Policy

The fact that these four cases all occurred between 1995 and 1997 is important because XL policies subsequently changed because of a shift in the political climate (less congressional antagonism toward environmental regulations) and as a result of several EPA evaluations.[4] State regulatory officials from across the country as well as business leaders encouraged EPA to give the states more authority to move forward with reform and experimentation. EPA itself was restructured and a separate Office of Reinvention was created to take charge of all the agency's reinvention efforts.

In October 2000, some five years after the program started, there were 48 signed agreements and six XL projects under development. Most of the organizations signing later XL agreements were public and nonprofit entities, with relatively few from the private sector. Six of the first seven agreements were with private companies, but out of the next 41 agreements, 16 were with private companies, 23 were with government entities, and two were with nonprofit organizations. Thirty-one new XL agreements were signed in the first 10 months of 2000, 19 of them with organizations in the public and not-for-profit sectors and 12 with businesses. Four of the 5 XL projects under development at that time were with entities in the public sector.

EPA tried to involve the private sector in changing the XL process by working with representatives from industry (Union Carbide and Dow Chemical), NGOs (Environmental Defense Fund, Environmental Law Institute, and Citizens for a Clean Environment), and other concerned stakeholder groups (EPA 1998c). EPA redrafted and revised many of the XL documents and created new ones. The revamped XL program differed from the earlier program in a number of important ways.

SEP and Flexibility

In 1997 EPA attempted to explain what it meant by SEP and regulatory flexibility in a Federal Register Notice (FRL-5811-7, April 1997). In defining SEP, EPA distinguished between tier 1 and tier 2 hurdles.

Tier 1 established an enforceable requirement that environmental performance, under an XL permit, would be *at least as good* as it would have been had the facility remained within the conventional regulatory framework. A tier 1 hurdle was a baseline composed of what would have happened under existing law if the proposal had not gone forward. According to the tier 1 criterion, an XL project had to result in environmental performance that was at least as good as the baseline. In determining the baseline, EPA allowed trade-offs—an increase in one kind of environmental impact for a decrease in another—provided there was a clear net environmental benefit.

Tier 2 was a determination of how much *better* environmental performance would be compared to the initial baseline environmental loading. Tier 2 was to include pollution prevention and other best practices, including quality of stakeholder involvement and reduced compliance costs. The Federal Register Notice categorized the project's tier 2 objectives as "voluntary commitments" that were spelled out in the Final Project Agreement (FPA). Although these commitments were not enforceable (as tier 1 requirements were), failure to achieve these commitments was "an appropriate basis for termination or modification of the XL project."

Neither of the newly defined criteria (tier 1 or tier 2) was particularly straightforward. Potential applicants, for instance, still had to ask themselves how much better than the baseline does a facility have to perform in order to gain approval? Moreover the more ambitious the commitment made by an applicant, the greater the risk of failing to meet the commitment and losing XL approval.

The EPA's 1997 Federal Register Notice explicitly linked the degree of regulatory flexibility EPA would grant a facility to the amount of SEP the project offered. The Notice strongly encouraged applicants to demonstrate a "factual link" between the two. EPA said that "the closer the factual link between the requested flexibility and anticipated environmental benefits, the more likely EPA is to approve the project." But again, as in the case of SEP, EPA was not very clear about what specifically it meant by regulatory flexibility. There was little certainty about what an applicant could obtain in exchange for the still nebulous SEP construct.

EPA also directly addressed the issue of obtaining credit for past voluntary controls in the Federal Register Notice (1997). It rejected the idea that preexisting voluntary measures should be credited to an XL project. Companies could not create "bank accounts" based on their existing voluntary controls, upon which they could potentially draw. EPA, however, allowed for exceptions to this general rule. If there were "other positive elements of superior environmental performance contained in the project," then it "would consider crediting the preexisting voluntary controls" (Federal Register Notice 1997). EPA seemed to be saying that facilities that had significant preexisting voluntary controls would not have to show as much environmental improvement as facilities that did not.

Best Practice's Guide

In October 1997 EPA began a process of reengineering that changed how XL decisions were made. The changes focused on a number of generic issues:

- The availability of information. EPA staff felt that they were not getting enough information from applicants.
- Coordinating the activities of the regulators. Communication within EPA needed to be improved and the way it made decisions clarified.

• Making stakeholder involvement more transparent and predictable. The process had been ad hoc and the role that the stakeholders were to play was ambiguous.

• Establishing a firmer management structure. A time frame was needed for tracking progress.

On June 3, 1998, EPA issued a *Best Practices Guide* (EPA 1998a) that laid out the kind of information EPA was looking for, what should be part of a proposal and why, and what else EPA was seeking. However, the revised process still depended on sequential decision making. An applicant first submitted an initial concept paper and obtained comments from the agency. It then submitted a formal proposal and the agency screened it for enforcement issues and notified the applicant. If these issues could be resolved, the applicant would then resubmit the proposal and EPA would review it again, have it examined by a special project team, and send it out to all relevant EPA offices.

Then the Agency would inform the applicant and stakeholders of all EPA's issues. The applicant would have to resolve the issues and take appropriate stakeholder process steps, as required. In this EPA-designed process, formal documents continued to flow back and forth many times between the agency and the applicant in an iterative, sequential cycle. In point of fact, the changes did not move EPA toward a system wherein all the parties, including state pollution control agencies and stakeholders, were part of an effective team-oriented process focused on attaining shared goals.

Transaction costs for XL pilots remained high, and since approval was by no means assured, business managers continued to be reluctant to allocate staff and financial resources to what might turn out to be an unsuccessful effort. If a company's XL proposal was accepted, there remained the considerable risk that the permit could be terminated if the voluntary commitments to achieve SEP under the tier 2 requirements were not met. Reversion back to conventional source-by-source regulation could entail additional costs to the company, making XL investments redundant or unnecessary. This latter risk was magnified by the fact that the tier 2 criteria were so vague and had to be fleshed out in each case through negotiation.

Conclusions

As the twenty-first century begins, a new generation of collaborative and cooperative policy mechanisms has emerged in the United States and elsewhere—sometimes as an alternative, sometimes as a supplement, and sometimes as a complement to conventional regulatory strategies. These new policy approaches require businesses and governments to forgo their traditional adversarial roles and adopt a more collegial, collaborative, and constructive posture.

This chapter has examined one such experiment, Project XL, which was initiated by the Clinton-Gore administration and implemented by the US EPA to test the feasibility of voluntary cooperative agreements for environmental protection. What unites the cases discussed here is that all four companies had significant pollution problems that they wanted to deal with in a creative and collaborative way that would be less complex, cumbersome, and bureaucratic.

The basic motivation for these companies to participate in XL was their desire to get products to market faster. They considered the existing rules to be a regulatory straitjacket and wanted greater flexibility to meet emission limits. EPA, for its part, wanted guarantees of superior environmental performance—better environmental quality than otherwise possible under current and anticipated rules—in exchange for granting flexibility. However, the companies participating in XL discovered that reaching an agreement with EPA was problematic because the agency was both uncertain and unclear regarding the definition of SEP (substance) and the procedures that would be used to arrive at a practical definition (process).

Three of the four companies—Intel, Weyerhaeuser, and Merck—are today relatively satisfied with the flexibility they achieved under XL, however hard won. The fourth company, 3M, failed in its bid to gain approval for an XL project, yet still managed to get virtually all of the flexibility it was seeking through a more traditional permitting process. Nevertheless, the frustration of the failed 3M-Hutchinson XL negotiation has left the parties involved generally dissatisfied and somewhat cynical about about EPA's reinvention efforts.

Though we have emphasized (1) the substantive issues related to defining SEP and (2) the process issues related to negotiating an uncertain quid pro quo as the primary causes of the 3M-Hutchinson stalemate, this outcome is certainly the result of multiple factors. For example, among the stumbling blocks that prevented an agreement was the fact that the 3M plant was so far below existing emission standards to begin with, which made it difficult to craft a suitable agreement that kept the XL allowables cap below actual emissions. Another factor was the predominant role of the EPA air office (Office of Air Quality Planning and Standards in North Carolina) in the 3M-Hutchinson case. Because the important issues were all air-related, the air office was often the key decision maker. There were few opportunities to trade off air emissions for reductions in water or land pollution (as was done in the Weyerhaeuser case). Furthermore, unlike the other three cases, there was no professional facilitator used in the 3M-Hutchinson negotiations, which made communication and compromise more difficult.

Perhaps most important, 3M and MPCA started from a false premise—they both assumed that MPCA had been delegated authority from EPA to negotiate an XL permit. Consequently they did not involve EPA Region V or EPA headquarters in the early discussions and presented EPA with what they thought was a fait accompli. However, it turned out that EPA had not delegated its authority to MPCA and took exception to both the substance and the process of the 3M-Hutchinson XL proposal.

Evaluating XL

In evaluating XL, it is important to consider the following questions: Did the government get better environmental performance? Did businesses get a system through which they could deliver better environmental performance in a more flexible manner? Making these determinations depends on how one defines superior environmental performance, which we have seen is a very difficult thing to do. It is still not clear how the various XL agreements defined SEP or whether they have achieved the agreed-on level of SEP. Was SEP defined in terms of pollution allowables or actuals? Was SEP to be measured in units of pollution-per-unit-of-production, technological feasibility, or some other metric? The definitions never were completely clear and they varied from case to case. Today the definition of SEP remains murky despite continuing EPA efforts to define the term more precisely.

Without a clear and consistent understanding of SEP, the controversy about what XL achieves will remain difficult to resolve. It will continue to be hard to say with any degree of certainty whether XL has been successful. What makes this conundrum even more difficult to grapple with is that it is not possible to say for certain if the same or even better results could have been achieved without XL. We do not currently have, and are unlikely to ever have the kind of hard, quantifiable data necessary to address these issues conclusively and credibly.

What we can say is that 3M seems to have stabilized its emissions at lower levels than the cap proposed in the XL agreement. Intel has enjoyed regulatory flexibility and has built a second lab on the Chandler site without having to go through additional permitting. Weyerhaeuser achieved almost all the environmental goals it set out for itself in its XL agreement, with the exception of perhaps the most important one, lowering of bleach flow effluent releases. Merck, although delayed, has constructed a natural gas boiler to replace a coal boiler, but it seems likely that Merck could and would have made a similar change even in the absence of XL.

The effort that went into XL was substantial. Do these gains justify the effort? In comparison to a conventional permit renegotiation, how much better off is the environment under XL and how much better off are the companies that participated? Have the companies achieved for themselves and for the environment results that are commensurate with their efforts?

Recommendations for Future Regulatory Reinvention Projects

We think that XL-like experiments should continue to be carried out, but only if they are structured in a way that will reduce the transaction costs substantially and make it more attractive for businesses to participate. Too many of EPA's recent XL agreements are with government entities and nonprofit organizations. Too few are with private businesses. Firms seem to have shied away from becoming involved in XL, yet we believe that it is from the business community that the most interesting XL experiments are likely to arise. However, as currently designed there simply is too much ambiguity in XL. Clear, consensus-based definitions are needed for critical and contentious issues, like SEP and regulatory flexibility.

We believe that regulatory experiments, like XL, must begin with an agreement on the definition of SEP. For example, a possibility might be that in attainment areas, SEP should be based on an emissions-per-unit-of-production standard, while in nonattainment areas SEP should be defined in terms of the actual amount of pollution emitted. Moreover, as a threshold for admission into the experiment, companies should be required to have a good compliance record and show real evidence of voluntary, beyond-compliance achievement. In addition to defining the vision and goals of pilot projects to test ideas, the federal goverment should downplay the precedent-setting nature of individual pilots by emphasizing their experimental nature and their potential to bring about system-wide change.

We agree with many other analysts that a new statute authorizing experiments in regulatory innovations would be helpful. In the absence of such a statute, the federal government should use its discretionary authority within existing laws to seek alternative solutions and approaches. In almost every case regulators have more statutory discretion than they either realize or care to admit.

In accord with the experimental nature of pilot projects, the government should take a broad-based approach encouraging the use of performance-based standards, multimedia pollution prevention, innovative environmental management systems, stakeholder accountability, and methods for achieving cost savings. When applicants meet certain minimum requirements, the government should grant them alternative permits that allow for operating flexibility. The government should then monitor and carefully evaluate the experiments it has authorized to determine whether granting greater flexibility results in the hoped for environmental and economic improvements. Experiments should be designed to test the hypothesis that environmental improvements will be achieved in exchange for greater operating flexibility.

In both designing and approving individual regulatory experiments, the federal government should rely on teams that consist of individuals from local communities, state pollution control agencies, regulated facilities, and relevant EPA offices (i.e., regional and headquarters). In order to facilitate communication and coordination, the government should insist that these teams meet together often and as a group, either in person or by means of teleconferencing technologies.

The government should avoid the serial passing of documents from one group to another because experience has shown that delay is likely and misunderstandings almost certain. Instead, it should promote parallel processes in which people from different groups are given the chance to work together collegialy and constructively.

The government should delegate final decision-making authority to the EPA groups that are part of these teams, thereby avoiding a consensus-based process that compels the agency to obtain agreement from each of its many units as well as from the team. In general, the government should empower the groups from EPA to take appropriate risks in order to move toward a new collaborative form of environmental management.

The governmental approval process should have explicit timetables, and the aim should be to complete the process, from application to design, in six months or less. For instance, failure to reach agreement within twelve months should be cause for automatic termination. EPA also should set a specific number of projects it hopes to approve within a certain time period. We believe such changes might go a long way toward ensuring that future experiments with voluntary environmental agreements will be illuminating, productive, and successful.

Acknowledgments

This is the fourth revised draft of a paper originally prepared for presentation at the Association for Public Policy Analysis and Management, 22nd Annual Research Conference, November 2–4, Seattle, Washington. The first revision was presented at a workshop at Harvard on Voluntary, Collaborative, and Information-Based Policies, May 10–11, Cambridge, Massachusetts. Alfred Marcus, Don Geffen, and Ken Sexton have published a book on Project XL in 2002 for RFF Press called *Reinvent-*

ing Environmental Regulation: Lessons from Project XL. This chapter draws on material contained in this book.

Notes

1. The authors were involved with the events in Project XL Minnesota as participants as well as observers. We helped to create a stakeholder group that advised the Minnesota Pollution Control Agency (MPCA). We also were a research team that monitored and evaluated the Minnesota project. We attended key project meetings, carried out interviews with MPCA, 3M, and EPA staff, collected documents and other archival evidence, and did a survey. Our work on Project XL was supported by an EPA contract to advise, monitor, and evaluate Project XL Minnesota.

2. Also see EPA (1997f, 1998d, 1999c).

3. It may be easiest to deal with "friends first," but this can be a mistake, if doing so excludes influential parties who have the capacity to block an agreement. By combining large numbers (a broad stakeholder coalition) and an "influentials first" approach (including the EPA region from the beginning), Weyerhauser, Merck, and Intel maneuvered better around a dilemma, to which the negotiation literature points. This literature suggests that a small numbers approach results in lower levels of cooperation, less equal outcomes, and a decrease in integrative trade-offs. Including everyone first, on the other hand, increases complexity which in turn leads to simplifying strategies and cognitive mistakes.

4. See *www.epa.gov/projectxl/guidexl.htm*.

References

Blackman, A., and J. Mazurek. 1999. *The Cost of Developing Site-Specific Environmental Regulations: Evidence from EPA's Project XL*. Discussion Paper 99-35. Washington, DC: Resources for the Future.

Clinton, B., and A. Gore. 1995. *Reinventing Environmental Regulation*. Washington, DC: US EPA, Office of Policy Analysis and Review, Office of Air and Radiation.

Dawson, E. 1998. Looking at voluntary participation programs: A case study of Project XL at the Weyerhaeuser Flint River Facility. Senior honors thesis. Natural Resources and the Environment, University of Michigan.

EPA. 1996a. Minutes to meetings held to develop final project agreement. May 15.

EPA. 1996b. Memorandum to paricipants of the Merck/NGO Project XL meeting. June 7.

EPA. 1996c. Minnesota XL permit? Final project agreement comments. V. Adamkus (Region 5 administrator) to C. Williams MPCA Commissioner, July 2.

Epa. 1996d. Letter from Gardiner to NRDC (van Loben Sels) regarding July 3 comment letter on the Intel XL project.

EPA. 1996e. Letter from McCabe (Region III administrator) to NRDC (Hawkins and van Loben Sels) regarding Merck XL project, August 15.

EPA. 1996f. *The XL Pipeline*, vol. 1, no. 2, October.

EPA. 1997f. "XL projects listed by EPA Region" and "Information on the various XL projects." epa.gov, 4/2.

EPA. 1998d. "XL process diagram." epa.gov, 11/4.

EPA. 1999c. *XL Project 1999 Comprehensive Report*. EPA 100-R-99-008.

Federal Register Notice (FRL 5197-9). 1995. Environmental Protection Agency. Regulatory Reinvention (XL) Pilot Projects, Solicitation of Proposals and Request for Comments. May 23, 27283–90.

Federal Register. 1997. XL Site Specific Rulemaking for Merck & Co., Stonewall Plant, Part III. Environmental Protection Agency 40 CFR. Parts 52, 60, 264, and 265.

Federal Register Notice (FRL-5811-7). 1997. Environmental Protection Agency. Regulatory Reinvention (XL) Pilot Projects, Notice of Modifications to Project XL, April 23, 19872–82.

Hirsch, D. 2001. Understanding Project XL. In E. Orts and K. Deketelaeare, eds., *Environmental Contracts*. London: Kluwer.

Kessler, E., and A. Chakrabarti. 1996. Innovation speed: A conceptual model of context, antecedents, and outcomes. *Academy of Management Review* 21(4): 1143–91.

Marcus, A., D. Geffen, and K. Sexton. 2001. The quest for cooperative environmental management. In E. Orts and K. Deketelaeare, eds., *Environmental Contracts*. London: Kluwer.

Mohan, T. 1997. The alternative compliance model: A bridge to the future of environmental management. *ELR News and Analysis* 27: 10345–56.

MPCA (Minnesota Pollution Control Agency). 1996d. Public Notice of Proposed Insurance of Minnesota XL Permit and Revocation of the MPCA Permit to be replaced by the Minnesota XL permit.

MPCA (Minnesota Pollution Control Agency). 1996e. Minnesota XL—Final Project Agreement (FPA) between MPCA, EPA, and 3M Draft. May 15.

Odell, J. 1999. The negotiation process and international economic organizations. Presented at annual convention of the American Political Science Association. September 3, Atlanta.

Permit No. 96-01 to 3M-company. Final draft. Charles Williams, commissioner, May 14.

Pilot Project Committee (PPC). 1996a. Notes from presentation of 3M's plans for XL permit at Hutchinson. D. Wefring, T. Zosel, C. Kedrowski, M. Nash.

Pilot Project Committee (PPC). 1996b. Notes from meeting to present the Minnesota XL project for 3M Hutchinson to US EPA Region 5, end of May Chicago Headquarters.

Sexton, K., A. Marcus, K. Easter, and T. Burkhardt, eds. 1999. *Better Environmental Decisions: Strategies for Governments, Businesses, and Communities.* Washington, DC: Island Press.

Sexton, K., B. S. Murdock, and A. Marcus. 2002. Cooperative environmental solutions: Acquiring competence for multi-stakeholder partnerships. In P. ten Brink, ed., *Environmental Agreements: Process, Practice, and Future Use.* Sheffield: Greenleaf.

Steinzor, R. 1996. Regulatory reinvention and Project XL: Does the emperor have any clothes? *ELR News and Analysis* 26: 10527–37.

12

Disclosure of Toxic Releases in the United States

Mary Graham and Catherine Miller

Under a 1986 federal law some categories of manufacturers in the United States are required to disclose to the public their releases of certain toxic chemicals, facility by facility and chemical by chemical. Releases must be reported annually in standardized formats. This Toxic Release Inventory (TRI) differs from existing national regulation of toxic chemicals in its emphasis on lawful releases of chemicals and in its multimedia approach. During the 1990s the federal Environmental Protection Agency expanded substantially both the chemicals for which reporting is required and the manufacturers required to report. Although the TRI was created as a modest "right-to-know" provision, the public disclosure it requires has come to be seen as an effective means of reducing toxic pollution. By 1998 it was credited with contributing to a 46 percent reduction in releases of listed chemicals over ten years.[1]

Behind this often-quoted decline, however, lies a more complicated story. Releases declined at a much more rapid rate in early years than later, with half of the total decrease registered in the first three years. Relatively few facilities cut releases by reducing waste at the source, the preferred method under national policy, but recycling increased substantially and releases of carcinogens have declined at a somewhat faster rate than overall releases. Trends in releases to air, water, and land vary significantly, as do trends in different industries. In addition reported releases have been concentrated in relatively few states, with Texas, Louisiana, and Ohio accounting for nearly a quarter of national releases.

The complex architecture of this seemingly simple disclosure requirement has influenced incentives created for businesses. The TRI's particular strengths lie in its unusual requirement that releases be reported annually, factory by factory, in standardized form. At the same time, some elements of the TRI's design placed some important limitations on its effectiveness. Because it targets releases only from manufacturing facilities, the TRI has had no effect on the nation's largest sources of toxic pollution: cars, trucks, and buses, and small businesses.

Structured disclosure is emerging as an important tool of risk regulation in the United States. Under the US system of government the development of new regulatory policies is inevitably incremental. The TRI provides early and valuable lessons about the role of disclosure as part of a complex web of changing political and economic forces that influence corporate decisions. Understanding the story behind the general decrease in toxic releases during the last decade and appreciating the influence of TRI's architecture on industry incentives will assist designers of future transparency systems. In addition the growing power of computers and the Internet offers hope for improving the effectiveness of such information-based regulation. The TRI creates the potential to bring specific trends in toxic pollution to the attention of interested members of the public and to provide each individual with the customized information that is most useful to them. The TRI also creates the potential for combining data from many sources in ways that overcome some architectural limitations and provide improving indicators of risk. Such progress should not be taken for granted, however. Disclosure systems, like other forms of regulation, can be manipulated to serve narrow political purposes, outdistanced by technology and markets and stymied by unintended consequences. Long-term improvements may depend heavily on the continuing presence of influential constituencies with a strong interest in improving the effectiveness of structured disclosure as a means of reducing risks.

This chapter consists of four sections. The first section provides an overview of the background, provisions, and evolution of TRI. The second section tells the interesting and complex story behind the decline in toxic releases frequently credited to TRI. The third section discusses the strengths and weaknesses of the requirement's structure and scope. A final section suggests some tentative conclusions about the usefulness of this novel requirement as an instrument of risk regulation.

A Novel Use of Structured Disclosure

The Toxics Release Inventory (TRI) provides an evolutionary bridge between familiar policies in the United States that treat information as a public right and emerging strategies that employ information as an instrument of risk regulation. The federal requirement that manufacturers disclose to the public their toxic releases was created in the mid-1980s as a modest right-to-know provision. But when industry leaders responded with major commitments to reduce toxic pollution and when releases of listed chemicals plummeted, the requirement came to be viewed as a leading example of the power of information to improve environmental protection.

The TRI is a disclosure provision that focuses primarily on routine toxic releases by manufacturers. But it was a last-minute addition to a law with a different purpose. The Emergency Planning and Community Right-to-Know Act of 1986 (EPCRA) aimed to protect the public against chemical accidents. It was a Congressional response to troubling recent incidents. Two years earlier a tragic escape of deadly gases at a Union Carbide plant in Bhopal, India, had killed more than 2,000 people. A number of less serious chemical leaks in the United States in the following two years, including one at a Union Carbide plant in Institute, West Virginia, which sent hundreds of people to the hospital, heightened public concern about the potential dangers of accidental pollution.[2]

The new law fostered creation of emergency response systems in communities around the country to respond to such incidents. The TRI was added after hearings were completed and aimed to inform the public about toxic releases that were routine and generally intentional. Its Congressional sponsors emphasized that the American people had a right-to-know about toxic pollution where they lived or worked. Such pollution could create risks to human health and the environment.

The idea of community right-to-know was by the mid-1980s a familiar if amorphous national goal in the United States. Since the 1960s a variety of federal and state laws had required broad public access to information that was collected by the government from corporations and other organizations, as well as to the government's own data. Because the TRI's initial emphasis was on right-to-know, questions about whether or how such information might create incentives for businesses to reduce such pollution were not debated in Congress or addressed in the law or regulations.[3]

Within a decade, however, the TRI had become a symbol of a newly prominent trend in US social policy. The idea was that systematic disclosure of factual information about environmental risks could itself create incentives for corporations and other organizations to reduce them. From 1988 to 1998, according to the TRI reports, facilities reduced releases of listed chemicals by 45 percent, from 3.40 billion pounds to 1.86 billion pounds, for chemicals reported in all years.[4] In 1997 the federal Environmental Protection Agency (EPA), which administered the law, declared that this simple disclosure system was considered "one of the most effective environmental programs ever legislated by Congress and administered by EPA."[5] In part because of this perceived success, structured disclosure of information began to be viewed not only as a right but also as an instrument of risk regulation.

The TRI law and implementing regulations required that manufacturers in certain Standard Industrial Classification (SIC) codes report releases into the environment of a government-provided list of chemicals each calendar year, facility by facility and chemical by chemical. Companies reported data to the administrator of the federal EPA. The EPA, in turn, created a national data base and issued an annual report that summarized the data. The law mandated use of a standardized form for reporting and required that data be made available to the public electronically as well as on paper and in other formats. Facilities that employed fewer than ten full-time employees, manufactured or processed 25,000 pounds or less, or otherwise used 10,000 pounds or less of listed chemicals were not required to report. Federal facilities also were not required to report. The law provided, however, that the administrator of EPA could add or remove chemicals from the initial list and broaden or narrow the categories of entities required to report. It also gave the administrator authority to change the prescribed thresholds for reporting for any chemical.[6]

The TRI complemented but did not replace established regulation of toxic chemicals. In 1976 Congress had given federal regulators broad authority to gather information about new and existing toxic chemicals, require testing, assess risks, and ban or limit their use as needed.[7] In addition more than a dozen national laws regulated the use, transportation, and cleanup of pesticides and other toxic chemicals in specific circumstances.[8]

The TRI differed in several fundamental respects from existing regulations. The mechanism by which disclosure influenced corporate decisions was public pressure, not sanctions. Conventional regulation aimed to create incentives for improved environmental performance by establishing minimum standards that were enforced by government action. The TRI influenced such performance by altering markets (i.e., altering choices by investors, employees or customers) or political action (i.e., altering choices by legislators, regulators or voters). The TRI also differed from other environmental regulation by adopting a multi-media approach. Before TRI, each of EPA's regulatory programs concentrated on a specific type of release. The Clean Water Act, for example, aimed only to reduce water pollution. The Resource Conservation and Recovery Act aimed to improve land disposal practices. Other laws focused on pesticides or common air pollutants. The TRI was the first law to require companies to add up toxic releases to air, water, and land.

The hallmark of this informational approach to reducing pollution was its flexibility. While the formats and intervals for reporting were centralized, decision making about reducing releases remained entirely decentralized. Each company could decide for each chemical and facility both *whether* and *how* to reduce releases.

Indisputably the new disclosure requirement inspired executives of some large companies to promise huge voluntary cuts in toxic pollution. Some took anticipatory action, announcing reduction goals many months before reports were made public. A prominent example was Monsanto. The day before manufacturers sent their first numbers to Washington in 1988 and nearly a year before the public would gain access to them, Richard J. Mahoney, then chief executive officer of the Monsanto Corporation, announced in a memorandum to his managers that Monsanto would eliminate 90 percent of its toxic air pollution in less than five years. "The public has spoken," Mahoney told *Newsweek* magazine, "and it's unmistakable they will no longer tolerate toxic emissions. Might as well get on with it."[9] In 1992 Mahoney announced that the company had met its goal.

Despite such bold commitments by some industry leaders, general implementation of the TRI was gradual and was characterized by contentious issues. In TRI's first two years as many as a third of firms failed to report, often because of confusion about the requirement's scope. In some states initial reporting rates were as low as 20 percent. Political conflicts and technical issues surrounding the law's requirement of electronic disclosure also slowed public access to industry reports. Manufacturers remained concerned that chemical-by-chemical disclosure might reveal trade secrets and that government-reported information might create an inaccurate or misleading picture of environmental performance, especially when combined with other sources of data. Right-to-know and environmental groups urged the government to create a system that was more user-friendly and gave the public a practical understanding of risks.[10]

For the first several years the EPA took the position that its job was to compile the data, not to suggest how pounds of releases might relate to health risks or offer other interpretations. Both industry and citizen groups had argued from the beginning that data should be placed in perspective for members of the public. Particularly as the Internet gained wider use, a variety of groups set up their own systems of tracking toxic pollution. Some of these systems combined the TRI data with other sources of information and interpretive material in an effort to provide the public with a more complete picture of corporate performance and/or of consequences of releases. In the late 1990s EPA itself also initiated a number of controversial projects to merge data from TRI and other sources for the purpose of creating more complete company profiles and estimations of risk. For example, the agency combined TRI data with other enforcement data to characterize pollution control efforts by facilities, companies, and industry sectors and created new methodologies to assess cumulative risks.

The perceived success of the TRI in reducing toxic releases created political pressures to expand its reach. During the 1990s the Clinton administration issued new rules that doubled the number of chemicals for which disclosure was required, called on several new industry sectors to report, and lowered the thresholds for reporting of some particularly harmful chemicals. Federal facilities also were required to report their releases. By the end of the decade the TRI had become a platform for a wide variety of government and private programs that aimed to monitor and reduce toxic pollution. Among them were government programs to recognize companies that were leaders in reducing toxic pollution and determine the environmental health of ecosystems, industry programs to encourage reduction of toxic wastes, and efforts by environmental and right-to-know groups to inform the public about toxic risks in their neighborhoods.[11]

Learning from Experience

The TRI is often cited as an example of the successful use of an informational strategy to improve environmental protection. Beneath the frequent assertion that the requirement contributed to a sharp decline in toxic releases nationally, however, lie more complex industry-specific, media-specific, and state-specific trends. Also beneath this seemingly simple mandate for disclosure lie architectural strengths and weaknesses that influence its usefulness. Both complex trends in releases of toxic chemicals and specific incentives created by the TRI's architecture provide insights into the effectiveness of this strategy and suggest avenues for future research.

The primary claim made for the TRI is that it has contributed to cutting releases of listed chemicals by nearly half in ten years (see figure 12.1). However, behind this overall decrease lies a complex story of increases in toxic waste, particularly dramatic decreases in releases of carcinogens, different trends in releases to air, water, and land, a variety of industry-specific patterns, and changes in geographical areas where toxic releases are concentrated.

As a starting point in understanding the broad decrease in toxic releases, it is important to summarize the relationship between toxic waste and releases. The TRI defines as "releases" only about 10 percent of total toxic waste reported. (Table 12.1 shows the categories of waste and releases as defined under TRI.) The term release was created by Congress to describe that portion of toxic waste that is discharged directly

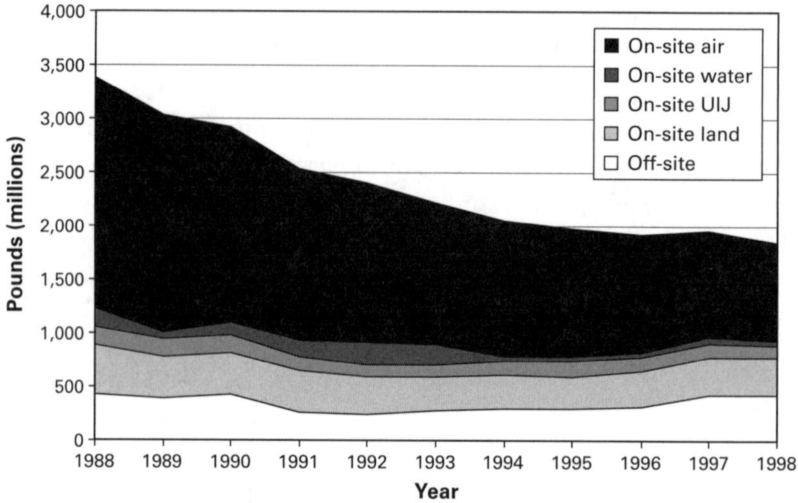

Percentage change 1988–1998

Total releases	−45%
Air	−58%
Water	−73%
UIJ	−29%
Land	−24%
Off-site	+0.2%

Figure 12.1
TRI on-site and off-site releases, 1988 to 1998. Not included are the delisted chemicals, chemicals added in 1990, 1991, 1994, and 1995, aluminum oxide, ammonia, hydrochloric acid, and sulfuric acid. The on-site releases from Section 5 of TRI Form R, off-site releases from Section 6 (transfers off-site to disposal) of TRI Form R, and off-site releases include metals and metal compounds transferred off site for solidification/stabilization and for wastewater treatment, including POTWs. Forms for industries newly reporting to TRI for 1998 are also not included. (Source: From TRI Public Data Release database, May 2000)

Table 12.1
TRI categories of waste and releases

Production-related waste
Recycled on site and off site
Energy recovery on site and off site
Treated on site and off site
Releases on site and off site
Air emissions
Surface water discharges
Underground injection
On site land disposal
Off site transfers to disposal (off site land disposal)

into the environment. The largest portions of toxic waste are recycled (about 45 percent) or treated (about 29 percent) or burned for energy recovery (about 16 percent) and are therefore not counted as releases. The 1998 TRI reported that total toxic waste of listed chemicals amounted to 24.1 billion pounds while total releases amounted to 2.4 billion pounds.[12]

During the 1990s, toxic waste reported under the TRI increased while releases decreased. From 1991 to 1998, total production-related waste reported by the manufacturing sector increased by 6 percent.[13] In the early years and again from 1994 to 1996 it decreased. However, since 1996 reports of toxic waste have increased again, growing from 1996 to 1998 by 8 percent. It is important to note that increases occurred in the context of a rapidly growing economy. Over the same time period, the manufacturing production index, an indicator of how US production levels in the manufacturing sector have changed, has risen by 40 percent. Thus toxic waste generated by manufacturing processes has increased at a much slower rate than production (see figure 12.2).

Facilities can achieve lower rates of waste generation, even in the face of rising production, in several ways. Processes may be more efficient in using the chemicals at higher rates of production. New uses for chemicals in the wastes may be developed. Or they can implement source reduction projects.

The federal Pollution Prevention Act of 1990 called for an environmental hierarchy that makes reducing waste at the source the preferred

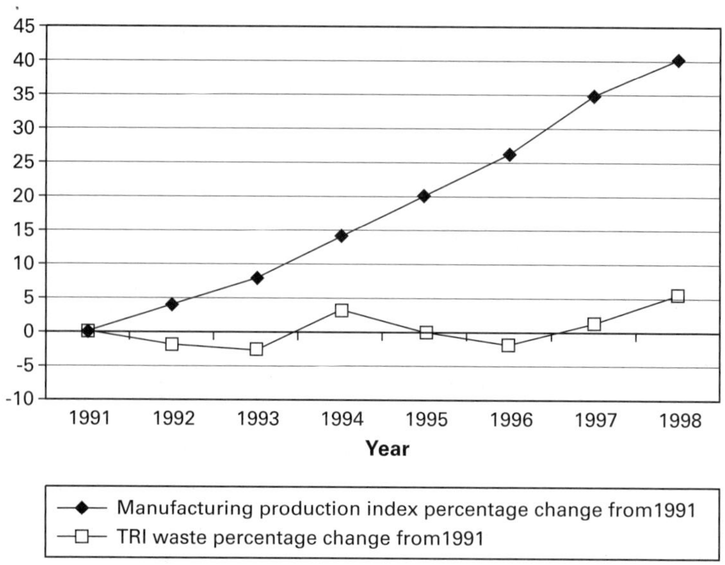

Figure 12.2
TRI production-related waste, 1991 to 1998, compared to manufacturing production index, cumulative change. Base 1991 = 100. Not included are delisted chemicals, chemical added in 1994 and 1995, ammonia, hydrochloric acid, and sulfuric acid. (Sources: Data from Section 8 of TRI Form R, TRI Public Data Release database, April 1999; production index from 1999 Statistical Abstract of the United States, No. 1240)

method of reducing environmental releases of these toxic chemicals. Source reduction includes substitution of raw materials, changes in maintenance procedures, product reformulation, and process efficiency changes. Only about 28 percent of all facilities reported that they undertook one or more source reduction projects. Source reduction is important not only because it prevents pollution but also because it reduces potential worker and community exposure and treatment, and disposal and liability costs. After source reduction, the preferred means of waste management is recycling.

While only about a quarter of facilities undertook source-reduction measures, recycling has increased by 27 percent (see figure 12.3). Trends in recycling or re-use of toxic wastes are influenced by economic as well as technological factors, including costs of raw materials, selling prices for by-products, and costs of off-site disposal versus costs of recycling.

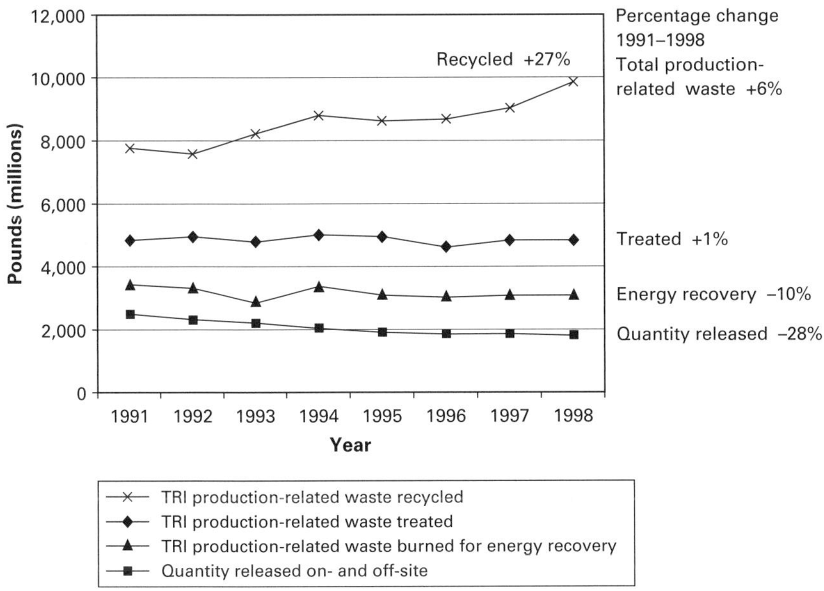

Figure 12.3
TRI production-related waste, 1991 to 1998. Forms from industries newly reporting to TRI for 1998 are not included. Not included are chemicals deleted or added since 1991, ammonia, hydrochloric acid, and sulfuric acid. (Source: Data from Section 8 of TRI Form R, TRI Public Data Release database, May 2000)

As these prices rise or fall for particular industry sectors, waste and releases (including off-site transfers to disposal) may decrease or increase in those sectors.

Reporting may also be influenced by the phenomenon of "paper changes." For example, a facility may one year report on-site recycling of a chemical but in another year consider that activity to be in-process recovery, which is not reportable to TRI.[14] One study of changes in TRI-reported waste found that fully half of the reductions reported were paper changes, with the majority being redefinition from on-site recycling to in-process recovery.[15] Of course, paper changes may also increase reported releases. Since reports can be based on estimates rather than measured emissions, changes in methods of estimation can increase or decrease releases. For example, the pulp and paper industry trade association published new estimating methods for 1994 that significantly increased reported air emissions of methanol from these plants, in some cases by a factor of ten.

Total TRI releases have declined dramatically, but the annual rate of decline in recent years has been much lower than in earlier years. Total releases on- and off-site declined by 45 percent in the ten years from 1988 to 1998, the figure often cited as an indication of TRI's success. Almost half of this decrease, however, was registered in the first three years. From 1988 to 1990, the annual decrease in total releases averaged 9 percent while the percentage decrease in the most recent three years, from 1996 to 1998, averaged 2 percent. Most of the decreases are found in reductions in on-site air emissions, which constitute over half of all reported releases. Some of these decreases have been offset, in recent years, by increases in disposal in landfills, both on and off site.

One possible explanation is that in the early years, as manufacturers saw the full picture of their releases for the first time, they were able to "pick the low-hanging fruit," finding ways to reduce waste through relatively easy techniques such as better maintenance, training in chemicals handling, and tweaking processes to be more efficient. In later years, industry had to use more costly and time-consuming methods, developing new processes and products to find additional ways to reduce remaining waste. Also some companies could have initially reported releases by using conservative estimates derived from published estimation factors based on the experiences of similar types of plants and processes. Since TRI was an "information" requirement, companies would be penalized for not reporting or willfully reporting incorrectly. So companies had the incentive to provide conservative estimates, if estimates were all they had. When they switched to actual measurements, it created an appearance of a sudden decrease in releases.

On-site releases to air, water, and underground injection have followed different patterns. Air releases have declined dramatically, from 2.2 million pounds in 1988 to under one million pounds in 1998, for a group of core chemicals reported in all years. Decreases have been reported each year and at a fairly steady pace, averaging 8 percent per year.

Changes in surface water discharges have varied sometimes quite significantly over the years, showing a 72 percent increase from 1988 to 1989 and a 79 percent decrease from 1993 to 1994. Relatively few facilities account for the majority of such releases. In 1998, 15 TRI facilities reported over half of all surface water discharges. Several facilities that dominate surface water discharges can have a large effect on these changes. For example, two facilities belonging to the same company accounted for the large decrease from 1993 to 1994 when they implemented source reduction measures by installing covers for their inactive gypsum stacks to reduce the amount of phosphoric acid run-off produced during rainstorms.[16]

Underground injection, like water discharges, has shown wide variation from year to year, rising 25 percent from 1994 to 1995 and declining by 13 percent from 1997 to 1998. In 1998 fewer than 0.5 percent of TRI facilities reported any on-site underground injection. Such disposal is permitted in only a few states where geographic configurations allow for such wells. When it is used, however, releases tend to be large, averaging almost 2.5 million pounds per facility. Variations in on-site surface water discharges and underground injection do not affect overall trends in total releases because they account for less than 10 percent of total releases.

Land disposal, on- and off-site, has been increasing in recent years. On-site land releases and off-site disposal (which is primarily disposal in off-site landfills) both experienced average decreases in the first three years, from 1988 to 1990, and experienced average increases in the last three years. Off-site transfers to disposal showed both the largest early decreases (average annual decrease of 13 percent from 1988 to 1990) and largest later increases (average annual increase of 15 percent from 1996 to 1998) of any type of release.

In 1998 metals and their compounds constituted over 75 percent of all land disposal on and off site. TRI requires reports on 18 metals and their compounds. Metals are particularly important because they do not degrade and are not destroyed by treatment. Some metals may be converted to less toxic forms. For example, hexavalent chromium (a known carcinogen) may be converted to the less toxic trivalent form. Or, some forms of metal may be treated so that they are less likely to be transported through soils, but such treatment does not destroy the metal. Metals can also present particular health risks; some, such as arsenic and inorganic arsenic compounds, beryllium, cadmium, hexavalent chromium compounds, cobalt, lead and inorganic lead compounds, and nickel, are known or suspected carcinogens. These and others can cause developmental defects in humans or adverse effects on aquatic and terrestrial organisms. They can bioaccumulate in fish and reach humans through the food chain.

Releases on and off site of metals decreased from 1988 to 1992 and then increased from 1992 to 1997, with the largest increases from 1995 to 1996 (14 percent) and from 1996 to 1997 (21 percent). Over 90 percent of all releases of metals are either on-site releases to land or off-site transfers to disposal with the majority being disposal in landfills, both on and off site. "Land disposal" of metals, the total of on-site land releases and off-site transfers to disposal, increased from 1995 to 1998, by 186 million pounds. On the other hand, there was a decrease of similar magnitude (160 million pounds) in off-site transfers of metals to recycling, even though prior to 1995 there had been substantial increases in off-site recycling of metals each year. Thus the trend for "land disposal" of metals is the reverse of the trend for off-site recycling of metals. In an effort to understand these shifts, the federal Environmental Protection Agency contacted some of the facilities with the largest off-site transfers of metals to disposal. They learned that a major recycler of metals had raised prices in the 1995 to 1997 time frame and the TRI facilities had switched to cheaper providers of disposal in landfills. The recycler subsequently lowered prices, and reduced levels of off-site disposal of metals were expected as the contracts for disposal expired and the facilities returned to recycling.

This example demonstrates another important point: A small fraction of the 20,000 facilities included in the TRI often reports a majority of the releases and therefore determines the trend. Annual reports that compare the data equivalent to TRI as reported by facilities in Canada regularly provide data for the 50 US facilities with the largest reported amounts.[17] For example, in 1997 just 50 facilities out of almost 20,000 facilities reported 37 percent of all on-site TRI releases, 50 facilities reported 40 percent of all off-site TRI transfers, and 50 facilities reported 27 percent of TRI total releases and transfers in that analysis.

This report also looked at TRI facilities reporting less than 100,000 kilograms of releases and transfers (including transfers to treatment as well as disposal) and compared them to TRI facilities reporting more than 100,000 kilograms in 1995. The change from 1995 to 1997 for these two groups of facilities differed considerably. The group of facilities with the relatively smaller amounts reported a 4 percent increase in total releases and transfers from 1995 to 1997, while the group of facilities with larger amounts reported a 7 percent decrease over the two years.[18]

Another report, using information from the New Jersey inventory, found similar results. New Jersey facilities using smaller amounts of the chemicals (less than 100,000 pounds per year) reported greater increases in releases and transfers as well as waste than the fewer facilities using the largest amounts. The New Jersey system also collects data on the amount of waste not generated due to source reduction projects that were implemented by the facilities as well as the amount chemical used. This study found that the smaller facilities reduced proportionally more waste through source reduction efforts than did the largest facilities. The waste reduced was equivalent to 3 percent of the waste reported for the year by the smaller facilities compared to 0.6 percent for the largest facilities. The smaller facilities did, however, have a longer way to go, since they reported on average that 35 percent of the amount of chemical used was generated as waste. The larger facilities reported from 10 to 25 percent of chemical use as waste generated.[19]

Trends in different industries also vary. All except one industry sector (the food and beverage industry) reported overall decreases. Significantly in 1988 the chemical manufacturing sector reported the largest amounts of total releases, but by 1998 it was ranked second to the primary metals sector. Chemical manufacturers reduced total releases by 57 percent from 1988 to 1998, including a 12 percent decline in the most recent year. On the other hand, the primary metals sector reported only a 2 percent decrease from 1988 to 1998, with a 3 percent increase from 1997 to 1998.

Releases are also concentrated in relatively few states. Texas, Louisiana, and Ohio were the states reporting the largest releases in each year from 1988 to 1998. In 1988 and 1998 these three states accounted for 23 percent of total releases in the United States. Texas had almost 1,250 TRI facilities and Louisiana had 315 TRI facilities in 1998. Ohio with almost 1,580 facilities had the highest number of TRI facilities reporting in 1998. In that year Louisiana reported, on average, over 550,000 pounds per facility while Texas reported over 200,000 pounds per facility, and Ohio had over 90,000 pounds per facility in 1998. The average nationwide was over 110,000 pounds per facility. Thus Louisiana and Texas appear among the states with the largest releases because facilities in these states report larger than average releases, while Ohio is ranked third for releases because it has an unusually large manufacturing base.

Again, a few facilities can account for the high rank of states and counties where they are located. The U.S. counties with the largest releases from manufacturing facilities in 1998 were Tooele, Utah, Ascension, Louisiana, and Harris, Texas, each reporting about 5 percent of all releases. A total of 4 facilities reported in Tooele, Utah, and 17 in Ascension, Louisiana, while the releases in Harris, Texas came from 275 facilities.[20] An encouraging trend is that releases of some of the most toxic chemicals have declined somewhat more rapidly than general releases in recent years. Among the more than 600 chemicals currently on the TRI list, EPA has identified about 165 known or suspected carcinogens that must be reported.[21] The designated carcinogens on the TRI list are subject to a lower reporting threshold when present in mixtures. From 1995 to 1998 total releases of this group of carcinogens decreased by 8 percent compared to 5 percent for all TRI chemicals. However, individual carcinogens showed significant variation. Styrene, dichloromethane, and formaldehyde, the carcinogens with the largest releases in 1998, accounted for almost 50 percent of all TRI carcinogen releases in 1998. While releases of dichloromethane decreased by 31 percent since 1995, both styrene and formaldehyde releases increased, by 26 percent and 13 percent, respectively (see figure 12.4).

To summarize, these complex trends provide a varied profile of a decade of trends in generation of toxic waste and releases of toxic chemicals required to be reported under the TRI. The data indicate that companies are decreasing toxic releases at a considerably slower rate in recent years than they did when TRI was first implemented and the most recent data suggest that only about a quarter of facilities cut releases by reducing waste at the source, the means of reducing pollution that is preferred under current national policy. On the other hand, recycling has increased substantially and releases of carcinogens have declined at a somewhat faster rate than overall releases.

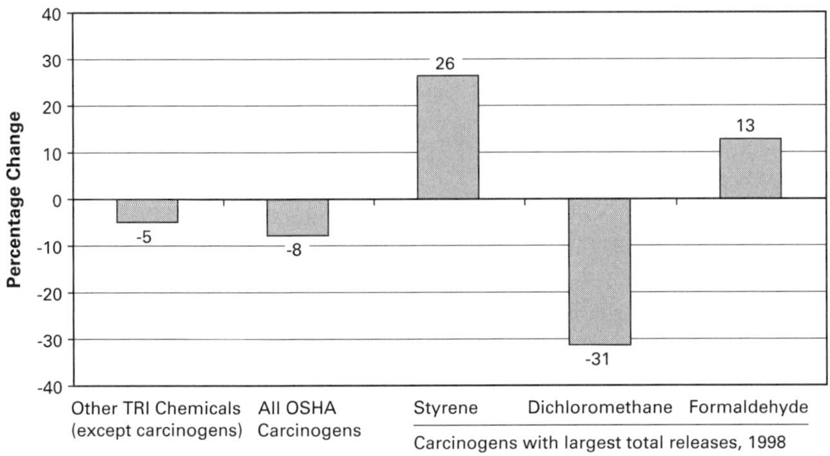

Figure 12.4
TRI on-site and off-site releases of carcinogens, percent change 1995 to 1998. Forms from industries newly reporting to TRI for 1998 are not included. Not included are chemicals deleted or added since 1991, ammonia, hydrochloric acid, and sulfuric acid. (Source: Data from Section 8 of TRI Form R, TRI Public Data Release database, May 2000)

Increases and decreases in some types of releases have varied significantly from year to year. While on-site air emissions have steadily declined year by year since 1988, land disposal decreased sharply in early years and increased sharply in recent years. Changes in surface water discharges and in underground injection have been erratic. The data suggest that one reason for such erratic changes is the small number of facilities that are sometimes responsible for large quantities of releases. Another reason is that company choices between such options as recycling and landfill disposal may be extremely sensitive to changing costs. Paper changes, such as re-definition of terms, can also result in large differences in reported releases from year to year. Trends in toxic releases have also varied by industry and tend to be concentrated in certain states.

TRI's Architecture Creates Strengths and Limits

In an open society the proposition that more public information is always better is often taken to be self-evident. But disclosure requirements

Table 12.2
Characteristics of mandatory disclosure systems

A public purpose	*Why* is disclosure required
A specific target	*Who* is required to disclose
A defined scope	*What* is required to be disclosed
An articulated structure	*How and when* is information communicated
An intended audience	*To whom* is information communicated

are not as simple as they may appear and often conflict with other public priorities. In creating this new system of public access to private-sector information in 1986, Congress struggled to resolve contentious issues and in so doing constructed the TRI with a specific architecture that in turn influenced incentives for target companies. The law and regulations specified particular purposes, targets, scope, structure, audience and enforcement, architectural elements that are common to most systems of mandatory disclosure (see table 12.2).

The character of each of these elements was created by political compromise. Compromise was needed because the perceived public interest in disclosure clashed with other enduring values, especially protecting trade secrets and minimizing regulatory burdens, and with powerful interests. The result was a disclosure system with particular strengths and weaknesses with regard to its potential for improving environmental protection.

Design Strengths

The TRI's particular strengths lay in its unusual structure. Reporting to the general public in standard formats, at regular intervals, by facility, and by chemical for all types of environmental releases made it possible to compare companies and track changes over time. Disclosure was structured to draw attention to both national and local levels of pollution and to limit claims of confidentiality. Because the law required executives to add up the numbers and to sign off on annual reports to Washington, managers were forced to focus on national levels of pollution from all their facilities, often for the first time. Because it also provided access to information about toxic releases of all types by chemicals and by factory, some managers responded locally as well. A number of companies, for example, began for the first time to communicate with people living and working in surrounding communities, in an effort to minimize public reaction. And because the magnitude of national releases was new to regulators in Washington and the magnitude of local releases was new to members of Congress representing industrial districts, the TRI created incentives for political action.

The initial totals were much larger than previous estimates. Congressman Henry A. Waxman recalled after the first reports that when he estimated in 1985 that 80 million pounds of toxic chemicals were released into the air each year "industry went haywire. They denounced the figure as environmental paranoia." When toxic releases reported in the TRI were added up, they initially showed releases of air toxins totaling 2.7 billion pounds.[22] These large numbers quickly added momentum to the debate about the need to revise national and state environmental laws to improve control of toxic pollution. EPA's past programs had focused mainly on a limited number of air and water pollutants and on improving landfills. The Clean Air Act Amendments of 1990, which emphasized the importance of reducing toxic emissions, were one example of a political action that derived strength from those early TRI revelations.[23]

The TRI produced surprising totals because in the past most environmental information collected by government had related directly to single-media compliance issues. That meant that seemingly simple questions about toxic pollution remained unanswered. No one knew what quantities of various chemicals were released into rivers or lakes, emitted from factory smokestacks, and buried or burned on land. No one knew which facilities released which chemicals. Even within companies, chief executives often did not have a complete picture of the toxic pollution their factories were creating. Information about costs of waste disposal tended to be subsumed in overhead and information about types of waste, when it was collected at all, often remained in the files of factory managers. Even in the mid-1980s the National Research Council concluded that the United States still lacked any coherent national picture of the movement of key toxic chemicals.[24]

The TRI also demonstrated the importance of viewing disclosure requirements as evolutionary. From its humble beginnings, the TRI grew in breadth and sophistication. Perceptions of its initial success as a regulatory measure prompted executive branch actions to expand its coverage, to improve estimating techniques, and to refine reporting formats. In its first formative decade the TRI benefited from a continuing interest on the part of government officials, citizen groups, and industry representatives in improving the quality of the data.

Design Limitations

At the same time some architectural features limited the TRI's usefulness as a tool for improving environmental protection. First, because its targets included only manufacturers, and a limited category of those, it could not create incentives for reduction of many of the nation's largest sources to toxic pollution, including mobile sources (cars, trucks, and businesses) and small businesses.[25]

Second, its scope was limited to a partial list of toxic chemicals. Initially its framers chose a politically expedient shortcut. They combined lists of toxic chemicals assembled by state officials in New Jersey and Maryland, both for state-specific purposes. While this list was later expanded significantly, for most of the TRI's first decade manufacturers had an incentive to substitute off-list toxic chemicals for listed chemicals, regardless of their relative toxicity.

Third, as has been noted, the TRI's structure did not require reporting of chemical *use*. That meant it created no incentives to reduce waste at the source, an emerging priority in national policy. Efforts to include chemical use in reporting were rejected repeatedly as industry groups argued forcefully for the need to minimize reporting and protect trade secrets.

Fourth, TRI's design placed important limitations on the data's timeliness and accuracy. Releases were reported to the public more than a year after they took place. While that pace was not unusual for government reports compiled from complex private-sector data, it limited the information's usefulness to community residents and businesses interested in avoiding exposure to particular chemicals, for example. In order to limit its cost to industry, the law also allowed reporting to be based on estimates rather than actual monitoring and permitted companies to choose from a variety of estimating techniques, thereby limiting the accuracy of the data and complicating year-to-year and company-to-company comparisons.

The requirement's structure also created special pressures for companies to come up with quick fixes. The annual reports on toxic releases inevitably produced national and local lists of top polluters. Efforts by government officials, environmental groups, and some journalists to explain that companies releasing the most pounds of listed chemicals were not necessarily those that created most serious health risks got lost in the general enthusiasm for ranking. Corporate executives who decided to take action in response to the TRI had one over-riding goal: to get off the list by the next time it was published. That meant that changes in cleaning procedures and maintenance and relatively simple substitutions of chemicals were appealing because they provided quick reductions, especially in the early years. More costly and more time-consuming modifications in products or processes were less likely to provide the quick relief that executives sought from media attention and other forms of public pressure. (Of course, staying off the "top polluter" list in later years could require more difficult modifications.)

A complicating factor was that the requirement's structure lacked a metric that was calibrated to risk. Toxic releases were reported only in total pounds, without adjustments for human exposure to chemicals or for their relative toxicity. Even if reporting had been calibrated to risks, estimates would have had limitations. The TRI was instituted at a time when little was actually known of the relative risk of most of these chemicals. The absence of any effort to take account of risks, however, meant that companies had no incentive to concentrate on reducing emissions that created the most serious threats to human health or the environment. It also meant that members of the public had no basis for taking action that was based on degree of risk. In the late 1990s the EPA, industry, and environmental groups initiated an ambitious program to expedite toxicity testing for a large number of chemicals in wide use, a costly and time-consuming effort.[26]

To recapitulate, Congress created the TRI with a set of architectural elements that influenced its potential effectiveness as an information strategy to improve environmental protection. The requirement's novel combination of structured disclosure to a broad audience at regular intervals of factual information about the environmental performance of identified companies and facilities created unusual incentives for some companies to reduce toxic releases. At the same time the requirement's narrow targets and scope limited its effectiveness as a means of improving environmental protection and structural characteristics such as reporting in a metric that did not include relative toxicity and exposure added further limitations.

Examining the complicated trends in toxic releases in the United States during the last decade and the strengths and weaknesses of the architecture of required disclosure is but one step toward improving understanding of this information strategy. Many important steps remain. To what degree reductions in toxic releases have been caused by government-required disclosure as opposed to other forces is a particularly difficult question. In addition to the economic forces at work, some of which are discussed above, the TRI was one among a multitude of public actions directed toward reducing toxic pollution in the late 1980s and 1990s. Increasing numbers of federal and state laws, international agreements such as the Montreal Protocol (which committed signatories to reducing use of solvents and other chemicals linked to ozone depletion) and actions by government entities to reduce their own use of such toxins could have contributed to industry incentives. Independent pressures on companies from industry trade associations and environmental groups could also have been significant factors. The picture that emerges from this analysis is one that features public disclosure as a part of a complex web of changing political and economic forces that influences corporate decisions about releases of toxic chemicals.

In the absence of more systematic direct evidence, most efforts to evaluate the TRI have relied instead on econometric studies, surveys, and other empirical research to assess the importance of disclosure about toxic releases to investors, employees, community residents, and other groups.[27] However, the answers to even basic questions remain elusive. For example, it is not yet possible to state with certainty that the TRI moves companies in the direction of reducing toxic pollution in a cost-effective manner. The requirement's emphasis on decreasing total pounds of releases as quickly as possible (rather than reducing risks) might skew incentives away from cost-effective reduction of toxic risks. As more complete data on the relative toxicity of these chemicals because available, research into this and other basic questions concerning risk will become feasible.

This analysis points to the need for further research. It is important to identify firms that are reducing toxic pollution in a cost-effective manner and assess the relative importance of the TRI and other factors in encouraging such reductions. The complexity of trends in toxic releases suggests that findings may vary by types of chemicals used, types of industrial processes, costs of raw materials and disposal, and the state and local regulatory climate as well as by the kind and degree of public pressure generated by the TRI. Important areas for research also include representative surveys of facilities to determine how and why they achieved reductions; assessing trends in toxic releases among subgroups of facilities, such as those with moderate levels of releases and those in different states or different industry sectors; and determining how estimation procedures have changed over time.

Conclusion

The growing power of computers and information technology may offer particular hope for the future role of the TRI in creating incentives to improve environmental protection. Such technology creates the potential to bring the nuances of trends in toxic pollution to the attention of the interested public. It also creates the potential for government agencies and private groups to combine data from the TRI with data from other sources and with indicators of risk as those become more sophisticated. Progress in improving the accuracy, timeliness, and completeness of data should not be taken for granted, however. Disclosure systems, like other forms of regulation, are difficult to reform. Long-term improvements may depend on such factors as the transparency of methodologies and the continuing presence of influential constituencies with a strong interest in improving the data. However, the TRI evolves. Its first decade has produced a number of positive lessons for information based regulation. In that respect it has already provided considerable public benefits by contributing to a national and international learning process about the emerging role of informational strategies in improving environmental protection.

Notes

Authors are listed alphabetically and contributed equally to this chapter. Mary Graham co-directs the Transparency Policy Project at the Kennedy School of Government and is associated with the school's Taubman Center for State and Local Government and Environment and Natural Resources Program. Trained as a lawyer, she is the author of *Democracy by Disclosure* (Governance Institute/Brookings Press 2002) and *The Morning After Earth Day* (Governance Institute/Brookings Press 1999) and numerous articles about the use of disclosure strategies to reduce health, safety and environmental risks. Catherine Miller is a senior researcher with the Hampshire Research Institute. She has worked with the TRI since before its passage and is the author of many reports analyzing the TRI data as well as other toxic pollutant release inventories worldwide.

1. TRI Public Data Release database (US EPA, April 2001).

2. 42 USC 11001–11050 (1994; supp. III 1997).

3. Some of the material in this chapter is drawn from Graham (2000, 2001a, 2002). For recent analysis of the Toxics Release Inventory, see also Karkkainen (2001), Fung and O'Rourke (2000), and Pedersen (2001). For a review of the TRI's implementation, see Greenwood and Sachdev (1999).

4. *1998 Toxics Release Inventory* (EPA 2000). Amounts in this chapter refer to data in this report for the original industry sectors and for chemicals that have been reportable since 1988, unless otherwise stated. In 1995, the TRI list of chemicals was almost doubled, but these new chemicals are not included in comparisons with 1988 or 1991. In 1998, several new industry sectors were added to the TRI reporting requirements, including metal and coal mines, electric generating facilities, and hazardous waste treatment and solvent recovery facilities, chemical wholesale distributors and petroleum terminal and bulk storage stations, but also are not included in data used in this chapter.

5. *1997 Toxics Release Inventory* (EPA 1999), pp. 1–7.

6. 42 USC 11023 (1994; supp. III 1997).

7. Toxic Substances Control Act, 15 USC S. 2601 et seq. (1976).

8. Davies and Mazurek (1998), pp. 11–26.

9. Quote appears in Newsweek magazine, July 24, 1989, p. 28; see also Graham (2000).

10. For early implementation of the TRI, see General Accounting Office (1991).

11. Executive Order 12,856 (August 3, 1993) required federal facilities to report toxic releases. The expansion of TRI to new industrial sectors is described in EPA (1999), pp. 1–3.

12. The above is true for all chemicals reported in 1998 by the manufacturing industry sectors that have reported to TRI since its beginning. Starting with the 1998 reporting year, several new industry sectors were also required to report, but they are not included in this analysis. The amount quoted above of 1.86 bil-

lion pounds represents releases of the chemicals reportable since 1988. In 1995, the TRI list of chemicals was almost doubled but these new chemicals are not included in comparisons with 1988 or 1991.

13. Comparisons of waste are made from the baseline year 1991 since that is the first year TRI required reporting on the elements of waste (in Section 8 of the TRI Form R).

14. In-process recovery is considered by the Pollution Prevention Act to be recycling that is integral to and necessary for the production of a product. However, TRI reporting instructions do not define in-process recovery so facilities are free to define their activities as they choose.

15. Natan and Miller (1998).

16. EPA (1996), p. 200. Phosphoric acid was deleted from the TRI list of chemicals for the 1999 reporting year.

17. The *Taking Stock North American Pollutant Releases and Transfers* series, published in 1994 to 1997 by the Commission for Environmental Cooperation, Montreal, Canada, selects the subset of TRI chemicals that are also reported under to Canadian National Pollutant Release Inventory. About 165 chemicals represent over 80 percent of the TRI reported amounts.

18. INFORM Inc., *Toxics Watch 1995*, pp. 457–59.

19. Data from TRI Explorer (www.epa.gov/triexplorer).

20. These chemicals are listed in at least one of three sources: the National Toxicology Program's "Annual Report on Carcinogens," the International Agency for Research on Cancer Monographs, or the Occupational Safety and Health Administration's list of Toxic and Hazardous Substances (29 CFR 1910, subpart Z).

21. Newsweek, July 24, 1989, p. 28. Initial totals included a number of chemicals that were later de-listed.

22. Graham (2001b).

23. See Shapiro (1993), Davies and Mazurek (1998), and National Academy Press (1984).

24. In 1998 EPA reported that 41 percent of air toxics derived from "mobile sources," 35 percent derived from small businesses and other diffuse sources, and 24 percent derived from manufacturers and on major sources (EPA 1998).

25. In April 1998 EPA announced a cooperative program with industry and environmental groups to collect more complete toxicity information on high production volume (HPV) chemicals. The program uses six internationally recognized testing protocols that together provide a basic picture of the toxicity of the chemical. Only 55 percent of TRI chemicals had been so tested. The primary objective of the program is to make the information available to the public, especially through the Internet. An example of a company using toxicity weighting to prioritize pollution reductions and demonstrate environmental performance is the ICI Group of international companies that calculates the "environmental

burden" of its air and water emissions based on such factors as the chemicals ozone depletion potential or potential to form acid rain.

26. See, for example, Hamilton (1999), Konar and Cohen (1997), and Arora and Cason (1996).

References

Arora, S., and T. N. Cason. 1996. Do Community Characteristics Determine Environmental Outcomes? *Southern Economics Journal* 65: 691.

Commission for Environmental Cooperation. 1994–97. *Taking Stock: North American Pollutant Releases and Transfers*. Montréal: CEC.

Davies, J. C., and J. Mazurek. 1998. *Pollution Control in the United States*. Washington, DC: Resources for the Future.

EPA. 1996. *1994 Toxics Release Inventory Public Data Release*. Washington, DC: EPA, p. 200.

EPA. 1998. *Taking Toxics Out of the Air*. Washington, DC: EPA.

EPA. 1999. *1997 Toxics Release Inventory*. Washington, DC: EPA.

EPA. 2000. *1998 Toxics Release Inventory*. Washington, DC: EPA.

Fung, A., and D. O'Rourke. 2000. Reinventing Environmental Regulation from the Grassroots Up, *Environmental Management* 25: 115.

General Accounting Office. 1991. *Toxic Chemicals: EPA's Toxic Release Inventory Is Useful but Can Be Improved*. Washington, DC: GAO.

Graham, M. 2000. Regulation by Shaming. *The Atlantic Monthly*.

Graham, M. 2001a. Information as risk regulation. Occasional paper. Innovations in American Government Program, John F. Kennedy School of Government. Harvard University.

Graham, M. 2001b. Mandatory disclosure as risk regulation. Occasional paper. Innovations in American Government Program, John F. Kennedy School of Government. Harvard University.

Graham, M. 2002. *Democracy by Disclosure*. Washington, DC: Brookings Press.

Greenwood, M. A., and A. K. Sachdev. 1999. *A Regulatory History of the Emergency Planning and Community Right to Know Act of 1986: Toxic Release Inventory*. Washington, DC: Chemical Manufacturers Association.

Hamilton, J. T. 1999. Exercising property rights to pollute: Do cancer risks and politics affect plant emission reductions? *Journal of Risk and Uncertainty* 18: 105.

INFORM. 1995. *Toxics Watch 1995*. New York: Inform, pp. 457–59.

Karkkainen, B. C. 2001. Information as environmental regulation: TRI and performance benchmarking, precursor to a new paradigm? *Georgetown Law Journal* 89: 257.

Konar, S., and M. A. Cohen. 1997. Information as regulation: The effect of community right-to-know laws on toxic emissions. *Journal of Environmental Economics and Management* 32.

Natan, T. E., Jr., and C. G. Miller. 1998. Are toxics release inventory reductions real? *Environmental Science and Technology* (August 1): 368–74.

National Research Council. 1984. *Toxicity Testing: Strategies to Determine Needs and Priorities*. Washington, DC: National Academy Press.

Newsweek. 1989. Air pollution: It's all legal. *Newsweek Magazine* (July 24), p. 28.

Pedersen, W. F. 2001. Regulation and information disclosure: Parallel universe and beyond. *Harvard Environmental Law Review* 25.

Shapiro, M. 1993. Toxic substances policy. In P. R. Portney, ed., *Public Policies for Environmental Protection*. Washington, DC: Resources for the Future, pp. 206–37.

13

Corporate Environmental Reporting in Norway: Beyond Emission Control?

Audun Ruud

In 1998 Norway introduced environmental reporting requirements as part of its financial legislation. The issues that must be reported extend far beyond the mandatory reporting schemes defined by the Pollution Control Act of 1981 and those included in the Norwegian Pollutants Release and Transfer Register (PRTR).[1] Like the Toxic Release Inventory (TRI) of the United States, the PRTR requires managers of hazardous manufacturing plants to submit specific environmental emission data to the environmental authorities. The environmental reporting requirements pursuant to the Norwegian Accounting Act of 1998 extend the focus beyond the concerns of the TRI and the PRTR. In the fifth paragraph of section 3.3, which refers to the annual report from the Board of Directors, the Accounting Act states: "Information concerning current activities including production inputs and products that may cause a not insignificant impact on the external environment shall be provided. The actual and potential environmental impacts of particular activities shall be specified and the firm shall specify efforts initiated to eliminate or reduce negative environmental impacts."[2] Through this provision, the Norwegian Accounting Act requires every commercial firm—regardless of whether it must submit data to the PRTR—to provide information concerning production inputs, production activities, and final products that may cause "a not insignificant" impact on the external environment. This more wide-ranging legislation applies to a significantly larger group of commercial firms than were previously affected. Each Board of Directors must include this environmental information in the company's annual report presented to the financial community.

Norway is the first country to introduce such a demanding environmental-reporting requirement as part of its financial legislation.[3] Firm-specific emission data are easily available from the PRTR, but this chapter questions whether Norwegian firms and particularly their Board of Directors are responding to the 1998 requirements. More specifically, the question is about the effectiveness and validity of the financial requirement on corporate environmental reporting. To answer this question, the chapter draws on a survey conducted among 112 large firms operating in Norway.

The Logic behind the Reporting Requirements

According to the OECD (1998) extending the corporate environmental perspective beyond plant-specific issues of pollution control will enhance competitiveness by stimulating similar concerns among its suppliers, by creating close collaboration with the consumers of the firm's products, by increasing cost awareness within the firm, and by attracting the goodwill of a variety of stakeholders with an interest in more complete environmental disclosure (WBCSD 1997; OECD 1998). This reasoning loosely parallels arguments for ecological modernization, particularly as it relates to consensual negotiations, partial self-regulation, and the use of market mechanisms and instruments (Mol 1996; Berger et al. 2001). As formulated by Hajer (1996: 248), "economic growth and the resolution of ecological problems can, in principle, be reconciled." However, some scolars have criticized this theory. Connelly and Smith (1999) argue that ecological modernization justifies the status quo and Western-style industrialization, as it hinders more radical environmental positions and fails to fully exploit the radical potential of the concept of sustainable development.

Despite disagreements about how to conceptualize and implement sustainable development, many agree that the total use of production inputs, the input's function in the production process, and the total environmental and economic outputs must be specified and integrated in greater detail (Stavins and Whitehead 1997; OECD 1998). References to improvements in resource and energy efficiency in line with the objectives of industrial transformation discussed in this book are also part of the logic behind the environmental reporting requirements of the Norwegian Accounting Act of 1998. The main purpose of this Act, however, is to enable judgments about whether a commercial firm is pursuing current product-related activities in an environmentally sound manner throughout the life cycle of a product.[4] Assuming that firms are complying, its reporting practices will reveal the difficulties of, and the potential for, strengthening corporate environmental awareness. The legislation explicitly states that the report should include the firm's environmental ambitions and objectives as well as the expectations of concerned external stakeholders such as government, customers, suppliers, and NGOs. Those advocating an extended approach of the reporting requirements argued for the need to address corporate environmental concerns beyond plant-specific issues. If firms respond by extending the reporting beyond emission control, industrial transformation toward more sustainable consumption and production patterns could also be achieved. These extended public concerns were reflected in the final legal text passed by the Norwegian Parliament in 1998 and were specified in accompanying accounting standards.

The Legislation: Environmental Data in Financial Reporting

In 1997 the government proposed a revised accounting act for commercial firms. The original legislation—the Accounting Act of 1977—merely requested Boards of Directors to include a vague environmental statement. Several argued that more specific environmental disclosures ought to be presented to the financial community. This was also supported by the Confederation of Norwegian Business and Industries (NHO).[5] However, the legislative text proposal suggested that only the Boards of Directors of firms that directly pollute the external environment should disclose more information specific to combating problems at specific plant sites. This proposal did not deviate significantly from the focus of the Pollution Control Act, which required emission reporting at the plant level.

Any legal proposal, however, requires public input. Concerned stakeholders may comment on a proposal before it is debated and passed in the Norwegian legislative assembly, the Stortinget. This "hearing process" is a democratic right defined in the Norwegian constitution. More than 50 organizations participated in the hearing process for the new accounting act. However, only the Ministry of Justice and Norsk Siviløkonom Forening (NSF)[6] objected to the proposed revision of the environmental reporting requirements. The legal department of the Ministry of Justice headed by Inge Lorange Backer, professor of environmental law, argued that the proposed environmental reporting requirement was too narrow. Citing the "cradle-to-grave" principle, the Ministry of Justice suggested that corporate environmental reporting requirements ought to cover any commercial firm and reflect inputs as well as outputs of its production processes, including the complete life cycle of its products.

It is important to keep in mind that the objections were coordinated by a scholar very familiar with the Pollution Control Act and its limited relevance to corporate environmental concerns beyond emission control. The NSF further argued that if the Board of Directors was forced to disclose detailed environmental consequences of the life cycle their products, environmental risks and liabilities would more easily be discovered. A hope was expressed that this process could move environmental issues into the core of the corporate decision-making process.

The Ministry of Finance, responsible for the law proposal, did not have any particular knowledge or experience of environmental reporting. It asked the Ministry of Environment to suggest a reformulation of the legal text as well as to propose the accounting standards that would provide the environmental data legally required. The Ministry of Environment had significant reporting experience pursuant to the Pollution Control Act of 1981. However, its expertise focused on plant-specific emission data—as illustrated with the PRTR. The cradle-to-cradle approach suggested during the hearing process geared the focus toward the entire life cycle of a firm's products.

While there is a considerable literature on life cycle analysis for environmental management and policy referring to life cycle analysis (e.g., Welford 1996; Roome 1998), as well as limited precedent for producers' environmental responsibility extending into the consumption stage,[7] mandatory reporting of consumption patterns had never been requested. Nevertheless, within a few weeks the Ministry of Environment presented a reformulated text that enabled a redrafting of the legal proposal that was subsequently approved by the Parliament. This new legislation represents a radical shift in the focus of legally driven corporate environmental disclosures in Norway.

Specific Rrequirements for Environmental Disclosures in the Annual Report

Due to the amendments the revised legal requirement requests each board to report "current activities including production inputs and products that may cause a not insignificant impact on the external environment." To facilitate this reporting, the Ministry of Environment composed a set of specific accounting standards. To operationalize the legislative changes, Norsk Regnskapsstiftelse (Norwegian Accounting Foundation) later proposed a set of (still preliminary) accounting standards.[8] They include:

1. the type and quantity of energy and raw materials consumed;
2. the type and quantity of discharges and emissions, including noise, dust, and vibrations;
3. the type and quantity of waste generated and deposited, the character of disposal sites, and actual or potential contamination or runoffs;
4. the risk of accident caused by current activities of the firm; and
5. the environmental loads caused by transportation of production input and output.

For those firms that manufacture material products, additional concerns need to be considered:

1. the type and quantity of hazardous chemicals included in the products;
2. the type and quantity of waste generated at disposal; and
3. environmental impacts during consumption of products.

These standards include plant-specific and both upstream and downstream environmental impacts. Firms must also report the environmental loads imposed by the transport of production inputs and outputs. To comply with the regulatory requirements the firms must report along the whole life cycle of the products. The central question is whether the

Accounting Act sets in motion a process of industrial transformation by extending the reporting beyond emission control.

Some Concerns Related to the Extended Reporting Requirements

The Confederation for Norwegian Business and Industry (NHO) and some branch organizations have expressed a concern about the clause "that may cause a not insignificant impact on the external environment." According to the NHO this statement may create confusion, as it differs from equivalent legal formulations. Usually, as in the Pollution Control Act, a reference is made to "significant [environmental] impacts." Those formulating the text responded that the phrase "a not insignificant impact" more easily facilitates the adoption of a product life cycle perspective.[9] The business community is also concerned that making environmental disclosures an explicit and specified part of the statements of Boards of Directors will make the focus on environmental issues in the annual report disproportionate compared to other mandatory issues.

Finally, other countries are not asking for similar disclosures. Some argue that foreign investors and other external concerned parties may get the impression that Norwegian firms have more environmental problems than their "nondisclosing" competitors.[10] The counterargument proposed by the Ministry of Justice and NSF during the legislative "hearing process" is that all relevant environmental concerns should be documented and reported. Firms operating in the same market as Norwegian companies complying with the law on environmental disclosure may be neglecting the impacts of their operations. Furthermore the broadened focus of environmental reporting might not necessarily create new expenditures, but rather enhance profits. By producing and presenting a more thorough and valid picture of all relevant environmental concerns, firms with good records will gain the goodwill and patronage of customers and other environmentally aware stakeholders. Other firms may discover opportunities to improve their environmental and economic performance.

The Legal Strength of the New Environmental Reporting Requirements

The Accounting Act was passed in 1998 and the specific standards were officially implemented later that year. All commercial firms and certified auditors were requested to follow the new reporting requirements beginning in the fiscal year 1999. While these legislative measures represent a significant environmental achievement, the legal strength of these reporting requirements remains questionable.

The Act quite clearly requires that corporate accounts must be audited by a certified external auditor. The law specifics how the accounts are set up, and the Auditing and Accountants Act of 1999 further identifies what is to be audited. A careful reading of this legal text, however, reveals that certain parts Board of Directors' statement are explicitly exempted from the auditors' concern and responsibility. This includes environmental reporting. According to section 5.1 of the Auditing and Accountants Act, the auditor can only consider the information included in the statement from the Board of Directors related to financial results, conditions for continued operations, and the suggested use of corporate profits or coverage of loss.

The reporting requirements of the Accounting Act are clear: those violating the requirements may incur fines and imprisonment up to three years. However, the Auditing and Accountants Act of 1999 doesn't specify that certified auditors should investigate whether firms are complying with the environmental reporting requirements of the Accounting Act of 1998. Thus, while reporting standards have been formulated, the law does not establish an authoritative body to control whether the firms are actually complying. The interpretation of what is to be disclosed, in what manner, and whether this should be part of the annual statement from the Board of Directors, is left to the reporting firm and its shareholders.

The Ministry of Finance, responsible for enforcing the intentions of both Acts, has not expressed any concern about this anomaly. The question is still whether the firms, given the loopholes in the auditing procedures, are complying with these requirements.

Compliance among Norwegian Firms

The Ministry of Finance has made no effort to control whether firms are complying with mandatory environmental reporting requirements.[11] During the summer of 2000 the Confederation of Norwegian Business and Industry (NHO) commissioned a study on the 1999 reporting practices of 219 Norwegian firms. The report is unpublished and was used only as an input to the work coordinated by the NHO related to the environmental reporting award of 1999. The study documents that 21 percent of the 219 firms did not include any environmental information or references in their annual reports. The study indicated that only 8 of the 219 firms fully complied with the specified reporting standards of the Accounting Act.[12]

Financial Requirements: Does It Make a Difference?

The NHO commissioned study focused mainly on corporate environmental reporting procedures, though the Accounting Act covers other issues. The explicit objective is to strengthen the environmental concerns of the Board of Directors—specified in section 3.3—through mandatory reporting on external environmental issues caused by commercial activities throughout the financial year.

This chapter questions whether the legal requirement affects corporate environmental disclosures included in the annual report signed by the Board of Directors. To answer this question a survey of reporting practices has been conducted for the financial year 2001. The survey was conducted among the largest firms operating in Norway (Ruud and Mosvold Larsen 2003). The changes in the environmental reporting requirements were introduced in 1999, and our survey reflected the third valid year of the revised and implemented Accounting Act. The term "largest" is understood as those firms having the largest turnover in the year 2000.[13] In addition to the largest firms, the most pollution-intensive firms were also included in the sample, no matter what their size. All the firms included in the sample had been requested to submit environmental reports on plant-specific environmental measures. Some of the largest firms are involved in banking, financial services, and accounting. However, we also included a number of smaller financing and consultancy firms to verify whether those asked to fund or assist firms to strengthen environmental cleanups are complying with the legal reporting requirements. The five plus three accounting standards proposed by the Norwegian Accounting Foundation—and referred to previously—are used as a criteria in our evaluation.

A total of 112 firms operating in Norway are included in the sample. The environmental reporting performance has been sorted into the following six categories:

Category 0: Missing reports. Firms included in this category fail to report or offer only one sentence stating they do not pollute the external environment.

Category 1: Very unsatisfactory. Firms included in this category briefly acknowledge that current activities cause an environmental impact, but the actual reporting is very unsatisfactory.

Category 2: Unsatisfactory. Firms included in this category acknowledge environmental impacts, and examples are included. However, the actual reporting is unsatisfactory as important and relevant aspects in the environmental reporting are omitted or neglected.

Category 3: Satisfactory, but wrong sender. This category includes firms that have published a satisfactory or very satisfactory environmental report, but have not included it in their annual report for the Board of Directors as required by the Accounting Act.

Category 4: Satisfactory. This category includes firms in which the Board of Directors has approved a satisfactory specification of environmental impacts or has explicitly referred to a separate environmental report.

Category 5: Very satisfactory. This category includes firms that quantitatively present environmental impacts and specific efforts made to improve the situation. These firms also report along the life cycle of the products produced and distributed. Those firms that refer to an approved separate environmental report also include a brief summary of this report in the annual report from the Board of Directors.

The six categories can be divided into two groups. The first represents those firms that can be classified as "violators," as they appear to contravene the legal reporting requirements of the Accounting Act. This refers to firms included in categories 0, 1, 2, and 3. The other group—categories 4 and 5—represents those firms that comply with the legal requirements. It is important, however, to underline that the regulatory requirements are somewhat unclear, and that making such a division is consequently problematic. The reporting practice among the firms included in our sample also varies. However, the presidents of all 112

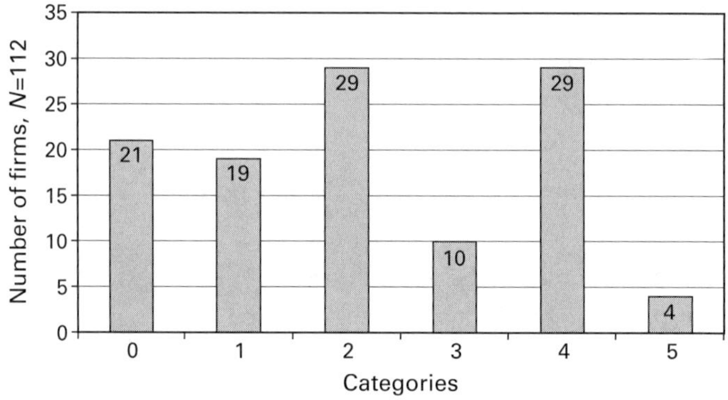

Figure 13.1
Classification of the environmental reporting by Norwegian films

firms were asked to comment on the relative classification. A total of 33 firms responded and provided additional information, but these new data failed to trigger any changes. The results are presented in figure 13.1.

As shown in figure 13.1, as many as 70 percent of the firms fail to fulfill the legal reporting requirements. Consequently the Boards of Directors of these firms appear to violate the Accounting Act of 1998. Still there are differences among the sample. The degree of "violation" varies significantly from those that bluntly neglect the legal requirements to those who misunderstand the legal text. The major finding nevertheless is that in the financial year 2001, only 30 percent of the firms surveyed complied with the 1998 legal requirements.

Category 0: Missing Reports = 21 Firms

As indicated above, this category includes firms that fail to refer to any external environmental issues in their annual report. A total of 21 firms were included in this category. These firms did not refer to any negative environmental impacts caused by their corporate or commercial activities. As many as 13 firms included only a statement that reflected the previous but obsolete reporting requirement defined in the Accounting Act of 1977. These firms provide an environmental statement that claims they are not polluting the external environment, but the report is missing. Among these 13 firms, one is actually among the potentially most pollution-intensive firms in Norway, and a discharge permit has been granted by the state pollution control board. One probable explanation for the missing reports is that the manufacturing operations of this firm are still in the planning stage. However, the legal reporting requirements also refer to potential environmental impacts. More important, the requirements demand the firm to specify the remedies taken or planned to combat negative impacts. This assessment is not provided, as the report is missing.

Category 1: Very Unsatisfactory = 19 Firms

A total of 19 firms were included in category 1. This refers to firms that do not make any thorough reference to the environmental challenges nor substantiate negative impacts caused by their commercial activities. In contrast to category 0, these firms have acknowledged they produce a negative impact. However, the specific reporting on particular activities is missing. Among the 19 firms included, two are involved in off-shore oil and gas activities. Neither of these firms possesses operational responsibility, but as partners in projects representing hazardous activities related to exploration, production, and transport of oil and gas, one questions whether the Board of Directors should have submitted an environmental report.

The Board of Directors of another firm involved in the import of fruits and vegetables reports that "the firm has an active environmental policy and works for the reduction in the use of nonrenewable resources." As an importing firm, however, transporting imported goods as well as producing and treating packaged materials cause significant environmental impacts. Despite the statements from the Board of Directors, these impacts remain unreported.

One pollution-intensive firm is also included in category 1. Their Board of Directors refers to the discharge permit issued by the state pollution control board. They also point to the installation of new measurement equipment to monitor atmospheric emissions. However, no information is provided on actual emissions generated by current production activities. Consequently this firm has produced a very unsatisfactory report.

Category 2: Unsatisfactory = 29 Firms

A total of 29 firms were categorized as unsatisfactory. This refers to firms that made significant attempts to deal with environmental challenges as well as the specific external impacts caused by their corporate activities. However, a large room for improvement exists, as the corporate reporting remains selective, fragmented, and limited to only a portion of the total range of actual or potential environmental concerns.

Based on a thorough reading of various annual reports written by the Boards of Directors of these 29 firms, it seems that efforts are being made to strengthen environmental disclosure. Still, however, there is work to be done. Firms in this category include some of the potentially most polluting firms in Norway. One common aspect of the environmental disclosures provided by these firms is that none of the firms disseminated any details on energy consumption and transportation use.

Category 3: Satisfactory, but Wrong Sender = 10 Firms

A total of 10 firms surveyed were included in this category: satisfactory but wrong sender. The term "wrong sender" refers to Boards of Directors that have failed to cite often impressive separate environmental reports. It is consequently called the wrong sender because the Boards of these 10 firms are not complying with the requirement of the revised Accounting Act. The revisions were introduced to make the Board more concerned with environmental issues. Category 3 refers to firms where this concern is lacking; therefore these firms are contravening the legal requirements.[14]

Category 4: Satisfactory = 29 Firms

A total of 29 firms are considered to have done satisfactory environmental reporting according to the requirements of the Accounting Act. In contrast to the fragmented and unsatisfactory reporting examples among the firms included in category 2, these firms present more thorough and inclusive reports. The majority are pollution-intensive firms with discharge permits, and they have already been asked by the environmental authorities to submit an annual environmental report on plant-specific measures. However, eight firms involved in trade and servicing activities are also included. This is particularly interesting as these firms are not subject to the same environmental reporting requirements.

Category 5: Very Satisfactory = 4 Firms

The four firms included in this top category have all produced environmental reports of high quality and the result is very satisfactory. These four firms have produced specific, quantitative data concerning all relevant business activities. Specific efforts for improvement are presented, and the reports also refer to the life cycle of the products manufactured or sold. The four firms are listed in alphabetical order:

- Møllergruppen
- Norske Skogindustrier ASA (Norske Skog)
- Tine
- Tomra Systems ASA

What Is Actually Reported?

Møllergruppen[15] has three business divisions: car sales, property, and investments. As with most of the firms that have produced a separate environmental report, the reporting included in the annual report for the Board of Directors is limited. However, an informative summary has been presented, and the impacts related to parts of the life cycle of the products are mentioned. The following is written: "The products sold by the firms and activities in workshops create an impact on the external environment. Consequently strong efforts are made with respect to hazardous waste management. Our workshops are equipped with oil settlers and suppressor tanks to handle oil and chemical spills. Over the last few years the technology for treatment of car emissions has been further developed, and our suppliers are well advanced in the development of fuel-efficient cars with limited atmospheric emissions during use." Møllergruppen is a relatively large firm in terms of property management, but related environmental concerns are not mentioned in the annual report. However, a separate environmental chapter of the financial report addresses the issue, and the Board explicitly refers to this chapter. In this separate chapter the total energy consumption during the last three years is presented. Kilowatt per hour per square meter is used to measure energy efficiency. Møllergruppen writes that relative energy consumption has decreased since the registration started in 1997. Despite this decrease, the commercial activities in the workshops have increased considerably —over the last five years, total energy consumption has decreased 20 percent.

Møllergruppen has presented solid and specific information on waste treatment procedures. The 2001 environmental objectives are discussed, and revised targets are defined for 2002. The reuse and recycling of cars is also treated, particularly with reference to the end-of-life vehicle directives that will be implemented in Norway in 2005.[16] For example, the report includes details of a car dealership in Stavanger that has been certified through the Eco-Lighthouse Program.[17] The general impression of the environmental reporting presented and approved by the Board of Directors of Møllergruppen is very satisfactory.

Norske Skogindustrier ASA[18] is a major transnational actor within the pulp and paper industry, and it is one of Norway's largest firms. In addition to an informative summary in the annual report, the separate environmental report of 53 pages is approved by the Board of Directors. This report provides a very satisfactory understanding of the environmental achievements and challenges of Norske Skog during the financial year of 2001. What is particularly impressive is the focus on challenges and unresolved matters. Firms still have a very strong tendency—even among those with satisfactory reporting—to present only achievements. In Norske Skog's annual report the environmental ambitions are presented: "The strategy of Norske Skog is to run a business in such a way that it supports sustainable development of the environmental and natural resources. The objective is to limit the environmental load to a minimum. The environmental strategy of the corporation includes all business units and Norske Skog will work for common environmental values for joint ventures and partly owned firms." Further, information on their demand for fiber is summarized in the annual report. It shows that 21 percent of their demand is supplied from recycled paper, a growth of 20 percent from the previous year. The summary also refers to current work on energy savings and waste minimization. In this section it is interesting to note that more than 80 percent of the organic waste at the European and South American factories are used as bioenergy in their pulp and paper factories. In a separate and approved environmental report, extensive reporting on a number of quantitative measures are presented, which are related to energy consumption, atmospheric emissions, liquid discharges, and waste treatment at all factories in Norway. In addition the report addresses the transportation of final products as well as raw materials. The quantitative data is further developed into eco-efficiency indicators. With explicit reference to the regulatory requirements of the Accounting Act, the environmental reporting of Norske Skogindustrier is very satisfactory. The separate report was also granted the Norwegian environmental reporting award of 2001.[19]

Tine Norske Meierier[20] writes in its annual report that the firm is "a coordinator of the environmental efforts made by Norwegian dairy cooperatives and a contact point vis-à-vis the authorities and other actors when it comes to compliance with rules and regulations." This claim is clearly reflected in the environmental reporting. A schematic overview of targets and achievements during 2001 is presented, and it includes the remaining number of farm tanks using CFCs as a cooling agent. The number has been reduced by more than 10 percent compared to the previous year. This achievement is in line with the targets set in the previous annual report. The activities of Tine Norske Meierier are also properly presented in the separate environmental report that was published together with the financial report. All share- and stakeholders interested in the financial report automatically received a copy of the environmental report. Tine's Board of Directors explicitly approved the content of the separate environmental report, and it is worth noticing that both reports are produced with eco-labeled paper. Interestingly only two other firms published their 2001 reports on eco-labeled paper.[21]

The separate environmental report produced by Tine is very solid and provides a comprehensive presentation of energy consumption split into various energy sources. Further it presents waste treatment procedures, packaging, and use of cooling agents for storage of dairy products as well as emission data. Detailed information on transportation is also included. As with Norske Skog, Tine presents quantified environmental data concerning the last three years, and this information is also developed into comparable eco-efficiency measures. Consequently Tine seems to have published a very satisfactory environmental report that is approved by the Board of Directors.

Tomra Systems ASA[22] manufactures reverse vending machines (RVM). As stated, "TOMRA's business concept is to offer cost-effective systems for recovering packaging for reuse and recycling. We deliver value through innovative technology and design, comprehensive operating support and services, and an integrated approach to recycling that encourages maximum efficiency and environmental benefit." The annual report states that the firm "makes an effective contribution toward closing the material cycle, thereby contributing to reduction of the negative effect on the environment." In a separate environmental chapter of the financial report—referred to by the Board—a quantified overview of the type and quantity of energy consumed (included the energy consumption of their fleet of vehicles), pollution, water consumption, and waste treatment is presented.

Tomra has promoted a recycling program for obsolete machines, and also provides an overview of specific production inputs used in each RVM. As stated, "Tomra RVMs are increasingly being designed for easy dismantling and recycling when they reach the end of their operating life." Tomra's business idea is to manufacture equipment that collects recycled packing or containers, but the firm also reports on the whole life cycle of its own product—the RVMs. This type of reporting corresponds very well with the intention of the reporting requirements. Consequently the environmental reporting of Tomra is very satisfactory.

Is the Business Community Complying with the Reporting Requirements of the Accounting Act?

The aim of the Accounting Act is to integrate firms' environmental concerns into the annual reports from the Boards of Directors. The findings are not very promising. To verify whether all relevant information is included, the presidents were asked to comment upon their specific classification. A total of 33 responses were received. Table 13.1 shows their reactions according to the extent to which they accepted the proposed classification and the extent to which they understood the regulatory requirement.

Table 13.1
Feedback by the firms on the classification

Approved the classification		Disapproved the classification			Did not understand the regulatory requirement		Total
Approved our classification, and expressed an interest in improving the environmental reporting	Approved our classification, and expressed a thanks for our the evaluation	Asked to be moved from category 4 to category 5	Referred to international environmental reports that did not refer to Norwegian activities	Did not approve the term "violater"	Did not understand/was not aware of the regulatory requirement—wanted further explanation	Not subject to these regulatory requirements	
7	1	2	3	8	10	2	33

Among the 33 firms that responded, 8 firms approved our classification and 7 expressed an interest in improving their environmental reporting practices. Two of these firms responded through their environmental representative, though these representatives complained that they continually struggled to place environmental concerns more centrally on the agenda of the Boards of Directors. Both representatives wanted to use this study to exemplify the need to strengthen the environmental concerns within the Board. A total of 13 firms did not approve the classification. Two firms wanted to be reclassified from category 4 to category 5. The firms in question had published extensive and solid reports, but they were separated from the annual report—not even a summary was offered. Only a brief reference was made, and while satisfactory, such a reference fails to merit an outstanding rating as exemplified in category 5. Three of the firms included referred to international environmental reports, but these reports did not focus on Norwegian activities. Eight firms did not approve the use of "violator," and wanted to be included in category 4 or 5. Among these, one firm did not distinguish between the external and internal environment. A total of 13 firms disapproved of the way this study interpreted the legal reporting requirements of the Accounting Act. Their arguments were considered seriously, but no reclassification was made. A total of 10 firms from various service industries expressed clearly that they did not understand the legal requirements and the relevance for their commercial activities. All wanted further information and specific explanation, but none asked for a reclassification. This indicates that even among the largest Norwegian firms, many explicitly acknowledged that they misunderstood the reporting requirements. If this fact is extended to those firms that rejected the proposed classification, the number becomes even larger.

Beyond Emission Control?

The general legal compliance is low, but—in line with the objectives of this book—attention must be drawn to whether the new regulatory requirements triggered reporting beyond emission control. This question is crucial, as extended reporting is a necessary step to promote industrial transformation to more sustainable consumption and production practices.

All 112 firms included in our sample should be capable of reporting on energy and raw materials, waste, and transport. Concerning the remaining five accounting standards—three of which are related specifically to the life cycle of products manufactured—as many as 57 firms are irrelevant (see table 13.2). Further it should be noted that all firms are included regardless of whether the reporting is included in the annual report or separately. Consequently category 3 (satisfactory, but wrong sender)—presented in figure 13.1—is omitted, as this category only fits into a general evaluation of legal compliance.

1. Energy and Raw Materials

Seventy percent of the firms in the sample did not comply with reporting requirements concerning energy and raw materials. This is equal to the percentage of firms not complying with the law in general. However, while only 19 percent of the firms in the overall study were classified in the lowest category, 46 percent of the firms were missing any reporting on consumption of energy and raw materials. Nevertheless, 10 percent reported very satisfactorily on energy and raw material consumption.

2. Pollution[23]

Only 5 out of the 72 relevant firms did not report on pollution. One of these is a maritime company whose emission of CO_2 and NO_x causes environmental impacts. Among the remaining four, two have been granted a discharge permit from the pollution control board. Production has not yet been initiated, but in accordance with the requirements, potential impacts should be reported. The two remaining firms are large importers of cars, yet they neglected to refer to any impacts that could be caused by their use.

3. Waste

Thirty-three percent of the firms did report in accordance with the legal requirements concerning type and quantity of waste. Further, waste reporting is incomplete for 38 percent of the firms. Our study documented variety in reporting, and seven of those with satisfactory or better waste reporting (37 firms) did not comply with the legal requirements on energy and raw material consumption.

Table 13.2
The reports evaluated against the eight criteria

Specific concerns	(0) Missing reports	(1) Very unsatis-factory	(2) Unsatis-factory	(4) Satisfactory	(5) Very satisfactory	Total firms evaluated	Non-relevant firms
1. Energy and raw materials	53 (46%)	13 (12%)	13 (12%)	22 (20%)	11 (10%)	112 (100%)	0
2. Pollution	5 (7%)	17 (24%)	15 (21%)	18 (24%)	17 (24%)	72 (100%)	40
3. Waste	43 (38%)	14 (13%)	18 (16%)	25 (22%)	12 (11%)	112 (100%)	0
4. Risk of accidents	47 (65%)	8 (11%)	12 (17%)	3 (4%)	2 (3%)	72 (100%)	40
5. Transport	79 (71%)	7 (6%)	14 (12%)	8 (7%)	4 (4%)	112 (100%)	0
a. Chemicals included	23 (42%)	12 (22%)	12 (22%)	8 (14%)	0 (0%)	55 (100%)	57
b. Impact when disposed	34 (62%)	4 (7%)	10 (18%)	4 (7%)	3 (6%)	55 (100%)	57
c. Impact during use	43 (78%)	0 (0%)	6 (11%)	6 (11%)	0 (0%)	55 (100%)	57

4. Risk of Accident[24]

In this category we also included general statements on accidents, and as many as 65 percent did not report properly. Only two firms reported very satisfactorily.

5. Transport

A total of 70.5 percent of firms did not report on transportation. Most of the firms included in the total sample, however, are heavily involved in various modes of transport. Still there are some promising exceptions. One telecommunication firm published data on the total energy consumption and pollution generated by their transportation activities.

From emission control toward the life cycle is a major concern for those firms involved in the manufacturing of material products. In our sample 55 firms are identified as relevant. Among these firms very few are reporting in a satisfactory way. The most striking finding is the fact that as many as 78 percent did not report on impacts during consumption or use. Referring to the first category "chemicals included" the information is missing among 42 percent of the 55 firms. Twenty-two percent have produced a very unsatisfactory report on chemical output. References are made to registers established, but the firms offered no inventory of the chemicals used in the products manufactured. Eight firms have provided some relevant information, and these are classified as satisfactory, but none of the firms in the sample have provided very satisfactory chemical reporting on products. Finally, 62 percent of the firms did not provide any information concerning product disposal, and only 13 percent reported in a satisfactory way. The findings indicate significantly lesser regulatory impacts of the Accounting Act on environmental reporting practices compared to the Norwegian PRTR as well as the Toxic Release Inventory.

Conclusion

Most firms in this survey are not complying with the environmental reporting requirements of the Accounting Act of 1998. Consequently it can also be questioned whether the legal efforts promote industrial transformation. The study does not provide empirical data to verify behavioral changes in actual production and consumption practices. However, to promote industrial transformation, the first necessary step must be documenting whether reporting practices are extended beyond emission control issues.

The particular legal intention of moving the focus beyond emission control toward the life cycle of the products or services has, so far, been unsuccessful. While promising exceptions exist, even among the four forerunners with a very satisfactory report, it can be questioned whether the legal requirements made a difference. In the case of Møllergruppen, extended reporting of the life cycle of the cars they sold was initiated in 1997, two years prior to the introduction of the legal requirements. Also, while the environmental reporting of Norske Skog has been developed in line with the proposed accounting standards, they make no explicit references to these legal requirements. Still the legal impact may be significant. The environmental performance of Norske Skog has improved. This is confirmed by studying PRTR data on specific pollution control issues. Yet it can be questioned whether the Accounting Act influenced the extended reporting on the type and quantity of hazardous chemicals included in their products, on the type and quantity of waste generated at disposal, as well as the reported environmental impacts during consumption of products provided by Norsk Skog.

Similar to Møllergruppen, Tine initiated extended environmental reporting prior to the introduction of the new Accounting Act. Further, as with Norske Skog, the improvements along the life cycle of the dairy products provided by Tine seem to have been pursued regardless of the Accounting Act. This was also the case of Tomra. Efficiency and environmental benefit was presented as an integrated dimension of their business philosophy, and this had been promoted many years prior to the legal reporting requirements of the Accounting Act.

As illustrated with the four top performers, some firms are extending the reporting requirements beyond emission control. This also includes some of the firms included in categories 3 and 4. However, the specified findings related to the accounting standards presented in table 13.2 do not present very promising findings. Compared to the TRI requirements in the United States, the Norwegian reporting requirements remain in limbo. Corporate environmental reporting is exempted from third-party verification. The limited legal strength of the Accounting Act may prevent the development of private sector mechanisms and networks for the implementation and enforcement of strengthened corporate environmental disclosures that support industrial transformation.

A large majority of the 112 firms did not comply with the reporting requirements, but does this imply that the firms are necessarily negligent? Forty-three of the firms evaluated have produced satisfactory reports, and one may question the usefulness of including mandatory environmental reporting in financial legislation, particularly if the reports are exempted from mandatory auditing requirements. In contrast to the specific reporting schemes enforced by the environmental authorities under the Pollution Control Act, as well as the TRI requirements in the United States, Norwegian firms are more or less left to decide whether they want to comply with the financial reporting requirements. Despite thorough data in their separate environmental report, even those firms classified as having produced a very satisfactory report do not fully disclose all environmental hazards within the life cycle of the products supplied to the market.

There are several reasons for noncompliance with the reporting requirements of the Accounting Act. One reason is that firms do not see any advantage in making detailed environmental disclosures about the company's environmental performance. Further the motivation is weakened when the Boards of Directors of foreign competitors are not required to include similar data, which may focus on negative aspects that can reduce general goodwill and competitiveness in the market. This debate must be taken forward, both by the business community as well as politicians. Extended environmental reporting has significant potential, and specific firms like the four best performers of this study are all contributing to industrial transformation through reported changes in production and consumption patterns. Still, as illustrated in table 13.2, most of the firms included in this survey fail to consider the life cycle of the products they produce.

In conclusion, the Accounting Act does not produce a great driving force for industrial transformation. Most of the companies that have more or less complied with the Act were already working toward the life cycle. They had begun to reduce their environmental impact because they had incentives outside the legislation to begin this process. Given the lack of legal enforcement and third-party verification, the potential of the Accounting Act is unfulfilled.

Notes

The material included in this chapter draws heavily on previous studies—in Norwegian only—done by Ruud and Larsen.

1. Further details of the design and content of the Norwegian Polluntants and Transfer register can be found at www.sft.no/bmi/main/english.asp.

2. My own translation. The original text is only available in Norwegian at www.lovdata.no/all/nl-19980717-056.html.

3. Sweden recently considered a similar request, but decided to limit the reporting requirements to those firms having a discharge permit granted from the pollution control board.

4. Stated on June 21, 2001, by Hege Andenæs at the Ministry of Environment, the Act is partly responsible for the work done at the Ministry of Environment.

5. According to Bjørn Sveen, environmental director of NHO.

6. NSF is a professional body for those with a Master of Business Administration (MBA) in Norway.

7. In Norway the Act controlling product and consumer services (*Lov om kontroll med produkter og forbrukertjenester*) was explicitly revised in 2000 to prevent health hazards and environmental impacts on eco-systems. As stated further in section 1, this Act promotes energy-efficiency in products. Further details in Norwegian only at www.lovdata.no/all/tl-19760611-079-0.html#1.

8. These accounting standards turned out to be identical to those initially proposed by the Ministry of the Environment.

9. According to Hege Andenæs, Ministry of the Environment; see note 4.

10. According to Bjørn Sveen; see note 5.

11. According to Espen Knudsen, Ministry of Finance.

12. The study analyzed compliance as related to the following four issues: environmental impacts caused by production or the use of energy or raw materials, environmental impacts caused by the use of products, potential environmental impacts, and planned environmental projects or initiatives. These four issues deviate slightly from the actual standards referred to in the legislation.

13. The selection of firms was made according to a classification provided by the Norwegian financial newspaper *Dagens Næringsliv* as of September 13, 2000.

14. In the subsequent analysis "beyond emission control," illustrated in table 13.2, category 3 is dissolved and the actual reports produced by these ten firms are treated equally to reports presented by those included in the other categories.

15. Further information on Møllergruppen is available at www.moller.no.

16. See chapter 4 of this volume for a more elaborate discussion on ELV policies.

17. The Eco-Lighthouse Program is a tailor-made program for environmental certification of small- and medium-sized companies and public administration in Norway. Further information available at www.eco-lighthouse.com.

18. Further information on Norske Skog at www.norske-skog.com.

19. Since 1995, an annual award has been granted to the company with the best environmental report in Norway. The Confederation for Norwegian Business and Industry (NHO) functions as the secretariat for the jury.

20. For further information on Tine, see www.tine.no/international/.

21. Further information is available at www.svanen.nu/Eng/default.asp.

22. Further information on Tomra is available at www.tomra.com.

23. Polluting emissions caused by heating could be included, but levels are limited, and the 40 office firms are excluded. Some of these refer to pollution in connection to transportation, but this reporting is included in the fifth category on transportation.

24. Firms with very limited risk for accidents are omitted.

References

Berger, G., A. Flynn, F. Hines, and R. Jones. 2001. Ecological modernization as a basis for environmental policy: Current environmental discourse and policy and the implications on environmental supply chain management. *Innovation* 14(1): 55–72.

Connelly, J., and G. Smith. 1999. *Politics and the Environment: From Theory to Practice*. London: Routledge.

Hajer, M. A. 1996. Ecological modernization as cultural politics. In S. Lash, B. Szersynshi, and B. Wynne, eds., *Risk, Environment and Modernity: Towards a New Ecology*. London: Sage.

Mol, A. P. J. 1996. Ecological modernization and institutional reflexivity: Environmental reform in the late modern age. *Environmental Politics* 5(2): 302–23.

OECD. 1998. *Eco-efficiency*. Paris: OECD.

Roome, N., ed. 1998. *Sustainable Strategies for Industry: The Future of Corporate Practice*. Washington, DC: Island Press.

Ruud, A., and O. M. Larsen. 2003. *Miljørapportering I større norske foretak: Fungerer Regnskapsloven etter intensjonen?* ProSus-report 5/03. Oslo: ProSus–University of Oslo.

Ruud, A., and O. M. Larsen. 2002. *Miljørapportering I årsberetningen: Følger norske bedrifter Regnskapslovens pålegg?* ProSus-report 8/02. Oslo: ProSus–University of Oslo.

Stavins, R., and B. Whitehead. 1997. Market-based environmental policies. In M. R. Chertow and D. C. Esty, eds., *Thinking Ecologically: The Next Generation of Environmental Policy*. New Haven: Yale University Press.

World Business Council for Sustainable Development (WBCSD). 1997. *Eco-efficiency: The Business Link to Sustainable Development*. Geneva: WBCSD.

Welford, R. 1996. *Corporate Environmental Management: Systems and Strategies*. London: Earthscan.

14

Conclusions: Lessons for the Design and Use of Voluntary, Collaborative, and Information-Based Approaches to Environmental Policy

Vicki Norberg-Bohm and Theo de Bruijn

Over the past decade the United States and many European countries have developed new approaches to environmental policy that are voluntary, collaborative, and information-based. These programs are attempts to engage industry in significant environmental improvements through dialogue, consensus-building, and voluntary action rather than the imperatives of direct regulation or the incentives of market-based approaches.

In chapter 1 we discussed the factors that determine the effectiveness of the new and innovative approaches. We identified three sets of factors that contribute to effectiveness, as well as three sets of factors that may limit effectiveness. The three major arguments for why they may achieve industrial transformation are, first, that they may build new relationships between stakeholders, leading to better solutions for the environment. Second, the programs may engage industry in a learning process that creates the capabilities within firms to engage in significant environmental improvements. Third, the programs may create first movers—firms willing to invest in business strategies, including managerial and technological innovations that can create substantial improvements in environmental outcomes. Their strengths notwithstanding, the new approaches may fail to be a force for industrial transformation because they have not succeeded in changing the competitive environment of the firm, suffer from complex implementation processes, or do not fit with the dominant legislative system.

In this chapter we take stock of the experiences by discussing the extent and ways in which the success and fail factors played out in the programs evaluated in this volume. The most effective programs, when measured against our demanding goals, were designed and implemented in ways that captured the potential strengths of voluntary, collaborative, and information-based approaches, while avoiding the pitfalls. Below we review the extent to which the programs in this book were able to do so. Based on this analysis, and the more detailed evaluations of each of the authors, we close the chapter with a set cross-cutting policy recommendations.

Experience with Innovative Approaches to Environmental Policy

The chapters in this book examined the effectiveness of voluntary, collaborative, and information-based policies, focusing particularly on superior environmental performance, radical technological innovation, industry leadership, and the involvement of other actors in the product chain. Table 14.1 gives an overview of successes and failures.

Taken as a whole, the programs examined in this book demonstrate more success than failure. Many of these programs have already improved the environment, as well as established long-term goals that hold up the prospect for more fundamental change in the future. There is also some evidence of beyond compliance behavior. However, when measured against the high standards for evaluation put forth in this book, our evaluation is more circumspect. While the programs have contributed to technology innovation, it was more often incremental than radical. And while there is evidence of private sector leadership, there is concern that it is most often one-off rather than ongoing, and focused on near-term opportunities rather than longer term and more difficult targets. When measured against the tall order of inducing changes in production and consumption systems (i.e., change process beyond the level of individual firms) there is rather limited evidence that voluntary, collaborative and information-based programs can contribute substantially to this.

Table 14.1 also indicates that sector-based programs seem to be more successful than facility-based programs. Below we address the differences in design, mechanisms and outcomes between industry sector approaches and firm-level approaches more specifically.

Table 14.1
Empirical evidence of industrial transformation

	Substantially improved environmental performance	Development and diffusion of environmental superior technologies	Private sector leadership	Patterns of changing behavior at levels beyond individual firms
Sector based				
Dutch Target Group	+	+/–	+	+/–
CSI	–	–	–	–
German ELV	+	–	+	+
Energy Star	+	+	+	+/–
R&D Collaboration	+	+	+	+/–
Danish CTP	+/–	+	+	–
Firm level				
Dutch EMS	+/–	–	+/–	–
EMAS in UK	+/–	–	+/–	–
StarTrack	+/–	–	+/–	–
Project XL	+/–	–	+/–	–
TRI	+	–	+	–
Norwegian Accounting Act	+/–	–	–	+/–

Note: + = effect is/will be notable; +/– = questionable whether effect has occurred/will occur; – = effect has/will not occur(red).

Industry Sector Approaches

As a group the sector-based programs in this book were more effective than facility-based programs in stimulating beyond compliance performance and technological innovation, but even within this group there were shortcomings and outright failures. There are several reasons for the relative effectiveness of sector-based approaches. By focusing on moving forward an entire industry sector, they create risks that nonparticipating firms will be left behind as well as lowering the risk of free-riding.[1] Furthermore they offer opportunities to develop strategies beyond the level and capability of individual firms.[2] In addition to these factors the sector-based approaches captured many of the aspects of the pathways of effectiveness. First, through sectorwide collaboration, they built new relationships that increased information flows among the firms in an industry, between industry and government, and in many cases also with nongovernmental actors. Second, by putting the onus on industry to propose solutions to environmental problems, these programs stimulated industry to build new capabilities and increased the steering capacity of governments. Third, many of the programs created adequate incentives for first movers, either within the program itself, or by linking with other policies and programs. Despite these many advantages, implementing sector-based programs is a complex and not always successful task. As shown by the examples in this book, some policy contexts facilitate the move to sectorwide collaboration better than others.

The "covenant" approaches, as captured in this book by two European programs, the Dutch Target Group Policy and the German end-of-life vehicles (ELV) program, hold the greatest promise for substantial long-term change. They not only captured the benefits of more creative solutions and increased steering capacity of governments through collaboration; both also changed the competitive environment of firms by combining stringent long-term requirements with flexibility in implementation. In the case of the Dutch Target Group Policy, government established long-term targets—50 to 90 percent emission reductions for specified pollutants—with industry and governments jointly developing plans for meeting these targets. In the case of the German ELV program, government challenged the automobile manufacturing and disposal industry to develop targets that were in line with public goals for the safe disposal of used vehicles, as well as plans for implementation, with government maintaining the right to impose targets and plans if the voluntary negotiations did not succeed. In this case capacity development across the supply chain was enhanced by the creation of a special institution (Arge-Auto) created for monitoring purposes.

For these programs many of the pitfalls during implementation have been avoided. In both cases strong ambitions were translated into relatively clear goals, thus facilitating effective implementation. These goals were either imposed (as in the Dutch policy program) or mediated (as in the case of the German ELV program) by government. Industry and governments then collaborated to find effective ways to reach these goals. Furthermore the programs had a good fit with the existing national policy style, which resulted in the programs being closely embedded in the larger policy system. In the Netherlands, the Target Group Policy brought environmental policy more in line with the Dutch national policy style, which is based on consensus-building and participation (Liefferink 1997; Bressers and Plettenburg 1997). Similarly the German ELV program could draw on the corporatist structure of German society, despite the historically strong preference for regulatory instruments in environmental policy (Jänicke and Weidner 1997).

Despite the many similarities in these programs, Hofman and Schrama (chapter 2) provide a more positive and cautiously optimistic assessment of the prospects for long-term effectiveness of the Dutch target group approach than Jörgens and Busch (chapter 3) provide for the German ELV program, which has now been superseded by what they view as a stronger and better designed ordinance prompted by EU regulation and informed by experience with the ELV program. In the Dutch Target Group Policy, links have been built with government-sponsored technology development programs as well as with the permit system. As Hofman and Schrama argue, concerns remain about the ability of the target group approach to reach long-term goals. In contrast, the ELV program, while providing clear long-term goals for auto producers, did not have adequate mechanisms to ensure that companies (or the government) made investments in R&D to reach these goals, nor did the agreement have enough specificity in the types of hazardous materials that should be phased out from use in automobiles. The auto companies, which were the politically and economically strongest actors in the supply chain, were able to shield themselves from such stringent requirements while dismantlers and return stations were required to take strict, detailed, and cost-intensive near-term actions.

In sum, these two examples of "covenant" approaches suggest that linking these programs more directly to other government policies that provide incentives for investment in new technologies—including government-sponsored R&D efforts, mid-term goals, and requirements that new technologies be specified for adoption in the implementation phase—may be necessary to ensure long-term success. Furthermore it is important to recognize that these approaches may be subject to industry capture. The relevant and yet unanswered question is whether the ability of dominant players to shape regulation is greater in these collaborative mechanisms than through more traditional regulatory processes.[3]

The US effort at a broad-based industry sector approach, the Common Sense Initiative (CSI), lacked the features that created success in the two cases above. This program challenged firms within an industry to join a collaborative process with government and NGOS to develop "cleaner, better, cheaper" solutions to environmental performance by tailoring environmental regulation to the specific circumstances of different industrial sectors. Coglianese and Allen (chapter 3) conclude that this voluntary, consensus-based approach was largely a failure, providing little incentive for first movers or for industry to reveal valuable information to improve government steering capacity. Several aspects of the program and the larger policy environment contributed to this failure. First, the program's lofty goals were not translated into clear targets. There were no government-imposed targets (or the threat to impose targets) and the voluntary, consensus-based approach was not able to achieve this. Second, without clear targets, the consensus-based decision making meant that the limited number of agreements that were reached tended to shift toward the lowest common denominator of the players involved, focusing more on the near-term interests of making environmental regulation more efficient, rather than the more difficult and long-term goal of increasing environmental protection. Furthermore this decision process meant that recalcitrant players held the process captive. Finally, CSI did not fit well with the adversarial and inflexible US policy regime. The adversarial system creates high risks for firms to voluntarily reveal information about their environmental behavior, and the limited flexibility for implementation meant that the program did not have legal authority to implement innovative solutions proposed during the collaborative process.[4]

Other more narrow US industry sector approaches were more effective, at least in stimulating near- and medium-term technological development for the environment. Energy Star, a labeling and marketing program focused on increasing the energy efficiency of equipment and buildings, contributed to the ability of firms to pursue a product differentiation strategy. However, it depended on links to other programs, specifically procurement and standards, to provide incentives for first movers. In the case of Energy Star computers, which pursued what Paton (chapter 5) calls a "converging" approach, the program focused on gaining the participation of all manufacturers. To accomplish this, computers qualifying for the Energy Star label initially only required the adoption of existing technology, and subsequent decisions about increasingly stringent requirements were also constrained by what was already technologically feasible. Furthermore the federal government and many large firms established procurement policies that required the purchase of Energy Star computers. This created a huge market for the computers, and an incentive for all manufacturers to meet the Energy Star standards. Once a process like this is in place, it can provide an incentive for firms to become first movers, as firms recognize that if they develop new energy-saving technology that does not require large cost increases, it is likely to become a requirement in a subsequent round of negotiation.

In the case of Energy Star washing machines, a "separating" strategy was feasible due to the combination of a voluntary labeling program and appliance standards. Washing machines are one of a number of home appliances that are governed by energy efficiency standards, which are revisited and revised, by statute, periodically. For washing machines, there were firms willing to become first movers. By gaining experience with higher energy efficiency in their more expensive models, a firm could not only position itself to capture market share based on experience with energy-efficient technologies, but a firms' technology could also become the basis for a future standard.

US R&D collaborations and the Danish Cleaner Technology Programs, although not a form of regulation, combined the same characteristics as the European covenant approaches—challenging goals and flexibility in meeting these goals—with the economic incentive of cost-sharing in order to reduce the risks that firms faced in pursuing technological innovation. The US R&D collaborations (for advanced gas turbines and solar PV technologies) focused on technological innovation for a next generation of energy technology that was beyond what industry would pursue on its own, with development timeframes of five to eight years. The Danish Cleaner Technology Programs focused on the development and deployment of pollution prevention technologies for a number of industries, with a dominant interest in incremental innovation and technology transfer between sectors, although it also supported some more radical innovation. Cost-sharing was an essential element in these programs, both for recruiting firms and in keeping firms at work on a goal that was a stretch technologically, even in the face of technological setbacks. Furthermore, through funding mechanisms and organization structures, these programs further developed existing networks of technological capability, bringing together manufacturers, suppliers, universities, national laboratories, and consultants to work together to reduce and spread the risks of technological development. Their support for technological development in some cases also contributed to strategies of product differentiation.

The ultimate success of these programs, however, is not simply technological but in the market for the widespread adoption of cleaner technologies. In this regard these technology development programs need to work synergistically with the existing environmental policy system, and often need to be linked to environmental regulation or other market creation policies. As discussed by Norberg-Bohm and Margolis (chapter 6), the US case, the Advanced Turbine Systems program, demonstrates the need for ongoing re-evaluation of environmental targets, and a close dialogue between those involved in technology innovation programs and those involved in environmental regulation. In this case, although at the inception of the program the turbines had ambitious goals for NO_x emission reduction, by the time the turbines were commercial, they could no longer meet NO_x emission requirements. This has serious implications for the market and cost of these technologies, and raises questions about whether increased examination of future NO_x emission scenarios would have led to a different trajectory of technological development. More generally, over the time frame of technology development, knowledge about environmental impacts will increase and may result in more stringent goals. Technology programs need to be designed with this possibility in mind.

In the Danish case, although the programs have had a measurable and significant impact on the availability of cleaner technologies, the diffusion and implementation of these technologies, while significant, has not reached its full potential. Jørgenson (chapter 7) argues that this is due to a lack of integration between the Cleaner Technology Programs and the dominant environmental policy system. The basic difficulty is that local authorities, which provide facility permits and have significant discretion in the Danish environmental policy system, have either not learned sufficiently about new technological options from the Cleaner Technology Programs or have not forced the implementation of newly developed cleaner technologies during their permit negotiations with industry. In short, the Cleaner Technology Programs have not succeeded in fundamentally penetrating and changing the traditional command-and-control approach to environmental regulation. The fact that a coherent practice has not emerged from the fifteen years of experience points to difficulties in transforming the entrenched regulatory regime through a voluntary program.

Firm-Level Approaches

In each of the firm-level programs, taking action for environmental improvement was voluntary, although in the case of the information disclosure programs, reporting information about the environment was not voluntary. The effectiveness of these programs in stimulating voluntary action varied quite a bit. The most effective of these approaches were able to develop capabilities and relationships that stimulated private-sector leadership for the environment and incremental and near-term investments in environmental improvement. By and large, however, these firm-level programs were unable to create first movers for radical and far-reaching technological innovation for the environment. Many of these firm-level programs faced difficulties in implementation, not only because of their complexity and the involvement of a large number of stakeholders but also because they did not fit with the existing regulatory system, which prevented creative deviations from current practices. Furthermore, because regulators had to interact with firms on an individual basis, these approaches suffered high transaction costs. Finally, these programs provided limited benefits to firms, and thus could not change their competitive environment. Voluntary, firm-level approaches therefore must depend heavily on incentives external to the program. In this sense the group of firm-level approaches in this book confirmed the limitations of relying solely on voluntary action, and refute the idea that one can rely on private "win–win" strategies as a road to sustainability. Rather, these firm-based approaches are best viewed as an important contributor in setting the stage for industrial transformation.

Two programs evaluated in this book promoted the adoption of environmental management systems (EMS), the Dutch Program on Environmental Management and the British adoption of the European EMAS Regulation. For both cases the adoption of EMS helped develop capacities for technological change. In this regard EMS programs may contribute to the effectiveness of the environmental policy system as a whole. EMS accomplish this through development and dissemination of information and changes in organizational structures and procedures. However, the evaluations by De Bruijn and Lulofs on the Dutch program (chapter 8) and by Gouldson on the British program (chapter 9) show that EMS programs depend on external incentives and imperatives for action, particularly action that would go beyond one-off and incremental changes. Furthermore the reason that firms chose to participate in EMS oriented programs was to improve their capability to respond to increasingly stringent environmental requirements from governments, other firms, and the public.

As discussed by De Bruijn and Lulofs, there were two unique features of the Dutch program on Environmental Management—which focused on EMS adoption by small and medium enterprises—that contributed to its ability to engage firms. First, the government's steering capacity was enhanced by the use of a network approach. While a network approach can be effective in increasing the steering capacity of governments, it is dependent on identifying the right partners, as the analysis by De Bruijn and Lulofs shows. Second, the program is part of a long-term strategy through which firms are encouraged, in a stepwise fashion, to develop and improve their capabilities. This stepwise approach is compatible with the long-term strategies that are necessary for the development of sustainable enterprises.

In contrast to the two programs above, in the United States, EMS has found its way into government policy as part of voluntary programs to create "beyond compliance" and "superior" environmental performance. In this role, it has been used as one indicator of a firm's commitment to being an environmental leader. Two of the programs, StarTrack and Project XL, challenged firms to voluntarily improve environmental performance in exchange for flexibility and public recognition. As argued by Nash (StarTrack, chapter 10) and Marcus, Geffen, and Sexton (Project XL, chapter 11), despite incorporating the building of new relationships and the development of environmental information and capabilities, both programs showed limited results, encountering the full range of the weaknesses of these innovative approaches: inability to change the competitive environment of firms, complexity in implementation, and misfit with the dominant regulatory system. On the whole these programs did not provide adequate benefits to stimulate proactive responses by firms. They were designed to provide the benefits of flexibility in meeting existing environmental requirements, fast-track permitting, reduced monitoring, and recognition. In implementation they found it difficult to provide all but the recognition, which in and of itself was not of great value to firms. The inability to provide the other benefits was due in great part to a poor fit with the US regulatory system.[5] Furthermore there were no clear targets, as "beyond compliance" and "superior performance" were not defined at program inception and remained sources of controversy among stakeholders. These controversies over what defined performance and what types of regulatory relief the programs could provide, resulted in high transaction costs. The combination of high transaction costs and uncertain or limited benefits led firms to withdraw, choose not to participate, or propose only minor and secure changes in response to these voluntary challenges. On a positive note, these programs have been part of an ongoing effort to improve the environmental policy system in the United States. At both the state and federal level, there are new efforts that are trying to build in flexibility and provide incentives for superior environmental performance, while overcoming the shortcomings of these programs (NAPA 2001).

The final two programs evaluated in this book are information disclosure programs, the Toxics Release Inventory and the Norwegian Accounting Act. These programs stimulated the development of new information within firms by requiring that information be released to the public. They do not, however, require firms to take action to improve their environmental performance. The TRI, passed in 1986, requires that firms annually report on toxic emissions (currently of 602 chemicals), as well as on-site and off-site storage, treatment, disposal, recycling, and energy recovery. These are legal emissions and waste management activities, so action to reduce them is voluntary. Similar to EMS programs, the TRI built capacity as firms generated information about environmental performance. In many cases this was the first time that facilities and firms had such a thorough accounting of their emissions. In contrast to the EMS adoption programs, incentives to take action were built into the program, as the public release of this information resulted in external pressures to improve performance—no firm wanted to be on the "top ten polluters" list.

Graham and Miller (chapter 12) find that the TRI was effectively implemented for a number of reasons: information disclosure was required by law, the requirements for disclosure were clearly specified, and the database gained a reputation for accuracy and legitimacy. Furthermore the TRI fits well within the US policy system, which has a tradition of information disclosure and of protecting competitively sensitive information while facilitating public access to information. Furthermore the United States has a well-developed set of stakeholders, at national and local levels, that use information as part of their political strategy—creating economic and political pressure for change. Despite its clear role as a stimulus for improving environmental performance, Graham and Miller's analysis points to ongoing concern that TRI, and information disclosure more generally, has its greatest effect when first introduced, and is better at generating "win–win" short-term action to reduce negative publicity rather than longer term strategies and investments.

The Norwegian Accounting Act (NAA), passed in 1998, requires the Boards of Directors of all commercial firms subject to external auditing requirements to disclose environmental data on activities that may cause "a not insignificant impact on the external environment." The NAA goes beyond the requirements for reporting on plant specific pollution control (as found in TRI) by requiring firms to report on the life cycle environmental impacts of their products and future plans for improving environmental performance. While it is too early to draw strong conclusions about its effectiveness, Ruud (chapter 13) identifies two aspects of the NAA that are likely to hinder implementation. First, the targets for reporting for the NAA are not clearly defined, and thus firms are left to determine what to report and what to leave out. As Ruud discusses, initial results suggest great variation, even among firms considered to have excellent environmental reports. Second, although the NAA requires environmental disclosure as part of annual financial reporting, it does not require a third party auditor to validate the environmental portion of the report. Thus the engagement of a third party as educator, implementer, and enforcer is not part of this legislation.

On a final note, most of the firm-level programs were designed to increase information flows with external stakeholders, and thus may have a role in helping firms improve environmental risk management. The programs provide leading firms with an opportunity to communicate their superior environmental performance. This is particularly the case when performance is certified by government or third-party auditors, as was the case with EMAS, StarTrack, and TRI. But, based on the cases in this book, there is no evidence that verified, external information communication has created a strong motivation for beyond compliance behavior and technological innovation; rather, firms participated in these programs for other reasons, and may as a side benefit improve their relationships with stakeholders and through this mechanism better manage environmental risks.

Policy Lessons

In this section we draw a set of policy lessons that are broadly applicable to the innovative approaches we have examined in this volume. Based on our previous discussion, the effectiveness of these programs depends critically on changing the competitive environment of firms and creating a fit with the dominant regulatory system. In order to achieve this, we suggest three lessons: integrate these innovative approaches with the environmental policy system, use governmental authority to create challenging long-term goals, and engage the core objectives of the firm. Two other cross-cutting themes are the need to consider the high transaction costs of these approaches and to undertake more systematic monitoring and evaluation. Given the diversity across these policies and programs, as well as the importance of details in design and implementation, we recognize that a more detailed set of recommendations for specific policies are also valuable, and refer readers to the individual chapters for these.[6]

Integration with the Environmental Policy System

There are three reasons why a strong integration with other policies and programs is needed: (1) the need for incentives external to the programs, (2) the need for different approaches for leaders and laggards, and (3) the need to change the dominant regulatory system.[7]

Incentives External to the Program

Voluntary, collaborative, and information-based approaches are most effective when tied to incentives or imperatives for change. In a few cases strong incentives or imperatives are an integral part of the program. But more often, particularly for strictly voluntary programs, the imperatives and incentives for change have been external to the programs. Strictly voluntary programs can be effective in stimulating firms to take win–win actions that they would not have identified without the intervention of a voluntary program. Beyond this, and from the standpoint of more fundamental industrial transformation and technological innovation, voluntary programs will be most effective if they are used synergistically with or as a complement to other policies that provide incentives or imperatives for action (Gouldson and Murphy 1998; Caldart and Ashford 1999; Spence and Gopalakrishnan 2001; Ten Brink 2002). While the need for linkage to other programs is particularly pressing for voluntary approaches, even for nonvoluntary programs, employing other policies simultaneously can create a stronger impetus to action.

An often-cited example of incentives internal to these approaches is a credible threat of regulation if voluntary action is not taken or negotiated agreements are not reached (e.g., the End-of-Life Vehicles Program and the Dutch Target Group Policy). However, a threat of traditional "command and control" approaches is not the only complementary mechanism that can be internal to these programs. Others include cost-sharing, technical assistance (as was the case in the US R&D Collaborations and the Danish Cleaner Technology Program), and information disclosure (e.g., TRI).

Similarly incentives external to these innovative approaches can take a variety of forms, including regulations, procurement policies, and other economic incentives. For example, the Dutch Program on Environmental Management was quite successful for capacity building, but external pressures for improved environmental performance, as found in the Dutch Target Group Policy, created the need for increased capacity. In the case of Energy Star, incentives were created by pursuing procurement policies and mandatory standards along with the voluntary negotiations over the performances needed to receive the label and consumer education through labeling. These external incentives created a risk for non-participating firms of falling behind the technological frontier and losing market share.

Different Approaches for Leaders and Laggards

Voluntary, collaborative, and information-based approaches may be most effective in a dynamic system of regulation, in which the level of regulation is established by best practices at leading firms, and laggards are then brought forward by regulatory requirements. These innovative approaches are most appealing to pro-active firms that are oriented toward creating competitive advantage by distinguishing their firm and its products as environmental leaders. However, if there are not sufficient incentives for laggards to adopt newly developed approaches to environmental protection—including new technologies—the potential leaders may choose not to participate. Thus these innovative policies need to be well integrated with other policies, for instance, with direct regulation or market-based mechanisms, that over time will force lagging firms to improve their environmental performance and in the process create markets for the environmentally superior technologies developed by leading firms. The issue of free-riding may be a particular hazard for voluntary programs (Delmas and Terlaak 2001a; Segerson and Dawson 2001), as witnessed by the difficulties Project XL and StarTrack had in serving as incubators for environmental leaders, and the Danish Cleaner Technology Program had in bringing forward the laggards. In the case of voluntary programs, links to external sources of incentives and imperatives are essential.

Fundamental Change in the Dominant Regulatory System

For these innovative approaches to have a strong impact, they have to work in tandem with the dominant regulatory system. In almost all Western countries, these new innovations are small changes in the larger environmental policy system. Direct regulation remains the dominant approach to environmental policy. The programs and policies examined in this book represent efforts to overcome the limitations of this "command-and-control" system, many with ambitions of creating new regulatory regimes. But changing the existing approach to environmental regulation is a tall order. For example, the Danish Clean Technology Programs and Project XL demonstrate how hard it is to be innovative and flexible within the context of a fairly strong system of direct regulation.

Yet change is possible. In the Netherlands, over the course of a decade, the basic policy approach has been changed quite fundamentally (Bressers and Plettenburg 1997; Keijzers 2000). Through the late 1980s, the Dutch government relied almost exclusively on direct regulation supplemented by some taxes for water pollution. Today the Target Group Policy, with its emphasis on collaboration and negotiation, stands central. In the Netherlands the government was able to draw on the strong neocorporatist traits of the Dutch society when changing the core features of its policy system. Representatives both of industry and government were willing to look for a way out of the traditional regulatory system. Without a broader context of collaboration and both parties willing to work toward an alternative system, efforts to implement any single innovative program would have been less successful.

The United States had less success with its industry sector program, as well as many other voluntary programs that bumped up against the rigidity of the current laws and enforcement culture of the EPA (National Academy of Public Administration 2001). In contrast, information disclosure in the United States has created more environmental progress, as it fit well with a culture that values access to information and has well-developed interest groups that can use this information to press for change within the adversarial environmental policy system.

The key lesson is not that all countries should follow the Dutch example for industry sector collaboration, nor the US example for information disclosure. While these programs have contributed significantly to environmental improvement, they have all been implemented in ways that suggest both strengths and weaknesses. Rather, the lesson from this exercise is that there is a need for a careful examination of the ways in which any policy innovation can either work within or change the existing regulatory structure (EEA 1997; Hoffman, Riley et al. 2002; Meadowcroft 1998). Notwithstanding the potential advantages of voluntary, collaborative, and information-based approaches, these cannot be effective unless designed to work synergistically with the larger policy system. In some cases, this will require legislative changes; in others, a carefully design package of programs and policies that can build capability and provide incentives for action. We conclude therefore that there is not one way for environmental policies to stimulate the fundamental innovations necessary for industrial transformation. It is the environmental policy system as a whole that must respond to this enormous challenge.

Setting Challenging Long-Term Goals

Clear and challenging long-term goals are essential for stimulating radical innovation for the environment.[8] The ability to engage industry and other stakeholders in establishing such goals is one of the greatest strengths of the voluntary and collaborative approaches (Blowers 1998; Ramesohl et al. 2002). Two major lessons for goal-setting emerge from the analyses in this book: First, government must play a leading role in the goal-setting process. Second, challenging long-term goals need to be translated into clear and specific targets, preferably with interim targets, in order to guide implementation. There is one outstanding issue related to long-term goals that must be addressed to use these approaches effectively in the future: how to reduce the uncertainty facing firms while retaining flexibility to respond to new scientific information on environmental hazards.

Government Leadership in Goal-Setting

Government played a key role in goal-setting for the most successful programs examined in this book. The programs in this book that were most successful in stimulating substantial improvement in environmental performance and radical innovation were based on challenging goals that were established by governments (e.g., the Dutch Target Group Policy, the US R&D Collaborations, and Energy Star) or mediated by governments (e.g., the German End-of-Life Vehicles Program). While collaborative approaches were often involved in the development of these goals, the government's role as final arbiter was essential.[9] Efforts to use consensus-based decision making to set goals led to lowest-common-denominator solutions or stalemate, as evidenced by the relative failure of two of the bold US experiments in regulatory reinvention, Project XL and CSI.

Combine Specific Targets with Flexibility

Program effectiveness depends on translating long-term goals into clear and specific targets, while providing flexibility during implementation.[10] This proved to be more feasible for some of the sector-based programs, while remaining a challenge for many facility level programs. Goals need to be sufficiently stringent to bring the private sector to the negotiating table, and yet sufficiently flexible to give the private sector the opportunity to bring in its knowledge and creativity to accomplish such goals (Meadowcroft 1998). Facility-level approaches require more case sensitive goals in order to address the highly divergent capabilities and needs of different firms. The programmatic goals of facility-level programs therefore are often more of a general nature. To be effective, these goals need to be articulated with enough specificity to facilitate successful implementation. The limited successes of StarTrack and Project XL, which both aimed for "superior environmental performance," show the difficulty of making this translation, and the consequences for not doing so.

Balancing Long-term Planning with Flexibility to Respond to New Environmental Knowledge

In many of these programs the private sector has made long-term commitments in exchange for promises that no new requirements will be imposed. This creates a dilemma, especially in cases where programs result in legally binding agreements between government and industry. Future insights into the cause and nature of environmental degradation can lead to a need for more stringent environmental targets. As industry, government, and other stakeholders collaborate to reach ambitious environmental goals, firms are looking for some assurances that these goals will not change in midstream. Although this is understandable and perhaps even necessary, governments have an obligation to respond to evolving knowledge about environmental hazards. To address this concern, while negotiating over long-term commitments, governments will have to make clear that future demands are likely to be more stringent (Meadowcroft 1998).

Engagement of the Firm's Core Objectives

Voluntary, collaborative, and information-based approaches are most effective in stimulating private sector leadership and radical technological innovation when they influence the core objectives of firms and, in doing so, are able to engage business units and not simply the environmental function in firms. The cases in this book demonstrate that there are a variety of ways to accomplish this. Challenging long-term goals, discussed above, is clearly one, and perhaps the most effective approach. Economic incentives linked to challenging goals, as in the technology development programs (US R&D Collaborative and Danish Clean Technology Program), provide another model that can lead to firms undertaking risky environmentally enhancing innovations to their core technology.

Other programs attempted to provide benefits to firms in exchange for superior or beyond compliance behavior, including flexibility, information generation, and public recognition. These resulted in limited value and thus limited success, suggesting both the potential and difficulties that these approaches can have in providing significant value without a linkage to regulatory requirements or economic incentives. For example, although Project XL had difficulty in delivering its promise of flexibility, by and large the companies that did come forward were those that saw a business advantage in the flexibility that Project XL could provide. For environmental management systems and information generation, programs provided greater value and thus a stronger impetus for change when they required firms to develop information that could help them become more eco-efficient or help them communicate superior environmental performance to external stakeholders. Negative public recognition created a stimulus to improved environmental performance, at least in the short run, but few of the programs based on positive recognition (aside from labeling) provided the basis for competitive advantage.

Transaction Costs

High transaction costs is a critique that was aimed at nearly all of the programs examined in this book, and for these approaches more generally (Caldart and Ashford 1999; Delmas and Terlaak 2001b). The high transaction costs are of two types. The first are those that can be reduced through better management and better integration of these innovative approaches with the larger environmental policy system, thus clarifying the possibilities and roles of each of the players and reducing the time needed for dialog and negotiation. The second category is inherent to the processes and/or types of environmental problems that are being addressed. Collaborative processes, by nature, require time for discussion among a broad group of people who may begin the process with very different goals and interests. Furthermore some of these programs are trying to address areas where markets have not provided adequate incentives or information flows precisely because of the high cost of developing and transferring information to the end-user, namely the person who would act on the information. These transaction costs will be impossible to reduce beneath a necessary threshold.

Potential transaction costs should be thoroughly examined when deciding whether to implement voluntary, collaborative, and information-based programs, and reduced through program design where possible. Furthermore inherent transaction cost should be evaluated against the benefits of the program and all the transaction costs should be compared to alternative approaches for reaching the same environmental goals. Transaction costs are most pressing with facility-based programs when they require lengthy negotiations on a case-to-case base. The analyses in this book, especially regarding Project XL and CSI, show the difficulties in creating sufficient (environmental) benefits in return. Finally, implementing voluntary programs through existing networks (e.g., the Dutch Environmental Management Program) may not only be a way of improving effectiveness, but also of limiting transaction costs.

Monitoring and Evaluation

Inadequate attention has been given to the evaluation of voluntary, collaborative, and information-based programs, despite widespread endorsement for monitoring and reporting on innovations in environmental policy (i.e., Mazurek 1998; UNEP 1998). In order to identify effective policies and thus enhance policy learning, the new approaches should be designed with a plan for evaluation.[11] Evaluation can be challenging, as we often need to answer the "what if" questions, that is, try to compare results from the new policy with what would have happened without it (i.e., develop a baseline scenario). Nonetheless, this type of analysis is critical if we want to learn through experimentation. Thus more attention should be given to evaluation, including making evaluation an integral part of program design, collecting real-time data, and putting the funding for evaluation into the budget of new programs. It is essential to look beyond process variables (e.g., how many firms adopted EMS) and evaluate actions taken to reduce environmental impacts as well as the actual reduction of environmental impacts.

The Prospects for Voluntary, Collaborative, and Information Strategies

In both the United States and Europe there is recognition that government, industry, and the NGO community must share responsibility for the transition to a sustainable industrial society. In sharing responsibility, governments must continue to steer and enforce, but do so in ways that facilitate and stimulate the private sector to develop and adopt environmentally sound processes and products. In recognition that command-and-control approaches cannot create shared responsibility, over the past decade scholars and practitioners alike have focused on the design and use of new approaches to environmental policy. In this book we have tried to capture the strengths and weaknesses of a broad group of these innovative approaches—including voluntary, collaborative, and information-based programs—for creating shared responsibility, and through this, contributing to industrial transformation.

The overarching conclusion of the book is that voluntary, collaborative, and information-based programs can play a useful role in a comprehensive environmental strategy if they are carefully designed to fit with and complement the other elements of a nation's environmental policy system. The strength of these innovative approaches is in their ability to build new relationships, create new capabilities, and stimulate firms to become first movers. At their best these approaches provide the opportunity for government and industry to negotiate a common agenda for the future and provide opportunities and incentives for firms to pursue strategies in which environmental sustainability is a primary driver. Despite the many benefits of the programs in general, however, we have seen little evidence for changes beyond the level of individual firms, thus leading to broader patterns of change in production and consumption systems. When they occurred, it was more often through the sector-based programs than through the facility-based programs.

The new approaches are not a panacea for industrial transformation (Caldart and Ashford 1999; Hartman et al. 2002). While the value of specific approaches and the optimal mix of policies will vary among countries, synergistically employing multiple approaches provides the strongest possibility of guiding industry on the path toward sustainable development. There will remain a role for direct regulations and market-based approaches as part of an overall strategy—these mechanisms will be needed to create sufficient pressures to push industry along the path toward sustainability. In the end the real question therefore is not whether the new approaches should be used, but rather *how* they should be used.

Notes

1. For discussion of the costs of free-riding and the impacts of free-riding on the effectiveness of voluntary agreements, see Delmas and Terlaak (2001a) and Segerson and Dawson (2001).

2. Many argue for involving firms in different stages of a product chain or joining forces with firms in a specific sector of industry; see, for instance, Hart (1995), IHDP (1999), Roome and Cahill (2001), and Schot et al. (1997).

3. For further discussion of administrative capture in voluntary agreements and regulatory contracts, see Faure (2001), Harrison (1998), Maxwell and Lyon (2001), Rennings et al. (1997), and Spence and Gopalakrishnan (2001).

4. These issues have been raised in analysis of other efforts to use stakeholder collaboration, negotiation, and consensus-based decision making within the US environmental policy (Caldart and Ashford 1999; Coglianese 2001).

5. Susskind and Secunda (1998a, b) discuss the constraints on the US regulatory system that prevent more effective use of administrative discretion, with particular reference to Project XL.

6. Policy lessons for different types of programs are also summarized in the report from the workshop in which these papers were first presented to a group that included participants from industry, academe, government, and NGOs (De Bruijn and Norberg-Bohm 2001).

7. Many scholars suggest a complementary role for these new approaches; see, for instance, Blowers (1998), Dowd et al. (2001), EEA (1997), Gouldson and Murphy (1998), Krarup and Ramesohl (2000), Paton (1999), Ten Brink (2002).

8. Many analysts of voluntary and collaborative approaches identify ambitious goals as a key for effectiveness, see Delmas and Terlaak (2001b), Dowd et al. (2001), and Krarup and Ramesohl (2000).

9. This conclusion is consistent with recommendations of others; see, for instance, EEA (1997).

10. This general finding is echoed by many analysts; see, for instance, Davies and Mazurek (1996), and Peters (1993).

11. Many analysts of voluntary and collaborative programs identify this need; see, for instance, EEA (1997), Harrison (1998), Krarup and Ramesohl (2000), and NRC (1997).

References

Blowers, A. 1998. Power Participation and Partnerships. In P. Glasbergen, ed., *Co-operative Environmental Governance: Public Private Agreements as a Policy Strategy*. Dordrecht: Kluwer.

Bressers, H. T. A., and L. A. Plettenburg. 1997. The Netherlands. In M. Jänicke and H. Weidner, eds., *National Environmental Policies: A Comparative Study of Capacity-Building*. Berlin: Springer, pp. 109–32.

Caldart, C. C., and N. A. Ashford. 1999. Negotiation as a means of developing and implementing environmental and occupational health and safety policy. *Harvard Environmental Law Review* 23(1): 141–202.

Coglianese, C. 2001. Is Consesnus an Appropriate Basis for Regulatory Policy? In E. W. Orts and K. Deketelaere, eds., *Environmental Contracts: Comparative Approaches to Regulatory Innovation in the United States and Europe*. London: Kluwer Law International, pp. 93–133.

Davies, T., and J. Mazurek. 1996. *Industry Incentives for Environmental Improvement: Evaluation of U.S. Federal Initiatives*. Washington, DC: Resources for the Future, p. 70.

De Bruijn, T., and V. Norberg-Bohm. 2001. *Voluntary, Collaborative, and Information-Based Policies: Lessons and Next Steps for Environmental and Energy Policy in the United States and Europe*. Cambridge: Energy Technology Innovation Project. Belfer Center for Science and International Affairs, and Regulatory Policy Program. Center for Business and Government, Kennedy School of Government, Harvard University.

De Bruijn, T., and A. Tukker, eds. 2002. *Partnership and Leadership: Building Alliances for a Sustainable Future*. Dordrecht: Kluwer.

Delmas, M. A., and A. Terlaak. 2001a. A framework for analyszing environmental voluntary agreements. *California Management Review* 43(3): 44–63.

Delmas, M. A., and A. Terlaak. 2001b. Voluntary Agreements for the Environment: Institutional Constraints and Potential for Innovation. In E. W. Orts and K. Deketelaere, eds., *Environmental Contracts: Comparative Approaches to Regulatory Innovation in the United States and Europe*. London: Kluwer Law International, pp. 349–68.

Dowd, J., K. Friedman, and G. Boyd. 2001. How well do voluntary agreements and programs perform at improving industrial energy efficiency. *Proceedings of the 2001 ACEEE Summer Study on Energy Efficiency in Industry*. Washington, DC: American Council for an Energy-Efficient Economy.

Driessen, P. 1998. Concluding remarks. In P. Glasbergen, ed., *Co-operative Environmental Governance: Public-Private Agreements as a Policy Strategy*. Dordrecht: Kluwer, pp. 251–68.

EEA. 1997. *Environmental Agreements: Environmental Effectiveness*. Copenhagen: Environmental Environment Agency.

Faure, M. 2001. Environmental contracts: A Flemish law and economics perspective. In E. W. Orts and K. Deketelaere, eds., *Environmental Contracts: Comparative Approaches to Regulatory Innovation in the United States and Europe*. London: Kluwer Law International, pp. 167–77.

Glasbergen, P., ed. 1998. *Co-operative Environmental Governance: Public-Private Agreements as a Policy Strategy*. Dordrecht: Kluwer.

Gouldson, A., and J. Murphy. 1998. *Regulatory Realities: The Implementation and Impact of Industrial Environmental Regulations*. London: Earthscan.

Hajer, M. A. 1995. *The Politics of Environmental Discourse: Ecological Modernization and the Policy Process*. Oxford: Oxford University Press.

Harrison, K. 1998. Talking with the donkey: Cooperative approaches to environmental protection. *Journal of Industrial Ecology* 2(3): 51–72.

Hart, S. L. 1995. A natural resource based view of the firm. *Academy of Management Review* 20(4): 986–1014.

Hartman, C. L., P. S. Hofman, and E. R. Stafford. 2002. Environmental collaboration: Potential and limits. In T. De Bruijn and A. Tukker, eds., *Partnership and Leadership: Building Alliances for a Sustainable Future*. Dordrecht: Kluwer, pp. 21–40.

Hoffman, A. J., H. C. Riley, J. G. Troast Jr., and M. H. Bazerman. 2002. Cognitive and institutional barriers to new forms of cooperation on environmental protection. *American Behavioral Scientist* 45(5): 820–45.

IHDP. 1999. *Industrial Transformation Science Plan*. Bonn: International Human Dimensions Programme on Global Environmental Change.

Jänicke, M., and H. Weidner, eds. 1997. *National Environmental Policies: A Comparative Study of Capacity-Building*. Berlin: Springer.

Keijzers, G. 2000. The evolution of Dutch environmental policy: The changing ecological arena from 1970–2000 and beyond. *Journal of Cleaner Production* 8(3): 179–200.

Krarup, S., and S. Ramesohl. 2000. *Voluntary Agreements in Energy Policy: Implementation and Efficiency*. Copenhagen: AKF–Institute of Local Government Studies, p. 72.

Liefferink, D. 1997. The Netherlands: A net exporter of environmental policy concepts. In D. Liefferink and A. S. Anderson, eds., *The Innovation of EU Environmental Policy*. Copenhagen: Scandinavian University Press.

Maxwell, J. W., and T. P. Lyon. 2001. An Institutional Analysis of Environmental Voluntary Agreements in the United States. In E. W. Orts and K. Deketelaere, eds., *Environmental Contracts: Comparative Approaches to Regulatory Innovation in the United States and Europe*. London: Kluwer Law International, pp. 333–48.

Mazurek, J. 1998. *The Use of Voluntary Agreements in the United States: An Initial Survey*. Paris: OECD.

Meadowcroft, J. 1998. Co-operative Management Regimes: A Way Forward? In P. Glasbergen, ed., *Co-operative Environmental Governance: Public-Private Agreements as a Policy Strategy*. Dordrecht. Kluwer, pp. 21–42.

NAPA. 2001. *Leading Change: Advancing Effective Governance in the 21st Century*. Washington, DC: National Academy of Public Administration.

NRC. 1997. *Fostering Industry-Initiated Environmental Protection Efforts*. Washington, DC: Committee on Industrial Competitiveness and Environmental Protection.

Orts, E. W., and K. Deketelaere, eds. 2001. *Envirionmental Contracts: Comparative Approaches to Regulatory Innovation in the United States and Europe*. The Hague: Kluwer Law International.

Paton, B. 1999. *Voluntary Environmental Initiatives and Sustainable Industry*. Greening of Industry Network Conference, Chapel Hill, NC.

Peters, B. G. 1993. *American Public Policy: Promise and Performance*. Chatham, NJ: Chatham House.

Poncelet, E. 2002. In Search of "win–win": Multistakeholder environmental partnerships and the pursuit of sustainability. In T. De Bruijn and A. Tukker, eds., *Partnership and Leadership: Building Alliances for a Sustainable Future*. Dordrecht: Kluwer, pp. 41–60.

Ramesohl, S., and K. Kristof. 2002. Voluntary agreements: An effective tool for enhancing organizational learning and improving climate policy-making? In P. T. Brink, ed., *Voluntary Environmental Agreements: Process, Practice and Future Use*. Sheffield: Greenleaf, pp. 341–56.

Rennings, K., K. L. Brockman, and H. Berman. 1997. Voluntary agreements in environmental protection: Experiences in germany and future perspectives. *Business Strategy and the Environment* (6): 245–63.

Roome, N., and E. Cahill. 2001. *Sustainable production*—Challenges and objectives for EU research policy. Report of the Expert Group on Competitive and Sustainable Production and Related Service Industries in Europe in the Period to 2020. EUR 19880 Research Directorate General.

Schot, J., E. Brand, and K. Fischer. 1997. The greening of industry for a sustainable future: Building an international research agenda. *Business Strategy and the Environment* 6(3): 153–62.

Segerson, K., and N. L. Dawson. 2001. Environmental voluntary agreements: Participation and free riding. In E. W. Orts and K. Deketelaere, eds., *Environmental Contracts: Comparative Approaches to Regulatory Innovation in the United States and Europe*. London: Kluwer Law International, pp. 369–88.

Spence, D. B., and L. Gopalakrishnan. 2001. The New political economy of regulation: Looking for positive sum change in a zero sum world. In E. W. Orts and K. Deketelaere, eds., *Environmental Contracts: Comparative Approaches to Regulatory Innovation in the United States and Europe*. London: Kluwer Law International, pp. 305–32.

Susskind, L. E., and J. Secunda. 1998–1999a. The risks and the advantages of agency discretion: Evidence from EPA's Project XL. *UCLA Journal of Environmental Law and Policy* 17(67).

Susskind, L. E., and J. Secunda. 1998–1999b. Improving Project XL: Helping adaptive management to work within EPA. in *UCLA Journal of Environmental Law and Policy* 17(155).

Ten Brink, P. 2002. Epilogue. In P. T. Brink, ed., *Voluntary Environmental Agreements: Process, Practice and Future Use*. Sheffield: Greenleaf, pp. 490–94.

Ten Brink, P. 2002. *Voluntary Environmental Agreements: Process, Practice and Future Use*. Sheffield: Greenleaf.

UNEP. 1998. Voluntary initiatives. *Industry and Environment* 21(1–2).

Index